Alf : coef s b cuu

The Art of Paper-Making

A Uscduldcl I coebppk pgui f P covgcduvsf pgUcr f s

Alf : coef s b cш

The Art of Paper-Making
A Practical Handbook of the Manufacture of Paper

JVBS OF AS / 6896668667536

Uslouf e lo Fvspr f . a VA. Ccocec. Avtusclkc. Mr co

Cp, f s/ Hpup 2 bf shhf ktu008 O r k f lkp@f

P psf c, cKcblf bppkt cu **www.hansebooks.com**

THE
ART OF PAPER-MAKING

WORKS BY THE SAME AUTHOR.

Just ready. Fourth Edition, Revised and Enlarged. Crown 8vo, 7s. 6d. cloth.

THE ART OF SOAP-MAKING: A Practical Handbook of the Manufacture of Hard and Soft Soaps, Toilet Soaps, &c. Including many New Processes, and a Chapter on the Recovery of Glycerine from Waste Leys. With numerous Illustrations.

"Really an excellent example of a technical manual, entering as it does, thoroughly and exhaustively, both into the theory and practice of soap manufacture. The book is well and honestly done, and deserves the considerable circulation with which it will doubtless meet." — *Knowledge*.

Second Edition. Crown 8vo, 9s. cloth.

THE ART OF LEATHER MANUFACTURE: Being a Practical Handbook, in which the Operations of Tanning, Currying, and Leather Dressing are fully Described, and the Principles of Tanning Explained, and many Recent Processes Introduced. With numerous Illustrations.

"A sound, comprehensive treatise on tanning and its accessories.... The book is an eminently valuable production." — *Chemical Review*.

Just Published. Third Edition, revised and much enlarged. 600 pp., crown 8vo, 9s. cloth.

ELECTRO-DEPOSITION: A Practical Treatise on the Electrolysis of Gold, Silver, Copper, Nickel, and other Metals and Alloys. With descriptions of Voltaic Batteries, Magneto and Dynamo-Electric Machines, Thermopiles, and of the Materials and Processes used in every Department of the Art, and several Chapters on ELECTRO-METALLURGY. With numerous Illustrations.

"Eminently a book for the practical worker in electro-deposition. It contains minute and practical descriptions of methods, processes and materials, as actually pursued and used in the workshop. Mr. Watt's book recommends itself to all interested in its subjects." — *Engineer*.

Just Published. Ninth Edition, enlarged and revised, 12mo, 4s. cloth.

ELECTRO-METALLURGY: Practically Treated. Ninth Edition, Enlarged and Revised, with Additional Matter and Illustrations, including the most recent Processes.

"From this book both amateur and artisan may learn everything necessary for the successful prosecution of electro-plating." — *Iron*.

CROSBY LOCKWOOD & SON, 7, Stationers' Hall Court, London, E.C.

THE ART OF
PAPER-MAKING

A PRACTICAL HANDBOOK OF THE MANUFACTURE
OF PAPER FROM RAGS, ESPARTO, STRAW, AND
OTHER FIBROUS MATERIALS, INCLUDING
THE MANUFACTURE OF PULP FROM
WOOD FIBRE

With a Description of the Machinery and Appliances
used

TO WHICH ARE ADDED

DETAILS OF PROCESSES FOR RECOVERING SODA FROM WASTE
LIQUORS

By ALEXANDER WATT

AUTHOR OF "THE ART OF SOAP-MAKING," "LEATHER MANUFACTURE," "ELECTRO-METALLURGY," "ELECTRO-DEPOSITION," ETC., ETC.

LONDON
CROSBY LOCKWOOD AND SON
7, STATIONERS' HALL COURT, LUDGATE HILL
1890
[*All rights reserved*]

LONDON:
PRINTED BY J. S. VIRTUE AND CO., LIMITED.
CITY ROAD.

PREFACE.

In the present volume, while describing the various operations involved in the manufacture of paper, the Author has endeavoured to render the work serviceable as a book of reference in respect to the processes and improvements which have from time to time been introduced, and many of which have been more or less practically applied either at home or abroad.

The recovery of soda from waste liquors has been fully dealt with, and the details of several applied processes explained.

Special attention has also been directed to some of the more important methods of producing pulp from wood fibre, since it is highly probable that from this inexhaustible source the paper-maker will ultimately derive much of the cellulose used in his manufacture. Indeed it may be deemed equally probable, when the processes for disintegrating wood fibre, so largely applied in America and on the Continent, become better understood in this country, that their adoption here will become more extensive than has hitherto been the case.

To render the work more readily understood alike by the practical operator and the student, care has been taken to avoid, as far as possible, the introduction of unexplained technicalities; at the same time it has been the writer's aim to furnish the reader with a variety of information which, it is hoped, will prove both useful and instructive.

It is with much pleasure that the Author tenders his sincere thanks to Mr. Sydney Spalding, of the Horton Kirby Mills, South Darenth, for his kind courtesy in conducting him through the various departments of the mill, and for explaining to him the operations performed therein. To Mr. Frank Lloyd he also acknowledges his indebtedness for the generous readiness with which he accompanied him over the *Daily Chronicle* Mill at Sittingbourne, and for the pains he took to supply information as to certain details at the Author's request. His best thanks are also due to those manufacturers of paper-making machinery who supplied him with many of the blocks which illustrate the pages of the book.

BALHAM, SURREY, *January, 1890*.

CONTENTS.

CHAPTER I.
CELLULOSE.

PAGE

Cellulose—Action of Acids on Cellulose—Physical Characteristics of Cellulose— Micrographic Examination of Vegetable Fibres—Determination of Cellulose—Recognition of Vegetable Fibres by the Microscope   1

CHAPTER II.
MATERIALS USED IN PAPER-MAKING.

Raw Materials—Rags—Disinfecting Machine—Straw—Esparto Grass— Wood—Bamboo—Paper Mulberry   9

CHAPTER III.
TREATMENT OF RAGS.

Preliminary Operations—Sorting—Cutting—Bertrams' Rag-cutting Machine—Nuttall's Rag-cutter— Willowing—Bertrams' Willow and Duster—Dusting—Bryan Donkin's Duster or Willow—Donkin's Devil   19

CHAPTER IV.

TREATMENT OF RAGS (continued).

Boiling Rags—Bertrams' Rag-boiler—Donkin's Rag-boiler—Washing and Breaking—Bertrams' Rag-engine—Bentley and Jackson's Rag-engine—Draining—Terrance's Drainer   29

CHAPTER V.

TREATMENT OF ESPARTO.

Preliminary Treatment—Picking—Willowing Esparto—Boiling Esparto—Sinclair's Esparto Boiler—Roeckner's Boiler—Mallary's Process—Carbonell's Process—Washing Boiled Esparto—Young's Process—Bleaching the Esparto   40

CHAPTER VI.

TREATMENT OF WOOD.

I. Chemical Processes—Watt and Burgess's Process—Sinclair's Process—Keegan's Process—American Wood-pulp System—Aussedat's Process—Acid Treatment of Wood—Pictet and Brélaz's Process—Barre and Blondel's Process—Poncharac's Process—Young and Pettigrew's Process—Fridet and Matussière's Process   53

CHAPTER VII.

TREATMENT OF WOOD (continued).

Sulphite Processes—Francke's Process—Ekman's Process—Dr. Mitscherlich's Process—Ritter and Kellner's Boiler—Partington's Process—Blitz's Process—M'Dougall's Boiler for Acid Processes

—Graham's Process—Objections to the Acid or Sulphite Processes—Sulphite Fibre and Resin—Adamson's Process—Sulphide. Processes—II. Mechanical Processes— Voelter's Process for preparing Wood-pulp—Thune's Process   68

CHAPTER VIII.

TREATMENT OF VARIOUS FIBRES.

Treatment of Straw—Bentley and Jackson's Boiler—Boiling the Straw—Bertrams' Edge-runner—M. A. C. Mellier's Process— Manilla, Jute, &c.—Waste Paper—Boiling Waste Paper—Ryan's Process for Treating Waste Paper   80

CHAPTER IX.

BLEACHING.

Bleaching Operations—Sour Bleaching—Bleaching with Chloride of Lime—Donkin's Bleach Mixer — Bleaching with Chlorine Gas (Glaser's Process)—Electrolytic Bleaching (C. Watt's Process)—Hermite's Process— Andreoli's Process—Thompson's Process—Lunge's Process—Zinc Bleach Liquor—Alum Bleach Liquor—New Method of Bleaching   89

CHAPTER X.

BEATING OR REFINING.

Beating—Mr. Dunbar's Observations on Beating—Mr. Arnot on Beating Engines—Mr. Wyatt on American Refining Engines— The Beating

Engine—Forbes' Beating Engine—Umpherston's Beating Engine—Operation of Beating—Test for Chlorine—Blending   101

CHAPTER XI.

LOADING. — SIZING. — COLOURING.

Loading—Sizing—French Method of preparing Engine Size—Zinc Soaps in Sizing—Colouring —Animal or Tub Sizing— Preparation of Animal Size—American Method of Sizing—Machine Sizing—Double-sized Paper—Mr. Wyatt's Remarks on Sizing   114

CHAPTER XII.

MAKING PAPER BY HAND.

The Vat and Mould—Making the Paper—Sizing and Finishing   129

CHAPTER XIII.

MAKING PAPER BY MACHINERY.

The Fourdrinier Machine—Bertrams' Large Paper Machine—Stuff Chests—Strainers—Revolving Strainer and Knotter— Self-cleansing Strainer—Roeckner's Pulp Strainers—The Machine Wire and its Accessories—Conical Pulp-Saver— The Dandy-Roll—Water-Marking—De la Rue's Improvements in Water-Marks—Suction Boxes —Couch Rolls—Press Rolls— Drying Cylinders —Smoothing Rolls—Single Cylinder Machines   133

CHAPTER XIV.

CALENDERING, CUTTING, AND FINISHING.

Web-Glazing—Glazing Calender Damping Rolls—Finishing—Plate Glazing—Donkin's Glazing Press—Mr. Wyatt on American Super-Calendering—Mr. Arnot on Finishing—Cutting—Revolving Knife Cutter—Bertrams' Single-sheet Cutter— Packing the finished Paper—Sizes of Paper   154

CHAPTER XV.

COLOURED PAPERS.

Coloured Papers—Colouring Matters used in Paper-making—American Combinations for Colouring—Mixing Colouring Materials with Pulp—Colouring Paper for Artificial Flowers—Stains for Glazed Papers—Stains for Morocco Papers—Stains for Satin Papers   165

CHAPTER XVI.

MISCELLANEOUS PAPERS.

Waterproof Paper—Scoffern and Tidcombe's Process—Dr. Wright's Process for preparing Cupro-Ammonium—Jouglet's Process— Waterproof Composition for Paper—Toughening Paper—Morfit's Process—Transparent Paper—Tracing Paper—Varnished Paper— Oiled Paper—Lithographic Paper—Cork Paper—New Japanese Paper—Blotting Paper—Parchment Paper—Mill and Cardboard— Making Paper or Cardboard with two Faces by ordinary Machine—Test Papers   174

CHAPTER XVII.

MACHINERY USED IN PAPER-MAKING.

Bentley and Jackson's Drum-Washer—Drying Cylinders—Self-acting Dry Felt Regulator—Paper Cutting Machine—Single-web Winding Machine— Cooling and Damping Rolls—Reversing or Plate-glazing Calender—Plate-planing Machine—Roll-bar Planing Machine—Washing Cylinder for Rag Engine— Bleach Pump—Three-roll Smoothing Presses—Back-water Pump—Web-glazing Calender—Reeling Machine—Web-ripping Machine— Roeckner's Clarifier—Marshall's Perfecting Engine   184

CHAPTER XVIII.

RECOVERY OF SODA FROM SPENT LIQUORS.

Recovery of Soda—Evaporating Apparatus—Roeckner's Evaporator—Porion's Evaporator—Yaryan's Evaporator—American System of Soda Recovery   204

CHAPTER XIX.

DETERMINING THE REAL VALUE OR

PERCENTAGE OF COMMERCIAL SODAS, CHLORIDE OF LIME, ETC.

Examination of Commercial Sodas—Mohr's Alkalimeter—Preparation of the Test Acid—Sampling Alkalies—The Assay—Estimation of Chlorine in Bleaching Powder—Fresenius' Method—Gay-Lussac's Method—The Test Liquor—Testing the Sample—Estimation of Alumina in Alum Cake, &c. 221

CHAPTER XX.

USEFUL NOTES AND TABLES.

Preparation of Lakes—Brazil-wood Lake—Cochineal Lake—Lac Lake—Madder Lake—Orange Lake—Yellow Lake—Artificial Ultramarine— Twaddell's Hydrometer—Imitation Manilla from Wood-pulp—Testing Ultramarines—Strength of Paper 235

TABLES.—Dalton's Table showing the Proportion of Dry Soda in Leys of different Densities—Table of Strength of Caustic Soda Solutions at 59° F. = 150° C. (Tünnerman)—Table showing the Specific Gravity corresponding with the Degrees of Baumé's Hydrometer—Table of Boiling Points of Alkaline Leys— Table showing the Quantity of Caustic Soda in Leys of different Densities—Table showing the Quantity of Bleaching Liquid at 6° Twaddell (specific gravity 1·030) required to be added to Weaker Liquor to raise it to the given Strengths—Comparative French and English Thermometer Scales—Weights and Measures of the Metrical System—Table of

French Weights and Measures	241
LIST OF WORKS RELATING TO PAPER MANUFACTURE	246

THE ART
OF
PAPER-MAKING.

CHAPTER I.

CELLULOSE.

Cellulose.—Action of Acids on Cellulose.—Physical Characteristics of Cellulose.—Micrographic Examination of Vegetable Fibres.—Determination of Cellulose.—Recognition of Vegetable Fibres by the Microscope.

Cellulose.—Vegetable fibre, when deprived of all incrusting or cementing matters of a resinous or gummy nature, presents to us the true fibre, or *cellulose*, which constitutes the essential basis of all manufactured paper. Fine linen and cotton are almost pure cellulose, from the fact that the associated vegetable substances have been removed by the treatment the fibres were subjected to in the process of their manufacture; pure white, unsized, and unloaded paper may also be considered as pure cellulose from the same cause. Viewed as a chemical substance, cellulose is white, translucent, and somewhat heavier than water. It is tasteless, inodorous, absolutely innutritious, and is insoluble in water, alcohol, and oils. Dilute acids and alkalies, even when hot, scarcely affect it. By prolonged boiling in dilute acids, however, cellulose undergoes a gradual change, being converted into *hydro-cellulose*. It is also affected by boiling water alone, especially under high pressure, if boiled for a lengthened period. Without going deeply into the chemical properties of cellulose, which would be more interesting to the chemist than to the paper manufacturer, a few data respecting the action of certain chemical substances upon cellulose will, it is hoped, be found useful from a practical point of view, especially at the present day, when so many new methods of treating vegetable fibres are being introduced.

Action of Acids on Cellulose.—When concentrated sulphuric acid is added very gradually to about half its weight of linen rags cut into small shreds, or strips of unsized paper, and contained in a glass vessel, with constant stirring, the fibres gradually swell up and disappear, without the evolution of any gas, and a tenacious mucilage is formed which is entirely soluble in water. If, after a few hours, the mixture be diluted with water, the acid neutralised with chalk, and after filtration, any excess of lime thrown down by cautiously adding a solution of oxalic acid, the liquid yields, after a second filtration and the addition of alcohol in considerable excess, a gummy mass which possesses all the characters of *dextrin*. If instead of at once saturating the diluted acid with chalk, we boil it for four or five hours, the *dextrin* is entirely converted into grape sugar (*glucose*), which, by the addition of chalk and filtration, as before, and evaporation at a gentle heat to the consistence of a syrup, will, after repose for a few days, furnish a concrete mass of crystallised sugar. Cotton, linen, or unsized paper, thus treated, yield fully their own weight of gum and one-sixth of their weight of grape sugar. Pure cellulose is readily attacked by, and soon becomes dissolved in, a solution of oxide of copper in ammonia (*cuprammonium*), and may again be precipitated in colourless flakes by the addition of an excess of hydrochloric acid, and afterwards filtering and washing the precipitate. Concentrated boiling hydrochloric acid converts cellulose into a fine powder, without, however, altering its composition, while strong nitric acid forms nitro-substitution products of various degrees, according to the strength of the acid employed. "Chlorine gas passed into water in which cellulose is suspended rapidly oxidises and destroys it, and the same effect takes place when hypochlorites, such as hypochlorite of calcium, or bleaching liquors, are gently treated with it. It is not, therefore, the cellulose itself which we want the bleaching liquor to operate upon, but only the colouring matters associated with it, and care must be taken to secure that the action intended for the extraneous substances alone does not extend to the fibre itself. Caustic potash affects but slightly cellulose in the form in which we have to do it, but in certain less compact conditions these agents decompose or destroy it."—*Arnot.*[1]

Physical Characteristics of Cellulose.—"The physical condition of cellulose," says Mr. Arnot, "after it has been freed from extraneous matters by boiling, bleaching, and washing, is of great importance to the manufacturer. Some fibres are short, hard, and of polished exterior, while others are long, flexible, and barbed, the former, it is scarcely necessary to say, yielding but indifferent papers, easily broken and torn, while the papers produced from the latter class of fibres are possessed of a great degree of strength and flexibility. Fibres from straw, and from many varieties of wood, may be taken as representatives of the former class, those from hemp and flax affording good illustrations of the latter. There are, of course, between these extremes all degrees and combinations of the various characteristics indicated. It will be readily understood that hard, acicular[2] fibres do not felt well, there being no intertwining or adhesion of the various particles, and the paper produced is friable. On the other hand, long, flexible, elastic fibres, even though comparatively smooth in their exterior, intertwine readily, and felt into a strong tough sheet.... Cotton fibre is long and tubular, and has this peculiarity, that when dry the tubes collapse and twist on their axes, this property greatly assisting the adhesion of the particles in the process of paper-making. In the process of dyeing cotton, the colouring matter is absorbed into the tubes, and is, as will be readily appreciated, difficult of removal therefrom. Papers made exclusively of cotton fibre are strong and flexible, but have a certain sponginess about them which papers made from linen do not possess."

Linen—the cellulose of the flax-plant—before it reaches the hands of the paper-maker has been subjected to certain processes of steeping or *retting*, and also subsequent boilings and bleachings, by which the extraneous matters have been removed, and it therefore requires but little chemical treatment at his hands. "Linen fibre," Arnot further observes, "is like cotton, tubular, but the walls of the tubes are somewhat thicker, and are jointed or notched like a cane or rush; the notches assist greatly in the adhesion of the fibres one to another. This fibre possesses the other valuable properties of length, strength, and flexibility, and the latter property is increased when the walls of the tubes are crushed together under the action of the beating-engine." From this fibre a very strong, compactly felted paper is made; indeed, no better material than this can be had for the production of a first-class paper. Ropes, coarse bags, and suchlike are made from hemp, the cellulose or fibre of which is not unlike that of flax, only it is of a stronger, coarser nature. Manilla[3] yields the strongest of all fibres. Jute, which is the fibre or inside bark of an Indian plant (*Corchorus capsularis*), yields a strong fibre, but is very difficult to bleach white. Esparto fibre holds an intermediate place between the fibres just described and those of wood and straw.... The fibre of straw is short, pointed, and polished, and cannot of itself make a strong paper. The nature of wood fibre depends, as may readily be supposed, upon the nature of the wood itself. Yellow pine, for example, yields a fibre long, soft, and flexible, in fact very like cotton; while oak and many other woods yield short circular fibres which, unless perfectly free from extraneous matters, possess no flexibility, and in any case are not elastic.

Micrographic Examination of Vegetable Fibres.—The importance of the microscope in the examination of the various fibres that are employed in paper manufacture will be readily evident from the delicate nature of the cellulose to be obtained therefrom.[4] Amongst others M. Girard has determined, by this method of examination, the qualities which fibres ought to possess to suit the requirements of the manufacturer. He states that absolute length is not of much importance, but that the fibre should be slender and elastic, and possess the property of turning upon itself with facility. Tenacity is of but secondary importance, for when paper is torn the fibres scarcely ever break. The principal fibres employed in paper-making are divided into the following classes:—

1. *Round, ribbed fibres*, as hemp and flax.
2. *Smooth*, or *feebly-ribbed fibres*, as esparto, jute, phormium (New Zealand flax), dwarf palm, hop, and sugar-cane.
3. *Fibro-cellular substances*, as the pulp obtained from the straw of wheat and rye by the action of caustic ley.
4. *Flat fibres*, as cotton, and those obtained by the action of caustic ley upon wood.
5. *Imperfect substances*, as the pulp obtained from sawdust.

In this class may also be included the fibre of the so-called "mechanical wood pulp."

Determination of Cellulose. For the determination of cellulose in wood and other vegetable fibres to be used in paper-making Müller recommends the following processes: [5] 5 grammes weight of the finely-divided substance is boiled four or five times in water, using 100 cubic centimètres[6] each time. The residue is then dried at 100° C. (212° Fahr.), weighed, and exhausted with a mixture of equal measures of benzine and strong alcohol, to remove fat, wax, resin, &c. The residue is again dried and boiled several times in water, to every 100 c.c. of which 1 c.c. of strong ammonia has been added. This treatment removes colouring matter and pectous[7] substances. The residue is further bruised in a mortar if necessary, and is then treated in a closed bottle with 250 c.c. of water, and 20 c.c. of bromine water containing 4 c.c. of bromine to the litre.[8] In the case of the purer bark-fibres, such as flax and hemp, the yellow colour of the liquid only slowly disappears, but with straw and woods decolorisation occurs in a few minutes, and when this takes place more bromine water is added, this being repeated until the yellow colour remains, and bromine can be detected in the liquid after twelve hours. The liquid is then filtered, and the residue washed with water and heated to boiling with a litre of water containing 5 c.c. of strong ammonia. The liquid and tissue are usually coloured brown by this treatment. The undissolved matter is filtered off, washed, and again treated with bromine water. When the action seems complete the residue is again heated with ammoniacal water. This second treatment is sufficient with the purer fibres, but the operation must be repeated as often as the residue imparts a brownish tint to the alkaline liquid. The cellulose is thus obtained as a pure white body; it is washed with water, and then with boiling alcohol, after which it may be dried at 100° C. (212° Fahr.) and weighed.

Recognition of Vegetable Fibres by the Microscope.— From Mr. Allen's admirable and useful work on "Commercial Organic Analysis"[9] we make the following extracts, but must refer the reader to the work named for fuller information upon this important consideration of the subject. In examining fibres under the microscope, it is recommended that the tissues should be cut up with sharp scissors, placed on a glass slide, moistened with water, and covered with a piece of thin glass. Under these conditions:—

Filaments of Cotton appear as transparent tubes, flattened and twisted round their axes, and tapering off to a closed point at each end. A section of the filament somewhat resembles the figure 8, the tube, originally cylindrical, having collapsed most in the middle, forming semi-tubes on each side, which give the fibre, when viewed in certain lights, the appearance of a flat ribbon, with the hem of the border at each edge. The twisted, or corkscrew form of the dried filament of cotton distinguishes it from all other vegetable fibres, and is characteristic of the matured pod, M. Bauer having found that the fibres of the unripe seed are simply untwisted cylindrical tubes, which never twist afterwards if separated from the plant. The matured fibres always collapse in the middle as described, and undergo no change in this respect when passing through all the various operations to which cotton is subject, from spinning to its conversion into pulp for paper-making.

Linen, or Flax Fibre, under the microscope, appears as hollow tubes, open at both ends, the fibres being smooth, and the inner tube very narrow, and joints, or *septa*,[10] appear at intervals, but are not furnished with hairy appendages as is the case with hemp. When flax fibre is immersed in a boiling solution of equal parts of caustic potash and water for about a minute, then removed and pressed between folds of filter-paper, it assumes a dark yellow colour, whilst cotton under the same treatment remains white or becomes very bright yellow. When flax, or a tissue made from it, is immersed in oil, and then well pressed to remove excess of the liquid, it remains translucent, while cotton, under the same conditions, becomes opaque.

New Zealand Flax (*Phormium tenax*) may be distinguished from ordinary flax or hemp by a reddish colour produced on immersing it first in a strong chlorine water, and then in ammonia. In machine-dressed New Zealand flax the bundles are translucent and irregularly covered with tissue; spiral fibres can be detected in the bundles, but less numerous than in Sizal. In Maori-prepared phormium the bundles are almost wholly free from tissue, while there are no spiral fibres.

Hemp Fibre resembles flax, and exhibits small hairy appendages at the joints. In Manilla hemp the bundles are oval, nearly opaque, and surrounded by a considerable quantity of dried-up cellular tissue composed of rectangular cells. The bundles are smooth, very few detached ultimate fibres are seen, and no spiral tissue.

Sizal, or Sisal Hemp (*Agave Americana*), forms oval fibrous bundles surrounded by cellular tissue, a few smooth ultimate fibres projecting from the bundles; is more translucent than Manilla, and a large quantity of spiral fibres are mixed up in the bundles.

Jute Fibre appears under the microscope as bundles of tendrils, each being a cylinder, with irregular thickened walls. The bundles offer a smooth cylindrical surface, to which the silky lustre of jute is due, and which is much increased by bleaching. By the action of hypochlorite of soda the bundles of fibres can be disintegrated, so that the single fibres can be readily distinguished under the microscope. Jute is coloured a deeper yellow by sulphate of aniline than is any other fibre.

CHAPTER II.

MATERIALS USED IN PAPER-MAKING.

Raw Materials.—Rags.—Disinfecting Machine.—Straw.—Esparto Grass.—Wood.—Bamboo.—Paper Mulberry.

In former days the only materials employed for the manufacture of paper were linen and cotton rags, flax and hemp waste, and some few other fibre-yielding materials. The reduction of the excise duty, however, from 3d. to 1½d. per lb., which took effect in the first year of Her Majesty's reign—namely, in 1837—created a greatly increased demand for paper, and caused much anxiety amongst manufacturers lest the supply of rags should prove inadequate to their requirements. Again, in the year 1861 the excise duty was totally abolished, from which period an enormously increased demand for paper, and consequently paper material, was created by the establishment of a vast number of daily and weekly papers and journals in all parts of the kingdom, besides reprints of standard and other works in a cheap form, the copyright of which had expired. It is not too much to say, that unless other materials than those employed before the repeal of the paper duty had been discovered, the abolition of the impost would have proved but of little service to the public at large. Beneficent Nature, however, has gradually, but surely and amply, supplied our needs through the instrumentality of man's restless activity and perseverance.

The following list comprises many of the substances from which cellulose, or vegetable fibre, can be separated for the purposes of paper-making with advantage; but the vegetable kingdom furnishes in addition a vast number of plants and vegetables which may also be used with the same object. We have seen voluminous lists of fibre-yielding materials which have been suggested as suitable for paper-making, but since the greater portion of them are never likely to be applied to such a purpose, we consider the time wasted in proposing them. It is true that the stalks of the cabbage tribe, for example, would be available for the sake of their fibre, but we should imagine that no grower of ordinary intelligence would deprive his ground of the nourishment such waste is capable of *returning to the soil*, by its employment as manure, to furnish a material for paper-making. Again, we have seen blackberries, and even the pollen (!) of plants included in a list of paper materials, but fortunately the manufacturer is never likely to be reduced to such extremities as to be compelled to use materials of this nature.

Raw Materials.

Cotton rags.
Cotton wool.
Cotton waste.
Cotton-seed waste.
Linen rags.
Linen waste.
Hemp waste.
Manilla hemp.
Flax waste, etc.
Jute waste, etc.
China grass.
Bamboo cane.
Rattan cane.
Banana fibre.
Straw of wheat, etc.
Rushes of various kinds.
New Zealand flax.
Maize stems, husks, etc.
Esparto grass.
Reeds.
Woods of various kinds, especially white non-resinous woods, as poplar, willow, etc.
Wood shavings, sawdust, and chips.
Old netting.
Sailcloth.
Sea grass (*Zostera marina*).
Fibrous waste resulting from pharmaceutical preparations.
Potato stalks.
Stable manure.
Barks of various trees, especially of the paper mulberry.
Peat.
Twigs of common broom and heather.
Mustard stems after threshing.
Buckwheat straw.
Tobacco stalks.
Beetroot refuse from sugar works.
Megass, or "cane trash"—refuse of the sugar cane after the juice has been extracted.
Fern leaves.
Tan waste.
Dyers' wood waste.
Old bagging.
Old bast matting.
Hop-bines.
Bean stalks.
Old canvas.
Old rope.
Gunny bags.
Waste paper.
Binders' clippings, etc.
Silk cocoon waste.
Oakum.
Flax tow.
Rag bagging.
Leather waste.
Tarpaulin. Etc., etc.

Rags.—Linen and cotton rags are imported into Great Britain from almost all the countries of Europe, and even from the distant states of South America, British South Africa, and Australasia. The greater proportion, however, come from Germany. The rags collected in England chiefly pass through the hands of wholesale merchants established in London, Liverpool, Manchester, and Bristol, and these are sorted to a certain extent before they are sent to the paper-mills. By this rough sorting, which does not include either cleansing or disinfecting, certain kinds of rags which would be useless to the paper-maker are separated and sold as manure. Woollen rags are not usually mixed with cotton rags, but are generally kept apart to be converted into "shoddy." The importance of disinfecting rags before they pass through the hands of the workpeople employed at the paper-mills cannot be over-estimated, and it is the duty of every Government to see that this is effectually carried out, not only at such times when cholera and other epidemics are known to be rife in certain countries from which rags may be imported, but at all times, since there is no greater source of danger to the health of communities than in the diffusion of old linen and cotton garments, or pieces, which are largely contributed by the dwellers in the slums of crowded cities.

Respecting the disinfecting of rags, Davis[11] thus explains the precautions taken in the United States to guard against the dangers of infection from rags coming from foreign or other sources. "When cholera, or other infectious or contagious diseases exist in foreign countries, or in portions of the United States, the health officers in charge of the various quarantines in this country require that rags from countries and districts in which such diseases are prevalent shall be thoroughly disinfected before they are allowed to pass their stations. Rags shipped to London, Hull, Liverpool, Italian, or other ports, and re-shipped from such ports to the United States, are usually subjected to the same rule as if shipped direct from the ports of the country in which such diseases prevail. It is usually requisite that the disinfection shall be made at the storehouse in the port of shipment, by boiling the rags several hours under a proper degree of pressure, or in a tightly-closed vessel, or disinfected with sulphurous acid, which is evolved by burning at least two pounds of roll sulphur to every ten cubic feet of room space, the apartment being kept closed for several hours after the rags are thus treated. Disinfection by boiling the rags is usually considered to be the best method. In the case of rags imported from India, Egypt, Spain, and other foreign countries where cholera is liable to become epidemic, it is especially desirable that some efficient, rapid, and thorough process of disinfecting should be devised. In order to meet the quarantine requirements, it must be thorough and certain in its action, and in order that the lives of the workmen and of others in the vicinity may not be endangered by the liberating of active disease-germs, or exposure of decaying and deleterious matters, and that the delay, trouble, and exposure of unbaling and rebaling may be avoided, it must be capable of use upon the rags while in the bale, and of doing its work rapidly when so used."

Disinfecting Machine.—To facilitate the disinfecting of rags while in the bale, Messrs. Parker and Blackman devised a machine, for which they obtained a patent in 1884, from which the following abstract is taken.

Formerly rags and other fibrous materials were disinfected by being subjected to germ-destroying gases or liquids in enclosed chambers, but in order to render the disinfecting process effectual, it was found necessary to treat the material in a loose or separated state, no successful method having been adopted for disinfecting the materials while in the bale. "This unbaling and loosening or spreading of the undisinfected material is absolutely unsafe and dangerous to the workmen, or to those in the vicinity, because of the consequent setting free of the disease germs, and the exposing of any decaying or deleterious matters which may be held in the material while it is compressed in the bale. The unbaling and necessary rebaling of the material for transportation also involves much trouble and expense and loss of time. Large and cumbrous apparatus is also necessary to treat large quantities of material loosened or opened out as heretofore."

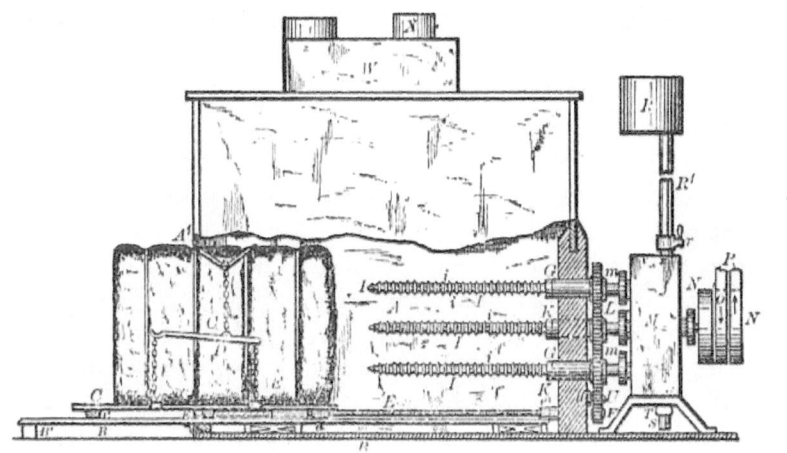

Fig. 1.

It is specially necessary that rags coming from Egypt and other foreign countries should be thoroughly disinfected by some rapid and effectual means, which, while not endangering the health of workmen employed in this somewhat hazardous task, will fully meet all quarantine requirements. The apparatus devised by Messrs. Parker and Blackman,[12] an abridged description of which is given below, will probably accomplish this much-desired object.

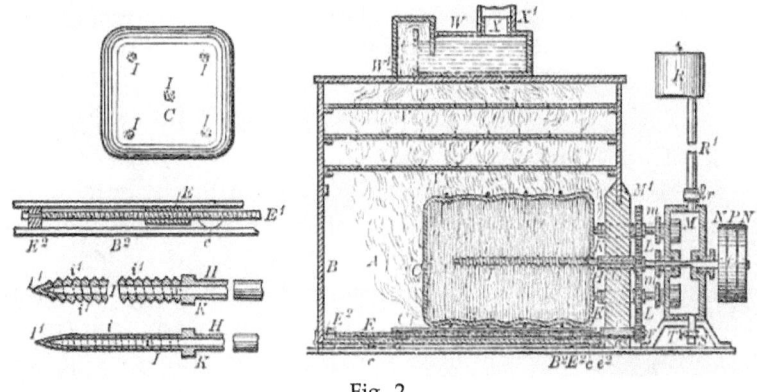

Fig. 2.

In the illustration, Fig. 1, A is the disinfecting chamber. At

31

one end is an opening A^1, and a door B, hinged at its lower edge and adapted to be swung up, so as to close the opening tightly. For supporting and carrying the bale C of material to be placed in the chamber is a carriage C^1, consisting of a platform supported upon wheels or castors $c\ c$. While the carriage is wholly within the chamber A, as shown in Fig. 2, these wheels rest upon the false bottom B^2; when the carriage is rolled back and out of the chamber, as shown in Fig. 1, they roll upon the upper face of door B swung down. The carriage is provided with a clamping device D to hold the bale firmly and immovably. To cause the carriage to move into and out of the chamber, the inventors provide upon the under side of the platform a fixed sleeve E, interiorly threaded to fit the screw E^1, journalled at one end near the opening in the chamber end in a stationary block E^2 fixed upon the false bottom B^2. From this end the screw extends along under the carriage through the screw sleeve and to the other end of the chamber. A collar e^2 on the screw bears against the inner end of this journal-bearing, and upon the end of the shank e bearing against the other end of the journal is fixed a pinion F, which is to be driven in either direction as desired. Above this journal-bearing is a series of similar bearings (five being shown), G G, passing through the wall of the chamber. Of these the middle one is in a line with the centre of the bale, supported and held on the carriage. The others are arranged at the corners of a square. Journalled in these bearings are the hollow shanks H H of the hollow screws I I pointed at $I^1\ I^1$. Each screw is perforated, $i\ i$, between the threads $i^1\ i^1$ from the fixed collar K K. Upon the tubular shanks H H of the screws are fixed the gear-wheels L L. At a short distance from the end of the chamber, A is the hollow chamber or receptacle M, into which is to be forced the disinfectant liquid or gas. The tubular shanks H H of the screws project through the wall M, passing through stuffing-boxes $m\ m$, and their bores communicate with the interior of the chamber, the shank of the middle screw being continued through the opposite wall and a stuffing-box, its solid or projecting end being provided with two fixed pulleys, N N, and a loose pulley O.

When a gaseous disinfectant is used, it can be forced by any desired means through the pipe S into the chamber. Where a liquid disinfectant is used, an elevated tank R containing the fluid may be used. As most fibrous materials, and especially rags, are baled so as to be in layers, it is preferable so to place the bale upon the carriage that the perforated screws may penetrate the material at right angles to the layers by which the gas or liquid issuing through the holes in the screws passes in all directions throughout the mass within the bale.

In the upper part of chamber A are perforated shelves V V, upon which, if desired, the material can be spread out and subjected to disinfecting gas or vapour. On the top of the chamber is a tank W nearly filled with disinfecting liquid. A passage W^1 extends from upper part of the chamber up into the tank above the level of the liquid therein, and is then carried at its end down below the surface of the liquid. At its other end the tank is provided at its top with a discharge opening X and a suitable pipe X^1, forming a continuation of the opening; by this means all foul and deleterious vapours or gases passing out of the closed chamber A through the passage W must pass through the disinfecting liquid in the tank before escaping through the opening X and stack X^1 into the air, and are thus rendered harmless.

When a sufficient amount of the disinfectant has been forced into and through the bale, the disinfectant is turned off, and cold dry air can be forced through chamber M, and out through the nozzles and bale, whereby the material within the bale becomes cooled and dried, and all the foul air from the chamber A driven out, so that it may be opened and entered with safety. Any suitable disinfectant may be used with this apparatus, as, for example, sulphurous acid, in gas or solution, superheated steam, carbolic acid, or any solution or vapour containing chlorine.

Straw.—Very large quantities of this material are used in the manufacture of paper, but more especially for newspapers, the straw from wheat and oats being mostly employed. Although the percentage of cellulose in straw is about equal to that of esparto, the severe treatment it requires to effectually remove the silicious coating by which the fibre is protected, and to render the knots amenable to the action of the bleach, greatly reduces the yield of finished pulp. Many processes have been introduced for the treatment of straw for paper-making, but the most successful of them appear to be modifications of a process introduced in 1853 by MM. Coupier and Mellier.

Esparto Grass.—This important fibrous material is largely imported from Algeria, Spain, and other countries, and constitutes one of the most valuable fibre-yielding materials with which the manufacturer has to deal. Some idea of the amount of esparto and other fibres which find their way to our shores may be gleaned from the fact that while the import of cotton and linen rags in the year 1884 was 36,233 tons, of the value of £487,866, that of esparto and other fibres amounted to 184,005 tons, of the value of £1,125,553.

Wood.—As a paper-making material, the fibre obtained from various kinds of wood now holds an important position, since the sources of supply are practically inexhaustible. The first practical process for manufacturing pulp from wood fibre was perfected and introduced by the author's father, the late Mr. Charles Watt, who, in conjunction with Mr. H. Burgess, obtained a patent for the invention on August 19th, 1853. The process was afterwards publicly exhibited at a small works on the Regent's Canal, when the Earl of Derby (then Lord Stanley), many scientific men and representatives of the press, were present, and expressed themselves well satisfied with its success. Specimens of the wood paper, including a copy of the *Weekly Times* printed thereon, were exhibited, as also some water-colour drawings which had been produced upon paper made from wood pulp. Failing to get the process taken up in England, an American patent was applied for and obtained in 1854, which was subsequently purchased; but with the exception of an instalment, the purchase-money was never paid to the inventor! Thus the process "got" into other hands, the original inventor alone being unbenefited by it.

It has been repeatedly stated,[13] no doubt unwittingly, that a person named Houghton first introduced the wood paper process into this country; but considering that his patent was not obtained until 1857, or four years after the process above referred to was patented and publicly exhibited in England, it will be seen that the statement is absolutely without foundation. The first knowledge Mr. Houghton received concerning wood as a paper-making material was from the author's father, and he (Mr. Houghton), in conjunction with Mr. Burgess, introduced the Watt and Burgess process into America in the year 1854. These are the facts.

Bamboo (*Bambusa vulgaris*).—The leaves and fresh-cut stems of this plant are used for paper material, but require to pass through a preliminary process of crushing, which is effected by suitable rolls, the second series of crushing rolls being grooved or channelled to split or divide the material, after which the stems are cut to suitable lengths for boiling.

Paper Mulberry (*Broussonetia papyrifera*).—The inner bark of this tree, and also some other basts, have long been used by the Japanese and Chinese in the manufacture of paper of great strength, but of extreme delicacy.

CHAPTER III.

TREATMENT OF RAGS.

Preliminary Operations.—Sorting.—Cutting.—Bertrams' Rag-cutting Machine.—Nuttall's Rag-cutter.—Willowing.—Bertrams' Willow and Duster.—Dusting.—Bryan Donkin's Duster or Willow.—Donkin's "Devil."

Preliminary Operations.—Before the rags are submitted to the various processes which constitute the art of paper-making, they are subjected to certain preliminary operations to free them from dirty matters, dust, and even sand, which is sometimes fraudulently introduced into rags to increase their weight. This preliminary treatment may be classified under the following heads, namely:—Sorting; Cutting; Willowing; Dusting.

Sorting.—The rags being removed from the bags or bales in which they are packed, require first to be sorted according to the nature and quality of the fabrics of which they are composed; thus linen, cotton, hemp, wool, &c., must be carefully separated from each other; the thickness of the substance, its condition as to the wear it has undergone, and the colour of the material, all these considerations are taken into account by the women and girls who are employed in the operation of sorting. The finer qualities are set aside for writing-paper, inferior sorts being used separately, or mixed, according to the requirements of the manufacturer. Blue rags are generally separated from the rest and kept for the manufacture of blue paper, but most of the other coloured rags require bleaching. In sorting rags, a good deal of judgment and skill are required to avoid mixing the better qualities with those of an inferior class, which would occasion loss in the manufacture. It is also important that those of inferior colour should not be mixed with the finer qualities, which would be liable to affect the colour and deteriorate the quality of the paper. Paper manufacturers generally classify the rags obtained from home sources, that is, from different parts of the United Kingdom, under the following heads:—

Home Rags.

New cuttings.
Linen pieces.
Cotton pieces.
Fines (whites).
Superfines (whites).
Outshots (whites).
Seconds (whites).
Thirds (whites).

Colours or prints.
Blues.
Gunny, clean.
Gunny, dirty.
Rope (white).
Rope (hard).
Rope, bagging, etc.

Foreign rags are distinguished as below:—

Belgian Rags.

White linens.
Mixed fines (linens and cottons).
Grey linens.
Strong linens.
Extra fine linens.
Blue linens.
Superfine white cottons.
Outshot cottons.
Seconds.

Half jute and linen.
Light prints.
Mixed prints.
Blue cottons.
Fustians.
Black calicoes.
White hemp, strings, and rope.
Tarred hemp, strings, and rope.
Jute spinners' waste.
Jute waste.

New.

White linens.
Grey linens.
Blue linens.
Unbleached cottons.
White linens and cottons.

Print cuttings (free from black).
Blacks.
Fustians.

French Rags.

French linens.
White cotton.
Knitted cotton.
Blue cotton.
Coloured cotton.

Black cotton.
Marseilles whites.
Light prints.
Mixed prints.
New white cuttings.

German Rags.

S. P. F. F. F.
S. P. F.
F. F.
F. G.
L. X. F.

L. F. R. blue.
C. S. P. F. F. F.
C. F. B. blue.
C. F. X. coloured.

Trieste.

P.P. white linen (first).
P. white linen (second).

S. fine greys.
X. coloured cottons.

Leghorn.

P. L. linens.　　　　　　　　S. C.
P. C. cottons.　　　　　　　　T. C.

Turkey and Beyrout.
Bright reds.

Alexandria.
Whites.　　　　　Blues.　　　　　Colours.

Baltic and Russian.

S. P. F. F.　　　　　　　　　F. F.
S. P. F.　　　　　　　　　　B. G.
L. F. B.　　　　　　　　　　L. F. X.
F. G.

Woollen rags are only used to a very moderate extent in blotting and filtering papers and also in coarse papers and wrappers. Many attempts have been made to bleach woollen rags, but the severity of the treatment required invariably ended in a destruction of the fibrous substances mingled with them. It is customary to dispose of such material for re-making into common cloths, and for shoddy. Rags collected in large cities, in consequence of the frequent bleachings they have been subjected to, are considerably weakened in fibre, tearing easily, and are therefore subject to loss in process of manufacture into pulp. Country rags, being coarser and greyer because less bleached, are stronger in fibre and give a better body to the paper. In sampling rags it is necessary to take precautions against the fraudulent "tricks of the trade," which are often resorted to to cheat the manufacturer. Samples should be taken from the interior of the bags or bales, to ascertain if the material in the interior is equal in quality with that at the outside—that is to say, that the quality is fairly averaged throughout. It may also be found that the rags have been purposely wetted to increase their weight. If such is found to be the case, a few handfuls should be weighed, and then dried in a warm room, and afterwards re-weighed, when if the loss exceeds 5 to 7 per cent. it may be assumed that the rags have been fraudulently wetted. It is generally found, however, that the merchants in the principal towns transact their business honourably and are therefore reliable.

The sorting is generally performed by women, who not

only separate the various qualities of the rags, which they place in separate receptacles, but also remove all buttons, hooks and eyes, india-rubber, pins and needles, &c., and loosen all seams, hems and knots. The rags are next carefully looked over by women called *over-haulers*, or over-lookers, whose duty it is to see that the previous operations have been fully carried out in all respects. Usually there is one over-hauler to every eight or ten *cutters*.

Cutting.—In some mills it is preferred to have the rags cut into pieces from 2 to 4 inches square, but the actual size is not considered of much importance. The chief object is to have them in such a condition that they may be thoroughly cleansed in subsequent operations, and able to float throughout the water in the rag-engine, without twisting round the roller. If the rag pieces are smaller than is required to effect this it tends to create a loss of fibre in the operations of willowing and dusting.

The process of cutting is performed by hand or by machinery. When the rags are cut by hand, the operation, which is accomplished by women, is conducted as follows:—The cutter takes her place in front of an oblong box, as in Fig. 3, covered with coarse wire netting, containing three threads per inch, through which dust, &c., passes to a receptacle beneath; in the centre is fixed, in a slanting position, a large-bladed knife of peculiar form, with its back towards the operator, who is surrounded by a number of boxes, corresponding with the number of the different qualities of rags; these are lined at the bottom with coarse wire gauze. In the operation of cutting, if any foreign

Fig. 3.

substances, such as buttons, hooks, &c., which may have escaped the sorters are found, these are at once removed. The rags as they are cut are put into baskets to be conveyed to the rag-engine room. In some mills rags are cut by machinery, but hand cutting is usually adopted for the better kinds of paper, as it is obvious that the machine would not be able to reject, as is the case in hand cutting, unpicked seams and other irregularities which may have escaped observation by the sorters and overhaulers. Machine cutting is, therefore, generally adopted for the materials which are to be used for the coarser papers. There are several rag-cutting machines in use, of which one or two examples are given below.

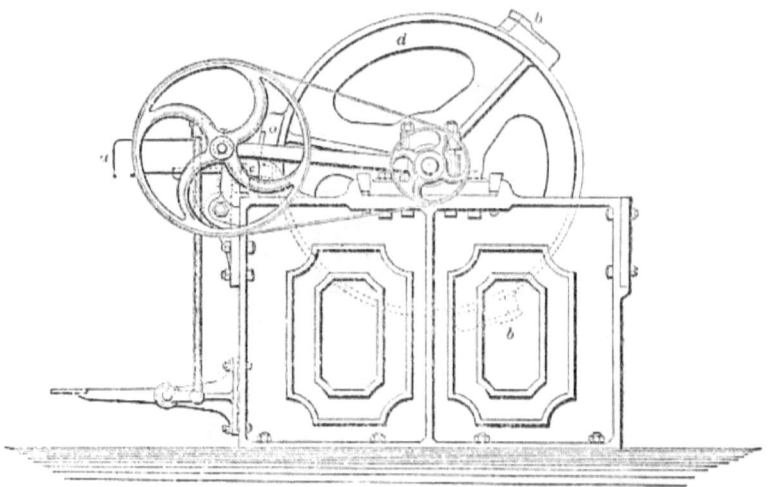

Fig. 4.

Bertrams' Rag-Cutting Machine.—The engraving, Fig. 4, represents a machine manufactured by Messrs. Bertrams, Limited, of St. Katherine's Works, Edinburgh, to whose courtesy we are indebted for this and other illustrations of their machinery, which have been reproduced in outline from their illustrated catalogue. The machine, which is suitable either for rags or ropes, has three revolving knives, and one dead knife, which is rendered reversible to four edges, and has self-acting feed gear, side frames, drum, and other connections of substantial construction; it is wood covered, and furnished with sheet-iron delivery spout. The material passes into the machine along the table at *a*, where it passes between the dead knife *c* and the knives *b* fixed to the revolving drum *d*. The cut rags fall into a receptacle beneath the drum.

Nuttall's Rag Cutter.—Another type of rag cutter, and which is also suitable for cutting bagging, sailcloth, tarpaulin, Manilla and other fibres, is Nuttall's Rag Cutter, a drawing of which is shown in Fig. 5. This machine is manufactured by Messrs. Bentley and Jackson, of Bury, near Manchester, and is generally known as the "Guillotine Rag Cutter," from the principle of its action, which is that of chopping the material. The machine is adopted at many mills, and a large-sized machine has recently been put down at the *Daily Telegraph* mills, Dartford. A medium-sized machine will cut about one ton of rags in an hour.

Fig. 5.

Willowing.—In some mills the cut rags are conveyed to a machine called the "willow," which in one form of machine consists of two cast-iron cylinders, 2½ feet in diameter and 3½ feet wide, provided with numerous iron teeth, which project about 4 inches. These cylinders are placed one behind the other, and beneath them is a semi-circular screw, and above them a cover of the same form. This cover is also furnished with teeth, and is so adjusted that the teeth in the cylinders pass those in the cover at a distance of ½ to ¾ of an inch. In front are a pair of rollers and revolving apron, which carry the rags into the cylinders, which rotate rapidly; and the rags, which are thrown by the first into the second cylinder, are allowed to remain in them for about 20 seconds, when a sliding door, which rises three times per minute, allows the rags to be discharged into a duster. Each time the sliding door opens the revolving apron moves forward and recharges the willow with a fresh supply. The rags, after being beaten and teazed in the willow, are considerably loosened in texture, and a good deal of dust and gritty matters fall through the screen beneath.

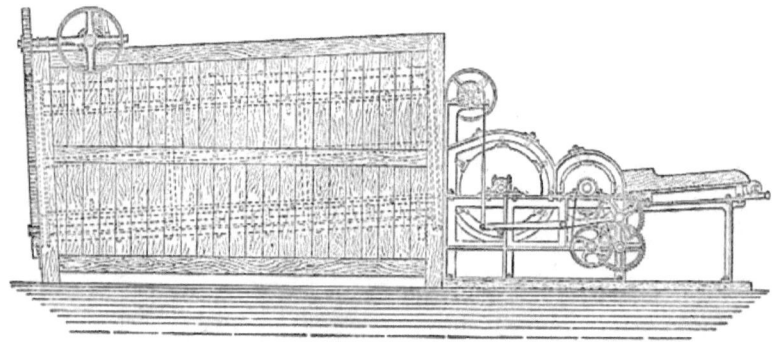

Fig. 6.

Fig. 6 represents a combined willow and duster, specially useful for waste rags and jute, but may be used for all fibres, manufactured by Bertrams, Limited, the main features of which are thus described:—"There are two drums, which have malleable-iron cross-bars and teeth, and malleable-iron harp motion below for escape of dust. The framework of the willow is of cast iron, and the sides are filled in with cast-iron panel doors, the top being covered in with sheet iron. The gear is arranged so that the willow will deliver to the duster or otherwise by self-acting motion continuously or intermittently. The feed to the willow can also be made continuous or intermittent. The drums, framework, panels, and casing being made of iron, the chance of fire from the friction of its working is reduced to a minimum. The duster, as a rule, is 12 feet long, about 5 feet in diameter, and has eight longitudinal bars of cast iron fitted between the front and end revolving rings. These bars are fitted with malleable-iron spikes, pitched and so arranged that the rags or fibres are delivered at the exit end automatically. The outside of the duster can be lined with wire-cloth, perforated zinc, iron, etc. It is driven by outside shafts and friction gear, so that there is no internal shaft to interfere with the delivery of the fibres."

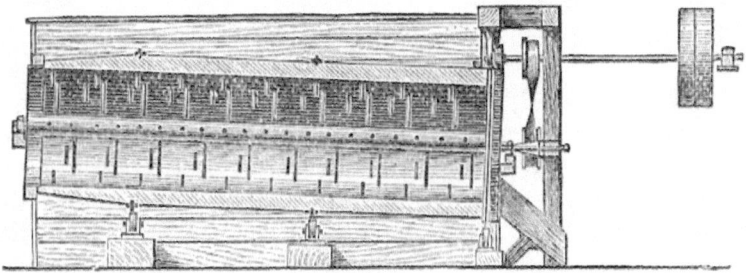

Fig. 7.

Dusting.—In Fig. 7 is shown a rag-dusting machine, manufactured by Messrs. Bryan Donkin and Co., of Bermondsey, London. The cylinder of this machine, which is conical in form, to enable the rags to travel from one end to the other, whence they are ejected, revolves, as also does a second cylinder of a skeleton form, but in the opposite direction. Each cylinder is fitted with knives, or spikes—those of the outer cylinder projecting towards the centre; the knives of the centre cylinder being attached to its exterior surface: when the machine is in motion the two sets of blades pass each other so that when the rags come between them the action is that of scissors. When the rags are ejected at the end of the cylinder, they pass into another cylinder of wire, through which the dust falls and leaves them in a fairly clean condition, when they are lowered through a trap-door to the boiling room below.

Fig. 8.

Donkin's "Devil."—For removing the dust and dirt from coarse and very dirty rags, oakum, rope, etc., the presence of which would seriously injure the quality of the paper, a still more powerful machine has been introduced, called the "devil," which is constructed on the same principle as the willow, but revolves at a lower speed. The revolving axle of this machine is conical, and is provided with teeth, arranged in a spiral form. The case in which it rotates is fed continuously, instead of intermittently; and although it facilitates the subsequent treatment of the fibre, it is said to be wasteful, while also consuming a considerable amount of power. A machine, or "devil," for cleaning rags or half stuff is manufactured by Messrs. Donkin and Co., a representation of which is shown in Fig. 8.

CHAPTER IV.

TREATMENT OF RAGS (continued).

Boiling Rags.—Bertrams' Rag Boiler.—Donkin's Rag Boiler.—Washing and Breaking.—Bertrams' Rag Engine.—Bentley and Jackson's Rag Engine.—Draining.—Torrance's Drainer.

Boiling Rags.—To remove greasy matters, and also to dissolve out the cementing substances from the stems of flax and shell of the cotton, the rags are next boiled in a solution of caustic soda, caustic lime, or a mixture of carbonate of soda and lime. The boiling has also the effect of loosening the dirt contained in the rags, whereby the colour of the material is greatly improved, while at the same time it is rendered more susceptible to the action of the bleaching agent. Strong linen rags will sometimes lose from one-third to one-fifth of their weight by the process of boiling. The vessels for boiling rags are of various construction, and have been the subject of numerous ingenious patents. These boilers are either cylindrical or spherical, and are also stationary or rotary—the latter form being devised for the purpose of keeping the caustic alkali solution freely diffused throughout the mass of fibre during the boiling.

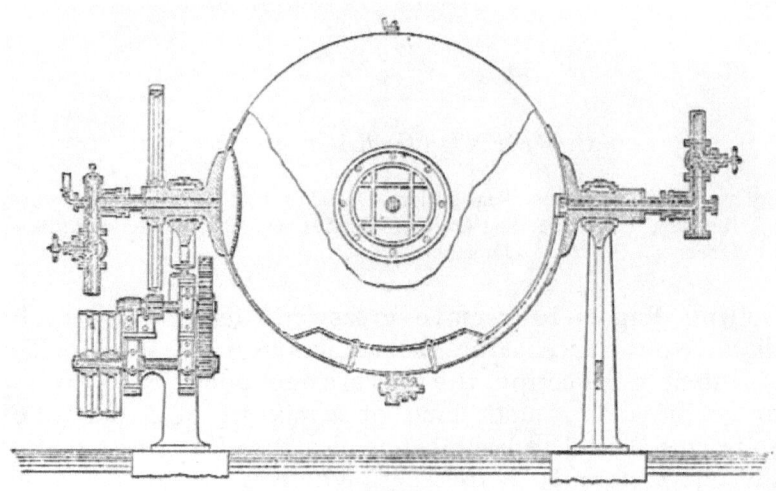

Fig. 9.

Bertrams' Rag Boiler.—An illustration of a spherical boiler, as manufactured by Bertrams, Limited, of Edinburgh, is given in Fig. 9. The shell of this boiler is made from malleable iron, is 8 feet in diameter and 9 feet deep. The boiler is constructed on what is termed the "vomiting" principle, by which a free circulation of the alkaline liquor is constantly maintained. These boilers are made to withstand any pressure of steam, but the size given is usually worked at from 35 to 45 lbs. pressure, and carries about 30 cwt. of dry esparto.

Fig. 10.

Donkin's Rag Boiler.—The spherical boiler of Messrs. Bryan Donkin and Co. is shown in Fig. 10. Being of a spherical form, it is twice as strong as a cylindrical boiler of the same diameter and thickness. The plates used are, notwithstanding, of the usual substance, thus rendering it perfectly safe, durable, and suitable for high-pressure steam. The spherical shape also allows the rags to fall out by themselves when the boiler is revolving with the cover off. Within the boiler are strainers to carry off the dirt, and lifters to agitate the rags during the process of either boiling or washing. To avoid cement, or even lead joints, the gudgeons and the boiler are turned true in the lathe to fit each other, the joints being simply made with red lead. These boilers are usually about 8 feet in diameter, and are capable of boiling from 20 cwt. to 25 cwt. of rags. The idea of giving motion to the boiler, so as to insure a perfect mixture of the rags and the caustic liquor, is of American origin, and was first introduced into this country by Messrs. Bryan Donkin and Co. It is usual to fix the boiler so that it can be fed with rags through a trap in the floor above, while the boiler is in a vertical position and the lid removed. The trunnions are hollow, to admit the introduction of steam, alkaline ley, or water, and its rotary motion, which is about three times in two minutes, is given by the gearing on the left of the illustration.

The alkalies used for boiling rags are either caustic soda, soda ash, slaked lime, made into a cream and sifted, or a mixture of slaked lime and carbonate of soda. A description of the preparation of caustic soda ley will be found in another chapter. It has been customary at most of the larger paper-mills to purchase their caustic soda direct from the alkali manufacturers, who supply it in a solid form enclosed in iron drums, hermetically closed, which are broken and the contents removed and dissolved when required for use. As to the strength of caustic soda liquor to be used for boiling rags, this is regulated according to the nature and condition of the material, and the quality of the paper it is intended for (see p. 34). For the finest papers the caustic soda should be perfectly pure, and as there are various grades of this chemical substance sold by the alkali makers, only the purer qualities are used for the better kinds of paper. The proportion of caustic soda per cwt. of rags varies to the extent of from 5 to 10 per cent. of the former to each cwt. of the latter, the coarser materials, of course, requiring more alkali than those of finer quality. In cases where rags are boiled in an open boiler—as was formerly the case—a much larger proportion of caustic soda would be required than when the boiling is conducted under high pressures, as is now very generally the custom. In boiling the finer qualities of rags, less pressure of steam is required than for the coarser qualities, and the heat being proportionately lower, there is less destruction of the fibre. Some paper-makers prefer to boil the rags with caustic lime only, in which case the lime, after being slaked in the usual way, is mixed with water until it attains a milky consistence, when it is passed through a sieve to separate any solid particles which may be present. About the same percentage of lime may be used as in the former case.

When a mixture of lime and carbonate of soda is used, a method much adopted on the Continent, the lime should be well screened from lumps before being mixed with the soda. The usual method of preparing this mixture is as follows:— A wooden tank, 15 feet long, 5 feet wide, and 4 feet deep is divided into three compartments, each of which has a false bottom perforated with ½-inch holes to keep back lumps, stones, pieces of coal, etc., which frequently abound in the lime. The fresh lime is put into the first compartment, where it is slaked with water in the usual way; the resulting powder is then put into the next compartment together with sufficient water, where it is agitated until converted into what is technically termed "milk of lime." In the partition which separates the second from the third division is a movable sluice, through which the milk of lime flows into the third compartment; in this is fitted a revolving drum, similar to the drum-washer of the breaking-engine, through which the milk of lime which flows from the sluice becomes strained, and is lifted in the same way as water is lifted by the drum-washer of the breaking-engine, and is thence discharged through a pipe into the rag boilers; an additional straining can be effected by placing a fine wire strainer over the mouth of this pipe leading to the boiler, which will prevent objectionable particles from entering the boiler. Each compartment is provided with a large waste pipe, through which, by the aid of a sufficient supply of water, all impurities which have been rejected by the drum are carried away. The soda solution is prepared by dissolving the required proportion in water, and the resulting liquor, after careful straining, is introduced into the boiler to which the charge of rags has been given; the head of the boiler is then fixed in its position and steam turned on, until a pressure of about 20 to 30 lbs. to the square inch is attained, and the boiling kept up for two to six hours, according to the quality of the rags. By the Continental system of boiling rags, for No. 1 stuffs, 216 lbs. of lime and 114 lbs., of 48 per cent., soda ash are used for every 4,000 lbs. of rags; for Nos. 3 and 5 stuffs, 324 lbs. of lime and 152 lbs. of soda ash are used; and for No. 4 stuff 378 lbs. of lime and 190 lbs. of soda ash, and the boiling in each case is kept up for twelve hours, under a pressure of 30 lbs., the operation being conducted in boilers which revolve horizontally.

In boiling the finest qualities of rags, it is considered preferable to boil with lime alone, which is believed to be less injurious to delicate fibres than caustic soda. Dunbar[14] gives the following proportions of 70 per cent. caustic soda per cwt. of rags:—

S. P. F. F. F. is boiled with lime alone, then washed in the boiler, and again boiled with 2 per cent. of soda ash.

				lbs. of	(70 per cent.)	caustic soda	per cwt.
S. P. F. F. is	boiled	with	12				
S. P. F	"	"	14	"	"	"	"
Fines	"	"	7	"	"	"	"
Seconds	"	"	6	"	"	"	"
L. F. X.	"	"	20	"	"	"	"
C. L. F. X.	"	"	27	"	"	"	"
C. C. L. F. X.	"	"	30	"	"	"	"
F. F.	"	"	15	"	"	"	"

These are all boiled at a pressure of from 20 to 25 lbs. for 10 hours, in stationary boilers without vomit, and also in boilers revolving horizontally. In some mills, where the best qualities of paper are made, iron boilers are objected to, as small particles of oxide of iron are apt to become dislodged from the interior of the boiler, and produce discolouration of the paper. In such cases wooden vats, with mechanical stirrers, are employed; sometimes a jacketed boiler is used.

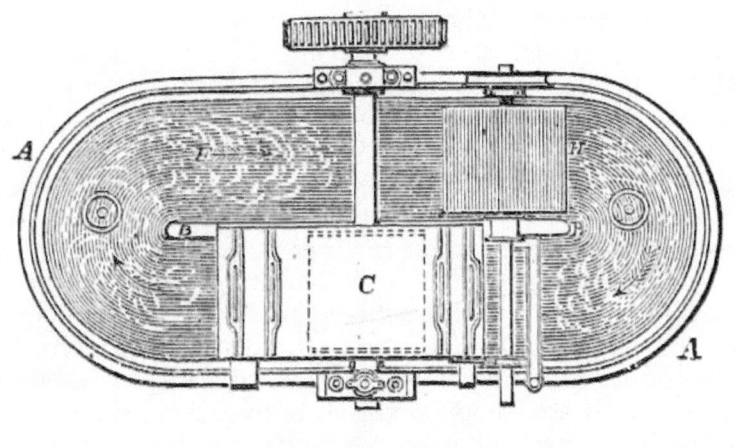

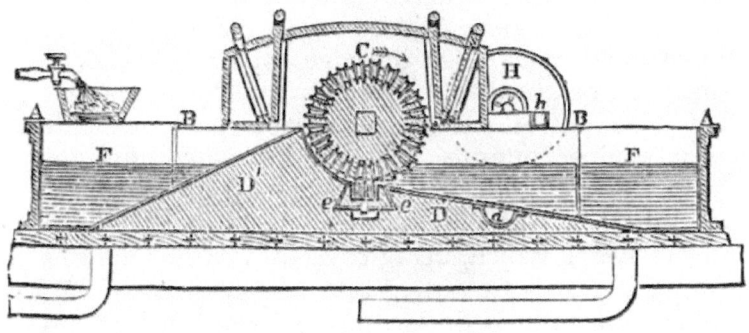

Figs. 11 and 12.

Washing and Breaking.—The removal of the dirty water resulting from the boiling is effected in the washing and breaking engine, or "rag engine," as it is commonly called, which is constructed on the same principle as the beating engine, but is provided with an extra drum, called the *drum-washer*, which, being covered with wire gauze, allows the washing waters to escape without permitting the fibrous stuff to pass through. The rag engine, having been invented by a Dutchman, acquired, and still retains, the name of the *Hollander*, and although it has been considerably improved upon, its principle is still retained in the modern engines, of which there are many different forms. The ordinary rag engine, Figs. 11 and 12, consists of a cast-iron trough A, about 10 feet long, 4½ feet wide, and 2½ feet deep, and rounded at the ends, and is firmly bolted to a wooden foundation. It is provided with a partition termed the *midfeather* B, of such a length as to have the trough of uniform width round it. A cylinder, or *roll*, C, furnished with a series of steel knives, rotates in one of the divisions formed by the midfeather, and the floor of the trough in this division is inclined in such a manner as to cause the pulp, as it travels, to pass under the roll. Beneath the roll is the *bed-plate*, which is fitted with a series of steel knives $c\,c$ similar to those on the exterior of the roll. The distance between the knives of the roll and the bed-plate is regulated by levelling screws, which are so adjusted that both ends of the roll are raised at the same time, which is a great improvement upon the older types of breaking engines in which only one end of the roll was raised, whereby the knives became unequally worn. By the present method of regulating the distance between the respective sets of knives, any required degree of fineness can be given to the fibrous substances treated. The roll is generally caused to rotate at a speed of about 230 revolutions per minute, causing the water and rags to circulate in the engine and to be constantly under the action of the knives. In the other division F F of the trough is the drum-washer H, which, being covered with fine gauze wire, allows the water to enter, but keeps back the fibrous material. The ends of the drum are formed of two discs of wood, generally mahogany, upon which the coarse gauze is fastened as a backing, and this is covered with the fine wire gauze. The interior of the drum is sometimes furnished with a series of buckets, which conduct the water to a trough in the axis of the drum, by which it is led away. This is also accomplished by dividing the interior of the cylinder into compartments by means of a partition. The drum-washer is so arranged that it can be wholly raised out of the trough, which is necessary in certain parts of the operation, when the removal of the liquid is not required; or it can be partially raised, or otherwise, according to requirement. The floor of the compartment containing the roll C is inclined at D, so as to cause the pulp to pass directly under the roll, and at D' is the *backfall*, over which the pulp travels to the opposite side of the midfeather.

In working the rag engine, it is first partly filled with water, and then set in motion; the boiled stuff is then gradually put in, and a constant supply of clean water is run in from a cistern provided with means of preventing sand or other impurities from finding their way into the engine. It is of the utmost importance that the water should be abundant and of good quality, more especially as the material (rags) is mostly required for making the finer qualities of paper. In this respect the county of Kent and a few other localities on the chalk formation are considered specially suitable for this particular manufacture.

With respect to the driving of the engines, this was formerly effected by what is called *toothed gearing*, but cog-wheels were afterwards replaced by iron spur-wheel gearing, which enabled manufacturers to drive four or more engines from one source of power, by continuing the line of shafting and spur-wheels; but even with small rolls the wear and tear on this system was considerable, while it was quite inadequate to the driving of a number of large rolls of 30 inches in diameter, such as are now used. The introduction of belt-gearing, by Messrs. G. and W. Bertram, proved to be a great improvement on the older system, and it is found that the rags are broken not only more uniformly, but in less time, as the rolls work more steadily on the plates than with any system of wheel-gearing, while the various working parts of the engine last longer than when subjected to the vibrating action of wheel-gearing.

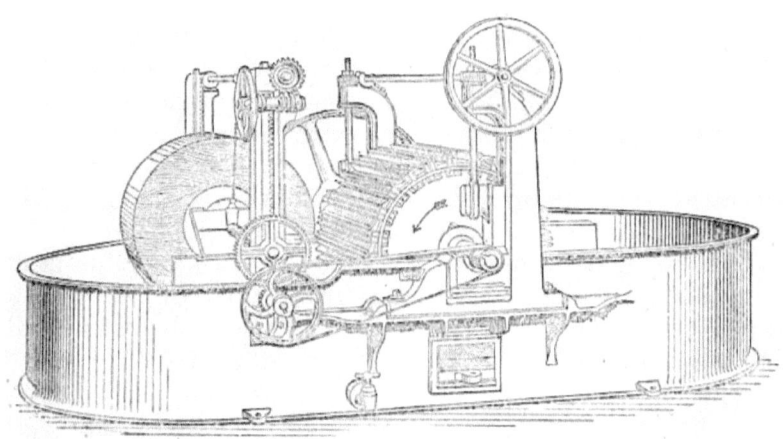

Fig. 13.

Bertrams' Rag Engine.—This engine, of which a drawing is shown in Fig. 13, may be used either as a washing and breaking engine, potcher, or beater. It is provided with double lifting gear, and has "all sweeps, curves, and angles" of the most improved design to save lodgments and ensure steady and thorough travelling of the pulp. The drum-washer is shown lifted by rack and pinion and worm gear, and empties down the midfeather direct to mouthpiece. The emptying can be done by spout and pipe, or by a chamber cast on the engine, down back or front side, as well as through the midfeather; but it is not advisable that it should be emptied down the midfeather if the rag engine is to be used as a beater.

Fig. 14.

Bentley and Jackson's Rag Engine.—This form of engine is shown in Fig. 14. The trough is of cast-iron, and made whole, and the engine can be obtained of any required dimensions. The trough is provided with a sand-well, cast-iron grate, and cock in front of the roll, and a sand-well, cast-iron grate, and brass valve on the back of the midfeather, a brass let-off valve and a brass waste-water valve. The bottom of the trough is "dished," to prevent the stuff from lodging. There are two movable bridge trees, fitted with pedestals and brass steps, and wrought-iron lifting links and screws, worm-wheels, worms, cross-shaft and hand-wheel for simultaneously lifting the roll on both sides. The roll is covered by a polished pitch-pine cover. The drum-washer may have either iron or wooden ends, has strong copper brackets, and is covered with brass backing and covering wires, mounted on a wrought-iron shaft, and carried by cast-iron stands, fitted with improved lifting gear, driving-wheels, and pulley.

When the engine is set in motion by the revolving shaft or spindle, the combined action of the knives of the roll and bed-plate causes the rags, which circulate in the water, to be gradually cut into small fragments, and the operation is kept up until the rags are converted into what is technically termed *half-stuff*. While this process is going on, fresh water is constantly supplied by a pipe at the end of the washing-engine; and when it is found that nothing but clear water escapes from the drum-washer, this is raised, and the spindle bearing the roll is lowered, so as to bring the respective knives closer together, to enable them to cut the reduced material still finer.

Draining.—When the material is sufficiently *broken*, as it is termed, the engine is then emptied by means of its valves, and the contents run into large vats or *drainers*, furnished with perforated zinc floors, in which it is allowed to drain thoroughly; and in order to remove the water more effectually, the pulp is afterwards pressed, either by an extractor or a centrifugal drainer, which dries it sufficiently for gas-bleaching, or for treatment in the *potcher* or *poacher*. This is a larger engine than the washer, and instead of the cylinder and bars, has a hollow drum which carries on its periphery a number of cast-iron paddles, which thoroughly agitate the pulp, and thus render it more susceptible of being freely and uniformly acted upon by the bleaching agent. The drum-washer of this engine should have a finer wire than is used for the breaker.

Torrance's Drainer.—This machine, which has been extensively used, is manufactured by Messrs. J. Bertram and Son, of Edinburgh. It consists of a perforated cylindrical box, enclosed in a fixed case, which revolves at about two hundred and fifty revolutions per minute. The machine is capable of treating about 4 cwt. of pulp per hour.

CHAPTER V.

TREATMENT OF ESPARTO.

Preliminary Treatment.—Picking.—Willowing Esparto.—Boiling Esparto.—Sinclair's Esparto Boiler.—Roeckner's Boiler.—Mallary's Process.—Carbonell's Process.—Washing Boiled Esparto.—Young's Process.—Bleaching the Esparto.

Preliminary Treatment: Picking.—Esparto is imported in bales or trusses, tightly compressed by hydraulic presses, and bound with twisted bands of the same material, much in the same manner as hay, except that which comes from Tripoli, which is bound with iron bands. The bands being cut, the loosened material is then spread out upon tables, partly covered with iron, or galvanised-iron, netting, to allow earthy matter or sand to pass through to a receptacle beneath. Here it is carefully picked by women and girls, who remove all roots, other kinds of grass, weeds, and heather. The material thus cleansed from impurities is transferred to the boiling-room. This careful preliminary treatment has been found necessary, since pieces of root and other vegetable matters which may be present are liable to resist the action of the bleaching liquor to a greater extent than the grass itself, and therefore produce specks, or "sheave" as they are termed at the mill, in the manufactured paper.

At some mills, however, as at the Horton Kirby Mills of Messrs. Spalding and Hodge, at South Darenth, for example, the cleaning of esparto is admirably effected by means of a willow, or esparto-cleaner, constructed by Messrs. Masson, Scott, and Bertram, which entirely supersedes the system of hand-picking. Having recently visited the mill referred to, we were enabled, through the courtesy of Mr. Sydney Spalding, to witness the action of this willow, which appeared to perform its functions with perfect uniformity, and to clean the grass most effectually. The *rationale* of the operation of willowing esparto may be thus described:—

Willowing Esparto.—A bale of the grass is unbound at a short distance from the machine, and the grass, which is in the form of small bundles or sheaves, tied with bands of the same material, is thrown by a woman on to a table or platform placed by the side of the willow, and a second woman, standing near the hopper of the machine, takes the bundles, a few at a time, and drops them into the hopper. The machine being in motion, in a few moments the grass, freed from its bands and dirty matters, appears in a perfectly loose condition at the wider end of the drum, and passes upward along a travelling-table to a room above, in the floor of which are the man-holes of a series of esparto boilers. During the passage of the loosened fibre, women standing on steps or platforms at the sides of the travelling-table are enabled to examine the material, and to remove any objectionable matters that may be present. Beneath the drum of the machine is a pipe, through which the dust and dirty matters are drawn away by means of a fan.

Boiling Esparto.—In the boiling-room at the mill referred to is a series of vertical stationary boilers, each about twenty feet high, and capable of holding about three tons of grass. The man-holes of these boilers pass through the floor of a room above, being nearly level with it, into which the cleaned esparto is conveyed, as described, by the travelling-table of the willowing machine. In this room is a series of compartments in which the willowed esparto is stored until required for boiling, when it is fed into the boilers by means of two-pronged forks provided for the purpose. The boiler being partially charged with caustic ley at 14° Twad., the esparto is introduced, and steam also, by which the esparto becomes softened, and thus a larger quantity of the fibre can be charged into the vessel. When the full charge of ley and esparto have been introduced the head of the boiler is securely fixed by means of its bolts, and steam then turned on until a pressure of about 20 lbs. to the square inch has been reached, which pressure is kept up for about three hours, when the steam is shut off and the blow-off tap opened. When the steam is blown off, the spent liquor is run off, and hot water then run into the boiler, steam again turned on, and the boiling kept up for about twenty minutes to half an hour, at the end of which time the steam is shut off and the blow-pipe opened. As soon as the steam has blown off, the washing water is run off by the bottom pipe, and the grass allowed to drain as thoroughly as possible. A door at the lower end of the boiler is then opened, and the grass emptied into trucks and conveyed to the washing-engines.

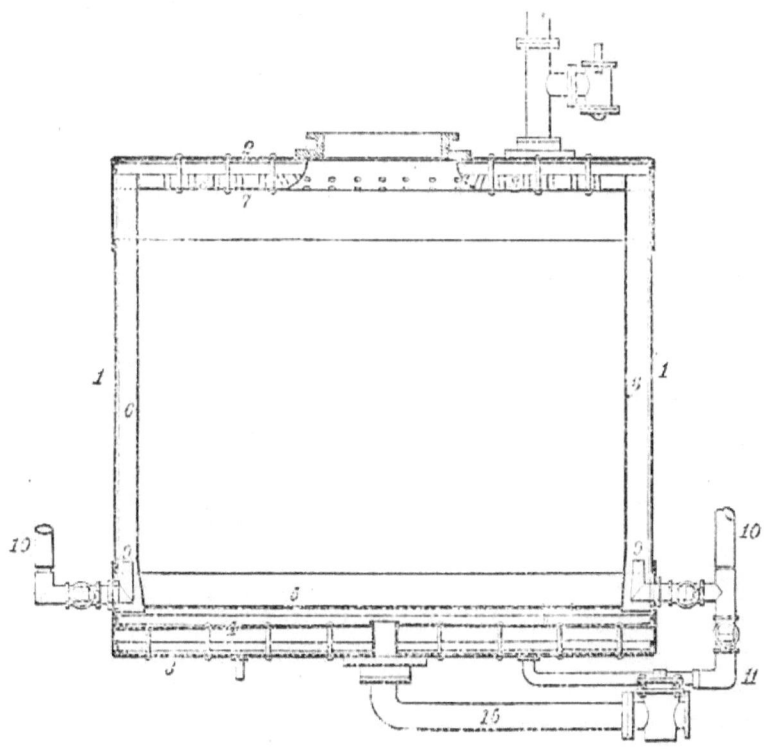

Fig. 15.

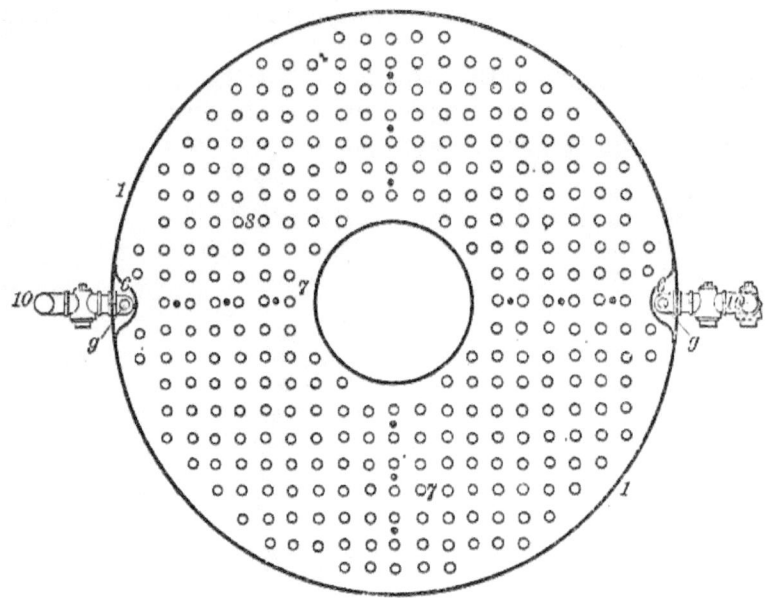

Fig. 16.

Sinclair's Esparto Boiler.—Another form of boiler, known as Sinclair's boiler, of the vertical cylindrical type, is shown in Figs. 15 and 16. It is constructed on what is termed the "vomiting" principle, but without the central vomiting-pipes generally used, and is fitted with one or more vomiting-pipes close to the side, two diametrically opposite pipes being used by preference. Steam jet pipes, with upwardly-directed nozzles, are fitted into the vomiting-pipes at points a little above the bends, between the vertical and horizontal parts. The liquid or ley thrown up the vomiting-pipes by the action of the steam is delivered from the upper ends of the pipes over a diaphragm or plate fixed near the top of the boiler, and the liquid is retained at a certain depth on the diaphragm by a number of small tubes fixed in it, and the liquid becomes well heated by the steam before overflowing down the tubes, which tubes also serve to distribute it uniformly over the fibrous materials in the boiler. A casing is formed at the bottom of the boiler, and in some cases extended more or less up the sides, and is supplied with steam, which should be superheated, or of high pressure.

With this arrangement the heat in the boiler is maintained without the excessive condensation of steam and consequent dilution and weakening of the liquors which occurs in ordinary boilers. Figs. 15 and 16 are horizontal and vertical sections of one form of this boiler. The boiler is made with a vertical cylindrical shell, 1; with a flat top, 2; and flat bottom, 3; and there is an inner or second bottom, 4; the space between it and the bottom, 3, being for steam to assist in heating the contents of the boiler. At a little distance above the inner bottom, 4, there is the usual perforated horizontal diaphragm, 5, down through which the liquid or ley drains from the fibre. Two diametrically opposite vertical vomiting-pipes, 6, are formed by the attachment of curved plates to the cylindrical shell, 1, and these vomiting-pipes, 6, have their upper ends above a horizontal diaphragm, 7, attached by stays to the boiler top, 2. This diaphragm is perforated, and short tubes, 8, are fixed in the perforations so as to project upwards, by which arrangement the liquid, rising up the vomiting-pipes, 6, lies on the diaphragm to the depth of the tubes, 8, and overflows down through them all equally, so as to be uniformly distributed over the materials in the boiler. Steam jet nozzles, 9, are fitted in the lower parts of the vomiting-pipes, being supplied with steam by pipes, 10, from one of which a branch, 11, supplies steam to the double bottom, 3, 4. The steam jets cause the liquid to be drawn from under the perforated diaphragm, 5, and thrown up the pipes, 6, whereby a constant circulation of the liquid through the fibre is maintained. The liquors are drawn off by the pipe, 15. In another form of boiler Mr. Sinclair employs vomit-pipes formed of thin steel plates riveted to opposite sides of the boiler, and the liquid which drains through the perforated double bottom is forced upward through the vomit-pipes to the perforated plates above, through which it distributes over the material in fine jets. The boiler is capable of holding from 2 to 3 tons of esparto, and under a pressure of from 40 to 50 lbs. the boiling occupies about two hours.

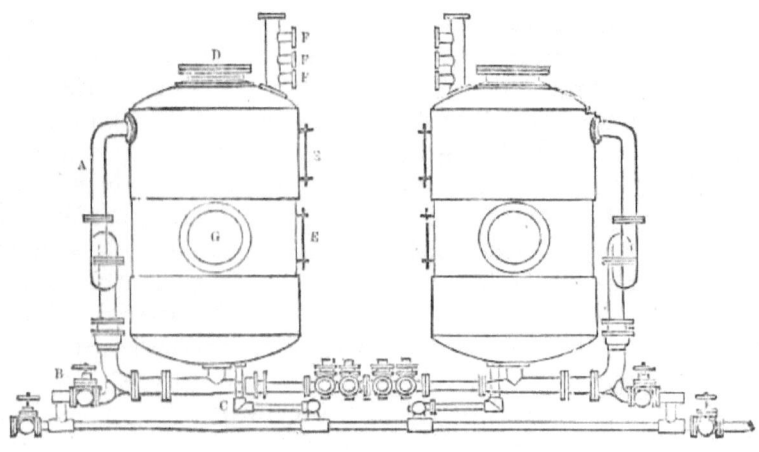

Fig. 17.

Roeckner's Boiler.—This boiler, of which an illustration of two in series is given in Fig. 17, has been extensively adopted by paper manufacturers. It will be noticed that the vomit-pipe A is placed outside the boiler, and the steam enters at the cock B, forcing the liquor up the vomit-pipe A and distributing it over the esparto. A pipe C is used for heating the liquor by means of waste steam at the commencement of the operation. The grass is fed into the boiler at the opening D. At E E are gauges for showing the height of the liquor in the boiler, F F F are pipes for the supply of steam, strong ley, and water, and the door G is for the discharge of the boiled grass. Each boiler is capable of holding 3 tons of esparto, and the boiling is completed in about two and a half hours, at a pressure of from 35 to 40 lbs. per square inch. It is said that the boiler effects a saving both in time and the amount of soda used.

Mallary's Process.—By this process the inventor says that he obtains the fibre in greater length, and gets rid of the gummy and resinous matters in a more economical way than by the present system. The materials used form a species of soap, with which and with the addition of water, the esparto is boiled. To carry out his process, he places in a boiler a suitable quantity of water, to which caustic soda, or a ley of the required strength to suit the nature of the fibre, is added; magnesite, or carbonate of magnesia, in the proportion of about 2 per cent. of the fibrous material, or a solution of sulphate of magnesia, is then added and mixed with the ley. He next adds "an improved saponaceous compound" to produce the required result, and when the boiling is completed, the stuff is treated as ordinary stock, to be applied for paper-making or other uses. The proportions are as follows:—2 gallons of petroleum or its products, 1 gallon of mustard oil, 10 to 15 lbs. of caustic soda, and 1 per cent. of boracic acid. These are placed in a copper and heated for 1 to 2 hours, until properly saponified. From 3 to 6 gallons of the "saponaceous compound" are added to the ley and magnesite, previously placed in the boiler with the fibre, and the boiling is kept up for the usual length of time, when the fibre will be found "beautifully soft, and the greater portion of the gum, silica, and resinous matters removed, or so softened as to be no hindrance to the perfect separation of the fibres, whilst the strength, silkiness, and softness are preserved in all their natural integrity." Considering that caustic soda ley "of the required strength" forms an essential part of this process, we should imagine that the auxiliaries mentioned would scarcely be necessary.

Carbonell's Process.—In this process, devised by M. Carbonell, of Paris, 200 lbs. of raw esparto are placed in a wooden vat furnished with a perforated steam-pipe, 20 lbs. of soda and 30 lbs. of quicklime being mixed with it: the vat is then supplied with cold water until the esparto is completely covered. Steam is then turned on, and the materials boiled for 4 hours. The spent liquor is then drained off, and the esparto submitted to hydraulic pressure. It is afterwards washed and broken in a rag engine, and in about 15 minutes is reduced to half-stuff. 20 lbs. of chloride of lime dissolved in water are then introduced, and the cylinder kept in motion as usual. In another vessel, lined with lead, 1¼ lb. of sulphuric acid is dissolved in 3 lbs. of water, and this gradually added to the pulp, which immediately assumes a reddish colour; but in the course of about three quarters of an hour it becomes perfectly white, when the pulp is ready for the paper-maker.

In the boiling of esparto, several important points have to be considered. The kind of esparto to be treated is the first consideration, since this grass differs materially in character in the different countries from which it is imported. Spanish esparto is considered the best for paper-making, as it is stronger in fibre and yields a whiter pulp than other varieties. Of the African espartos there are several varieties, which are known respectively as Oran, Tripoli, Sfax, Gabes, and Susa. Of these, the first-named (Algerian esparto) is held in highest estimation amongst paper-makers, since it more closely resembles Spanish esparto than the other varieties, though not so hard and stiff as the latter. These grasses usually have a length of about 10 to 12 inches. Tripoli esparto has an entirely different growth, being sometimes as long as 2½ or 3 feet, and proportionately stouter, and is also softer than Oran esparto, which is not so hard as the Spanish variety. Tripoli esparto does not yield a strong paper by itself, but in conjunction with Oran esparto gives more favourable results. Sfax and Gabes espartos have a closer resemblance to Oran than Tripoli, but are not so strong as Oran, being green and spongy, and not so dry as the latter variety. Susa esparto of good quality is said to equal Oran, but not to yield so high a percentage of fibre.

The next important consideration is to determine the percentage of caustic alkali which should be used per hundredweight of the particular variety of esparto to be treated, and we cannot do better than give the following proportions recommended by Mr. Dunbar.

		lbs. of	70 per cent.	caustic soda	per cwt.
Fine Spanish	18 to 20				
Medium Spanish	16 to 18	"	"	"	"
Fine Oran	18	"	"	"	"
Medium Oran	16 to 17	"	"	"	"
Susa	18	"	"	"	"
Tripoli	19 to 20	"	"	"	"
Sfax	20 to 21	"	"	"	"

Mr. Dunbar says that the above figures "insure a first-class boil, with the steam pressure of 25 lbs. and not exceeding 30 lbs., but are liable to alteration according to circumstances—such as the form of boilers, quality of the water for boiling purposes, and steam facilities, which ought at all times to be steady and uniform to get the absolute regularity required."

Respecting the strength of caustic ley used for boiling esparto, as indicated by Twaddell's hydrometer, this appears to range from 7° to 15°, some preferring to boil with stronger liquors than others. The time occupied in boiling also varies at different mills, and depends greatly upon the character of the boiler used. We are informed that a Sinclair boiler will turn out, on an average, three boils in twenty-four hours, including filling, boiling, discharging, &c., the boiling occupying about four hours for each batch of grass.

The boiling being completed, the liquor is run off into tanks, to be afterwards treated for the recovery of the soda, and the esparto is then subjected to a second boiling with water only for about 20 minutes. The liquor from the second boiling is sometimes thrown away, even when the soda from the first liquor is recovered; but a more economical method is to use this liquor, in lieu of water, strengthened with soda for a first boiling; or to mix it with the first liquors and evaporate the whole together. The second boiling being finished, the steam is turned off, and water then run in and steam again turned on for a short time, and the water then run off and the esparto allowed to drain thoroughly. The boiled grass is then discharged into trucks which convey it to the washing engines.

The liquor resulting from the boiling of esparto, which is of a dark brown colour, contains nearly all the soda originally used, but it also contains silicious, resinous, and other vegetable matters which it has dissolved out of the grass, the silica taking the form of silicate of soda. The esparto liquor, which was formerly allowed to run to waste, polluting our rivers to a serious extent, is now treated by several ingenious methods for the recovery of the soda with considerable advantage alike to the manufacturer and the public. The process consists essentially in boiling down the liquor to dryness, and incinerating the residue. During the process of incineration the carbonaceous matter extracted from the grass is converted into carbonic acid, which, combining with the soda, reconverts it into carbonate of soda, which is afterwards causticised with lime in the usual way, and the caustic soda thus obtained is again used in the boiling of esparto. Although one or other of the "recovery" processes is adopted at a good many of our paper-mills, the recovery of the soda is by no means universal as yet, but the time will doubtless soon arrive when the economical advantages of the process will be fully recognised. Indeed, we know it to be the fact that some manufacturers are watching, with keen interest, the progress of some of the newer systems of soda recovery, with the full intention eventually of adopting one or other of them.

Washing Boiled Esparto.—This operation is usually performed in engines similar to those used in washing rags, but in some mills the boiled grass is washed in a series of tanks, so arranged that water flows in at one end of the series, thence passing in succession through each batch of grass in the other tanks, and finally issues at the farthest end of the series as a very concentrated liquor. By this arrangement there is great economy of water, while at the same time no loss of fibre occurs. The concentrated washing liquors thus obtained may be evaporated, and the alkali recovered, which would be an undoubted saving, since these liquors obtained in the ordinary way by washing in the boilers are generally run off as waste. The engines used for washing esparto and converting it into half-stuff are generally of large size, and capable of treating a ton of boiled esparto. In this engine, however, there is no bed-plate, as the action of the roll alone is sufficient to reduce the boiled and softened esparto to half-stuff. A drum-washer is also furnished to the engine, which carries off the dirty washing water, while an equivalent proportion of clean water is kept constantly running into the engine from an elbowed pipe at its end. In charging the washing-engine, it is first about three parts filled with water, when the washing cylinder is lowered, and the esparto is then put in, care being taken not to introduce more of the material than will work freely under the action of the roll; if the mass be too stiff, portions of the material may be imperfectly washed. While the washing is in progress, the workman, armed with a wooden paddle, constantly stirs the esparto, clearing it away from the sides of the engine, so that none of the material may escape a perfect washing. At the bottom of the engine is a "sand-trap," covered with perforated zinc, through which any sand or other solid particles which may be present escape. When the washing is complete, the fresh water supply is shut off, and the drum-washer allowed to run until enough water has been removed to make room for the bleaching liquor.

Young's Process.—By this process the boiled and strained esparto is passed through elastic covered rollers, so adjusted as to split up and squeeze out the dissolved matters or liquid from the fibres, thus leaving them clean and open for the access of the bleaching liquor.

Bleaching the Esparto.—It is usual to bleach esparto in the washing engine, for which purpose a tank of bleaching liquor of the required strength (about 6° T. for Spanish) is placed close to the engine, which is provided with a pipe leading to the engine and another pipe proceeding from the tank in which the bleaching liquor is stored. The supply tank is furnished inside with a gauge, divided into inches— each inch representing so many gallons of liquor—by means of which the workman is enabled to regulate the quantity of bleaching liquor he is instructed by the manager or foreman to introduce into the engine. About half an hour after the bleach has become well incorporated with the fibre, sulphuric acid in the proportion of six ounces of the acid (which must be well diluted with water) to each hundredweight of the fibre. The dilute acid should be added gradually, and the proportions given must not be exceeded. The bleaching being completed, the half-stuff is next treated in a machine termed the *presse-pâte*, which not only cleanses the material from sand and dirt, but also separates all knots and other imperfections from the fibre in a most effectual and economical manner. Indeed, we were much struck with the excellent working of this machine at Messrs. Spalding and Hodge's mill, at South Darenth, and the remarkably fine quality of the finished pulp obtained through its agency. The presse-pâte was formerly used in the preparation of pulp from straw, but its advantages in the treatment of esparto are now fully recognised. The apparatus and method of working it may be thus briefly described:—

The machine is on the principle of the *wet end* of a paper machine, and consists of several stone chests for holding the bleached half-stuff, in which are fitted agitators to keep the stuff in suitable condition. From these chests the stuff is pumped into a mixing box, and from thence over a series of sand traps made of wood, and with slips of wood fixed in the bottom, in which any sand present is retained. The stuff then passes into a series of strainers, which, while allowing the clean fibre to pass through, retain all impurities, such as knots, &c., and the clean stuff is allowed to flow on to the wire-cloth in such a quantity as to form a thick web of pulp. A greater portion of the water escapes through the wire-cloth, but a further portion is removed by the passage of the pulp across two vacuum boxes, connected with four powerful vacuum pumps, which renders the half-stuff sufficiently dry to handle; but to render it still more so, it now passes between couch rolls, and is either run into webs, or, as is sometimes the case, it is discharged into boxes, the web of pulp thus treated being about an inch in thickness.

CHAPTER VI.

TREATMENT OF WOOD.

Chemical Processes.—Watt and Burgess's Process.—Sinclair's Process.—Keegan's Process.—American Wood Pulp System.—Aussedat's Process.—Acid Treatment of Wood.—Pictet and Brélaz's Process.—Barre and Blondel's Process.—Poncharac's Process.—Young and Pettigrew's Process.—Fridet and Matussière's Process.

The advantages of wood fibre as a paper material have been fully recognised in the United States and in many Continental countries, but more especially in Norway, Sweden, and Germany, from whence large quantities of wood pulp are imported into this country. There is no doubt that our home manufacturers have recently paid much attention to this material, and it is highly probable that wood, as an inexhaustible source of useful fibre, will at no distant date hold a foremost rank. Indeed, the very numerous processes which have been patented since the Watt process was first made known, indicate that from this unlimited source of fibre the requirements of the paper-maker may be to a large extent satisfied, provided, of course, that the processes for reducing the various suitable woods to the condition of pulp can be economically and satisfactorily effected. The great attention which this material has received at the hands of the experimentalist and chemist—the terms not being always synonymous—shows that the field is considered an important one, as indeed it is, and if successfully explored will, it is to be hoped, yield commensurate advantages both to inventors and the trade.

The object of the numerous inventors who have devised processes for the disintegration of wood fibre—that is, the separation of cellulose from the intercellular matters in which the fibres are enveloped—has necessarily been to dissolve out the latter without injury to the cellulose itself, but it may be said that as yet the object has not been fully attained by either of the processes which have been introduced. To remove the cellular matter from the true fibre or cellulose, without degrading or sacrificing a portion of the latter, is by no means easy of accomplishment when practised on an extensive scale, and many processes which present apparent advantages in one direction are often found to exhibit contrary results in another. The field, however, is still an open one, and human ingenuity may yet discover methods of separating wood fibre from its surrounding tissues in a still more perfect manner than hitherto.

The various processes for treating wood for the extraction of its fibre have been classified into: (1) chemical processes; and (2) mechanical processes. We will give precedence to the former in describing the various wood pulp processes, since the pulp produced by the latter, although extensively used, is chiefly employed, in combination with other pulps, for common kinds of paper. In reference to this part of our subject Davis says:—"Experience has dictated certain improvements in some of the details of those earlier methods, by which so-called 'chemical wood pulp' is manufactured very largely on the Continent of Europe.... It is possible to obtain a pulp of good quality, suitable for some classes of paper, by boiling the chipped wood in caustic soda, but when it is desired to use the pulp so prepared for papers having a perfectly white colour it has been demonstrated in practice that the action of the caustic soda solution at the high temperature which is required develops results to a certain degree in weakening and browning the fibres, and during the past five years much labour has been expended in the endeavour to overcome the objections named. The outcome of these efforts has been a number of patents, having for their object to prevent oxidation and subsequent weakening of the fibres." In several of these patents, to which we shall refer hereafter, bisulphite of lime is employed as the agent to prevent oxidation and consequent degradation of the fibres, and in other processes bisulphite of magnesia has been used for the same purpose. Davis further remarks: "Although a common principle runs through all these methods of preparing cellulose from wood, they differ in detail, as in the construction of the digesters employed, methods of treating the wood stock before boiling it in the sulphurous acid solution, and also as regards pressure, blowing off the sulphurous acid gas, etc., but all these processes present a striking similarity to the method patented by Tilghmann in 1867." There can be no doubt that the action of caustic soda, under high pressures, is highly injurious both to the colour and strength of the fibres, and any process that will check this destructive action in a thoroughly practical way will effect an important desideratum.

I. Chemical Processes: *Watt and Burgess's Process.*—This process, which, with some modifications, is extensively worked in America, consists in boiling wood shavings, or other similar vegetable matter, in caustic soda ley, and then washing to remove the alkali; the wood is next treated with chlorine gas, or an oxygeneous compound of chlorine, in a suitable vessel, and it is afterwards washed to free it from the hydrochloric acid formed. It is now treated with a small quantity of caustic soda in solution, which instantly converts it into pulp, which only requires to be washed and bleached, and beaten for an hour and a half in the beating engine, when the pulp is ready for the machine. The wood-paper process as carried out in America has been described by Hofmann, from whose work[15] we have abridged the following:—

The wood, mostly poplar, is brought to the works in 5-feet lengths. The bark having been stripped off by hand, it is cut into ½-inch slices by a cutter which consists of four steel knives, from 8 to 10 inches wide by 12 to 15 inches long, which are fastened in a slightly inclined position to a solid cast-iron disc of about 5 to 7 feet diameter, which revolves at a high speed, chopping the wood—which is fed to the blades through a trough—into thin slices across the grain. The trough must be large enough to receive the logs, usually 10 or 12 inches thick, and it is set at such an angle that the logs may slide down towards the revolving cutters; this slanting position only assists the movement of the logs, while a piston, which is propelled by a rack, pushes them steadily forward until they are entirely cut up. The piston, or *pusher*, then returns to its original position, fresh wood is put into the trough, and the operation repeated. In this way many tons of wood can be chopped up by one of these cutters in a day. The sliced wood is conveyed by trucks to an elevator by which it is hoisted up two storeys to a floor from which the boilers are filled. The boilers are upright cylinders, about 5 feet in diameter and 16 feet high, with semi-spherical ends, provided inside with straight perforated diaphragms, between which the chips from one cord of wood are confined. A solution of caustic soda, at 12° B., is introduced with the chips, and fires are started in a furnace underneath. At other works the boilers are heated by steam circulating through a jacket which covers the bottom and sides of the boiler.

The boiling is continued for about six hours, when the digestion is complete, and the contents of the boilers are emptied with violence, under the pressure of at least 65 lbs. of steam, which had been maintained inside. A large slide valve is attached to the side of each boiler for this purpose close to the perforated diaphragm, and connected by a capacious pipe with a sheet-iron cylinder of about 12 feet diameter and 10 feet high, which receives the contents— pulp, liquor, and steam. The object of these large chambers —one of which serves for two boilers—is to break the force of the discharging mass. The steam passes through a pipe on the top of each, and from thence through a water reservoir, while the liquid containing the pulp flows through a side opening and short pipe into movable boxes, or drainers, mounted on wheels, and each capable of holding the contents of one boiler; these boxes are pushed along a tramway up to the collecting chambers, where the pulp is received. In a building 132 feet long and 75 feet wide, ten digesting boilers are arranged in one straight line, and parallel with the boilers runs the main line of rails, side tracks extending from it to each of the chambers, and a turn-table is supplied at every junction. By this arrangement the drainer waggons can be pushed from the side tracks on to the main line, which leads to the washing-engines in an adjoining room. A system of drainage is established below the tramways, by which all the liquid which drains from the waggons is carried away and collected for treatment by evaporation; these carriers remain on the side tracks until the pulp is ready for the washing-engine.

When the greater portion of the liquor has drained off, warm water is sprinkled over the pulp from a hose for the purpose of extracting all the liquid which is sufficiently concentrated to repay the cost of evaporation—the most advantageous method of recovering the soda. The contents of the waggons—from the same number of boilers—are then placed in two washing-engines, each capable of holding 1,000 lbs. of pulp. After being sufficiently worked in these engines the pulp is transferred to two stuff-chests, and from thence conveyed by pumps to two wet-machines. The screens (strainers) of the wet-machines retain all impurities derived from knots, bark, and other sources, and the pulp, or half-stuff, obtained is perfectly clean and of a light grey colour. The pulp is bleached with solution of bleaching powder like rags, then emptied into drainers and allowed to remain therein with the liquid for twenty-four to forty-eight hours, or long enough to render the use of vitriol in the bleaching unnecessary. The portion of the white pulp which is to be worked up into paper in the adjoining mill is taken from the drainers into boxes running on tramways in the moist state, but all the pulp which has to be shipped to a distance is made into rolls on a large cylinder paper-machine with many dryers. The object being merely to dry the pulp, a very heavy web can be obtained, since the water leaves this pulp very freely. The wood pulp thus obtained is perfectly clean, of a soft, white spongy fibre, and a greater portion of it is mixed with a small proportion of rag pulp and worked into book and fine printing papers. Sometimes the wood pulp is used alone or mixed with white paper shavings for book paper. The fibres are rather deficient in strength, but as a material for blotting paper they are said to be unsurpassed, while the wood paper is much liked by printers.

The wood from poplar, which is generally preferred, furnishes a very white fibre, and is easily digested, but since the fibres are short it is sometimes found advantageous to mix them with longer fibres, as those of the spruce or pine, although the latter wood requires a much more severe treatment in boiling with alkali than the former. In reference to this process the following remarks appeared in *The Chemist*,[16] 1855:—"The process occupies only a few hours; in fact, a piece of wood may be converted into paper and printed upon within twenty-four hours." An interesting verification of this was published a few years since in an American paper, the *Southern Trade Gazette*, of Kentucky, which runs as follows:—"At a wood-pulp mill at Augusta, Ga., a tree was cut down in the forest at six o'clock A.M., was made into pulp, and then into paper, at six o'clock in the evening, and distributed amongst the people as a newspaper by six o'clock the next morning. From a tree to a newspaper, being read by thousands, in the brief round of twenty-four hours!" The wood-paper process referred to has given rise to many subsequent modifications, some of which we will briefly describe.

Sinclair's Process.—The wood is first cut into pieces about 1 inch broad, ⅛th inch thick, and from 2 to 3 inches long. It is then placed in a boiler and a solution of caustic soda, in the proportions of 600 gallons to 10 cwts. of dry wood, is poured over it. The boiler having been securely closed, the heat is raised till a pressure of 180 to 200 lbs. on the square inch is obtained, when the fire is withdrawn and the boiler allowed to cool, after which the ley is blown off, the top door removed, and the contents scalded. The discharge door is now opened and the pulp transferred to a poaching-engine to be washed with pure water, when the resin, &c., are easily removed and the clean fibres obtained, which, it is said, are longer and firmer than those obtained by other methods.

Keegan's Process.—By this method soft deal or pine is sawn up into pieces from 6 to 12 inches long and ½ inch thick, it being preferable that all the pieces should be of an equal size, but the smaller they are the more rapid, of course, will be the operation. The pieces of timber are placed in a cylindrical boiler, turning upon a horizontal axis while the digestion is progressing. In a second boiler is prepared a solution of caustic soda of about 20° B. (specific gravity 1·161), which is introduced through a pipe into the first boiler, this being afterwards hermetically closed, and the soda is forced into the pores of the wood by means of a pump. When the wood is not more than half an inch in thickness a pressure of 50 lbs. on the square inch is sufficient, and the injection of the caustic soda solution is completed in half an hour. The superabundant liquor is pumped back into the second boiler for the next operation.

The excess of liquor having been removed from the wood as stated, steam is introduced between the double sides of the first boiler, and the temperature of the wood raised from 150° to 190° C. (334° to 438° F.). The wood is next washed in the usual way until the liquor runs off perfectly limpid, and the half-stuff thus produced may be converted into pulp either before or after bleaching, according to the quality and colour of the paper to be produced.

American Wood-Pulp System.—Another method of carrying out the wood-pulp process has recently been described by Mr. E. A. Congdon, Ph.B.,[17] from which we extract the following:—"Poplar, pine, spruce, and occasionally birch, are used in the manufacture of chemical fibre. Pine and spruce give a longer and tougher fibre than poplar and birch, but are somewhat harder to treat, requiring more soda and bleach. Sticks of poplar, freed from bark, and cleansed from incrusting matter and dirt, are reduced to chips by a special machine having a heavy iron revolving disc set with knives, and are then blown by means of a Sturtevant blower into large stove chambers after passing over a set of sieves having 1¼-inch for the coarse and 1⅛-inch mesh for the fine sieves, from whence they pass to the digesters, which are upright boilers 7 by 27 feet, with a manhole at the top for charging the chips and liquor, and a blow-valve at the bottom for the exit of the boiled wood. A steam-pipe enters at the bottom, beneath a perforated diaphragm, and keeps the liquor in perfect circulation during the boiling of the wood by means of a steam-ejector of special construction."

Boiling.—The average charge of wood for each digester is 4·33 cords,[18] giving an average yield of 4,140 lbs. of finished fibre per digester. A charge of 3,400 gallons of caustic soda solution of 11° B. is given to each digester charged with chips, and the manhead is then placed in position and steam turned on. Charging the digester occupies from thirty to forty-five minutes, and steam is introduced until the gauge indicates a pressure of 110 lbs., which occupies about three hours. This pressure is kept up for seven hours, when it is reduced by allowing the steam to escape into a large iron tank which acts as a separating chamber for the spent liquor it carries, the steam entering in at one end and passing out at the other through a large pipe, the liquor remaining in the tank. The steam is allowed to escape until the pressure is reduced to 45 lbs., when the digester is blown. The blow-cap being removed, the blow-valve is raised and the contents of the digester are discharged into a pan of iron covered with a suitable hood. The contents strike against a dash-plate placed midway in the pan, which thoroughly separates the fibres of the wood. The time occupied in the foregoing operations is from eleven to eleven and a half hours. It takes from nine to ten hours to free the pans from alkali, when they are removed to washing-tanks with perforated metal bottoms, where the material receives a final washing before being bleached.

Washing.—Each of the three digesters has a pan into which its contents are discharged, and there are also four iron tanks used for holding the liquors of various strengths obtained from the cleansing of the pulp and a fifth tank is kept as the separating-tank before mentioned. When the digester is blown, the pulp is levelled down with a shovel, and the liquor from the separating-tank is allowed to flow into it. The contents of the next strongest pan are pumped upon it, while at the same time the strongest store tank flows into this pan. This flowing from the tank to the pan, pumping from here to the pan just blown, and from there to the evaporators, is kept up until the liquor is not weaker than 6° B. hot (130° F.). The second pan is now down to 4° B. hot, and the process of "pumping back" is commenced. The two weakest tanks are put upon this pan and pumped out of the bottom of it into the two tanks in which are kept the strongest liquors. The two weak tanks have been filled in the process of completing the cleansing of the third pan (the weakest) on which water was pumped until the last weak tank stood at only ½° B. This pan, now cleaned, is hosed and pumped over to the washing tanks. A fresh blow is now made in this pan, and the same treatment kept up as with the first pan.

The foregoing system is thus illustrated by Mr. Congdon:—

Pan   A.—   Just blown.

"	B.—	Partly cleaned.		
"	C.—	Almost cleaned.		
Tank	1.—	3½°	B.	hot.
"	2.—	2°	"	"
"	3.—	1°	"	"
"	4.—	½°	"	"

Separating tank, strong.

A is levelled down; contents of separating-tank allowed to flow upon it; B is pumped on to A; at the same time liquor from the two strong store tanks is put on it (B), and this continued to be sent from A to the evaporator until it is now weaker than 4° B. hot; the process of "pumping back" is then commenced. The two weakest are allowed in succession to flow on to it, and the liquor purified from the bottom of B into the two strong tanks, filling No. 1, the stronger, before No. 2. The weakest are filled in the process of completing the cleansing of C, on which water is pumped until the last tank from it tests only ½° B. C is now hosed and pumped over to the washing tanks. A fresh digester is blown in C, and the process repeated as with A.

The above system has been modified by having an extra pan into which the liquor from the last pan blown (after sending to the evaporators until down to 6° hot, and bringing down to 4° hot, by the stored liquor) is pumped. When the strength is reduced to 4° the pumping is stopped. The liquor from this pan is put in the next pan blown, after the liquor from the separating-tank has been put upon it, whereby an economy in time is effected.

The pulp, after being partially cleaned in the pans, still contains an appreciable quantity of soda. It is hosed over to the washing-tanks and receives a final washing with hot water. When the pulp is thoroughly free from alkali, and the water flowing from under the tank is colourless, the contents are hosed down by hot water into the bleaching-tanks. The superfluous water is removed by revolving washers, and about 1,000 gallons of a solution of chloride of lime at 4° B. are then introduced, and the contents agitated as usual. The bleaching occupies about six or seven hours, when the pulp is pumped into draining tanks, where it is left to drain down hard, the spent bleach flowing away. The stock is then hosed and pumped into a washing-tank, where it acquires the proper consistency for the machine. From here it is pumped into the stuff chest, whence it goes over a set of screens and on to the machine, from which the finished fibre is run off on spindles. The rolls are made of a convenient size to handle, averaging about 100 lbs. each. The fibre is dried on the machine by passing over a series of iron cylinders heated by steam. The finished product is a heavy white sheet, somewhat resembling blotting paper. The whole of the foregoing operations are stated to occupy forty-five hours.

Aussedat's Process.—By this method the wood is disintegrated by the action of jets of vapour. In one end of a cylindrical high-pressure boiler, about 4½ feet in diameter and 10 feet high, is fixed a false bottom, whereby the wood placed upon it may be removed from the liquor resulting from steam condensed in the chamber, the whole being mounted on lateral bearings which serve for the introduction of the vapour, and the wood is fed through a manhole at the upper end of the boiler. Taps are fixed at the upper and lower ends for the liquid and uncondensed vapour. The wood having been placed in the boiler, the jet is gradually turned on in such a way that at the end of three or four hours the temperature becomes about 150° C., the pressure being about five atmospheres, which point is maintained for an hour. As the slightest contact between the wood and the condensed water would at once discolour the former, it is essential that the liquid be removed from time to time by one of the outlets provided for the purpose.

The treatment above described is said to be suitable for all kinds of wood, and although it is the usual practice to introduce it in logs about a yard long, any waste wood, as chips, shavings, etc., may be used. It is preferable, though not necessary, to remove the bark, but all rotten wood may be left, as it becomes removed in the condensed water. The logs, after the above treatment, by which the fibre is disintegrated and the sap and all matters of a gummy or resinous nature are removed, are afterwards cut up by any suitable means into discs of about an inch, according to the nature of the fibre required. These are then introduced into a breaker, in which they become converted into half-stuff, which, after being mixed with a suitable quantity of water is passed through mills provided with conical stones, in which it becomes reduced to whole-stuff. The pulp thus prepared is principally used in the manufacture of the best kinds of cardboard, but more particularly such as is used by artists, since its light brownish shade is said to improve the tone of the colours. Bourdillat says that in the above process the vapour has a chemical as well as a mechanical action, for in addition to the vapour traversing the cellular tissues of the wood and dissolving a considerable portion of the cell-constituents, acetic acid is liberated by the heat, which assists the vapour in its action on the internal substance of the wood.

Acid Treatment of Wood.—A series of processes have been introduced from time to time, the object of which is to effect the disintegration of wood fibre by the action of acids. The first of these "acid processes" was devised by Tilghmann in 1866, in which he employed a solution of sulphurous acid; the process does not appear to have been successful, however, and was subsequently abandoned, the same inventor having found that certain acid sulphites could be used more advantageously. Other processes have since been introduced, in which wood is treated in a direct way by the action of strong oxidising acids, as nitric and nitro-hydrochloric acids, by which the intercellular matters of the wood become dissolved and the cellulose left in a fibrous condition.

Pictet and Brélaz's Process.—By this process wood is subjected to the action of a vacuum, and also to that of a supersaturated solution of sulphurous acid at a temperature not exceeding 212° F. In carrying out the process a solution of sulphurous acid is used, consisting of, say from $\frac{1}{5}$ to $\frac{1}{3}$ lb. avoirdupois of sulphurous acid to each quart of water, and employed under a pressure of from three to six atmospheres at 212° F. Under these conditions the cementing substances of the wood "retain their chemical character without a trace of decomposition of a nature to show carbonisation, while the liquor completely permeates the wood and dissolves out all the cementing constituents that envelop the fibres." In carrying out the process practically, the wood is first cut into small pieces as usual and charged into a digester of such strength as will resist the necessary pressure, the interior of which must be lined with lead. Water is then admitted into the vessel and afterwards sulphurous acid, from a suitable receiver in which it is stored in a liquid form until the proportion of acid has reached that before named, that is, from 100 to 150 quarts of the acid to 1,000 quarts of water. The volume of the bath will be determined by the absorbing capacity of the wood, and is preferably so regulated as not to materially exceed that capacity. In practice it is preferable to form a partial vacuum in the digester, by which the pores of the wood are opened, when it will be in a condition to more readily absorb the solution and thereby accelerate the process of disintegration. When disintegration is effected, which generally occurs in from twelve to twenty-four hours, according to the nature of the wood under treatment, the liquor, which is usually not quite spent in one operation, is transferred to another digester, a sufficient quantity of water and acid being added to complete the charge. In order to remove the liquor absorbed by the wood, the latter is compressed, the digester being connected with a gas-receiver, into which the free gas escapes and in which it is collected for use again in subsequent operations. The bath is heated and kept at a temperature of from 177° to 194° F. by means of a coil in the digester supplied with steam from a suitable generator. The wood, after disintegration, undergoes the usual treatment to convert it into paper pulp, and may thus be readily bleached by means of chloride of lime. The unaltered by-products contained in the bath may be recovered and treated for use in the arts by well-known methods.

Barre and Blondel's Process consists in digesting the wood for twenty-four hours in 50 per cent. nitric acid, used cold, by which it is converted into a soft fibrous mass. This is next boiled for some hours in water and afterwards in a solution of carbonate of soda; it is then bleached in the usual way.

Poncharac's Process.—In this process cold nitro-hydrochloric acid (aqua regia) is employed for disintegrating wood in the proportions of 94 parts of the latter to 6 parts of nitric acid, the mixture being made in earthen vessels capable of holding 175 gallons. The wood is allowed to soak in the acid mixture for six to twelve hours. 132 lbs. of aqua regia are required for 220 lbs. of wood. When it is desired to operate with a hot liquid, 6 parts of hydrochloric acid, 4 parts of nitric acid, and 240 parts of water are used in granite tubs provided with a double bottom, and it is heated by steam for twelve hours and then washed and crushed.

Young and Pettigrew's Process.—These inventors use either nitric or nitrous acids, and the acid fumes which are liberated are condensed and reconverted into nitric acid.

Fridet and Matussière's Process.—This process, which was patented in France in 1865, consists in treating wood with nitro-hydrochloric acid, for which purpose a mixture of 5 to 40 per cent. of nitric acid and 60 to 95 per cent. of hydrochloric acid is used, which destroys all the ligneous or intercellular matter without attacking the cellulose. After the wood (or straw) has been steeped in the acid mixture, the superfluity is drawn off, and the remaining solid portion is ground under vertically revolving millstones. The brownish-coloured pulp thus obtained is afterwards washed and bleached in the usual way.

It is quite true that cellulose can be obtained from wood and other vegetable substances by treatment with nitric acid alone, or with a mixture of nitric and hydrochloric acids, but it will be readily seen that the employment of such large quantities of these acids as would be required to effect the object in view on a practical scale, would be fraught with incalculable difficulties, amongst which may be mentioned the insuperable difficulty of obtaining vessels that would resist the powerful corrosive action of the acids. Moreover, since nitric acid forms with cellulose an explosive substance (*xyloidin*) of the gun cotton series, the risk involved in the drying of the cellulose obtained would be quite sufficient to forbid the use of processes of this nature.

CHAPTER VII.

TREATMENT OF WOOD (continued).

Sulphite Processes.—Francke's Process.—Ekman's Process.—Dr. Mitscherlich's Process.—Ritter and Kellner's Boiler.—Partington's Process.—Blitz's Process.—McDougall's Boiler for Acid Processes.—Graham's Process.—Objections to the Acid or Sulphite Processes.—Sulphite Fibre and Resin.—Adamson's Process.—Sulphide Processes.—Mechanical Processes.—Voelter's Process.—Thune's Process.

Sulphite Processes.—An important and successful method of treating wood has been found in employing sulphurous acid, combined in certain proportions with soda, lime, or magnesia, whereby a bisulphite of the alkaline or earthy base is obtained. One of the principal attributes of these agents is that in boiling wood at high pressures oxidation and consequent browning of the fibres is prevented. Of these sulphite, or more properly bisulphite, processes, several of those referred to below have been very extensively adopted, and vast quantities of so-called "sulphite pulp" are imported into this country from Norway, Germany, Scandinavia, &c., the product from the latter source being considered specially suited for the English market. Some of these processes are also being worked in this country, but more particularly those of Partington, McDougall, and Ekman.

Francke's Process.—In this process, which is known as the "bisulphite process," the active agent employed for the disintegration of wood is an acid sulphite of an alkaline or earthy base, as soda or potassa, lime, &c., but it is scarcely necessary to say that the process has since been modified by others. The invention is applicable to the treatment of wood, esparto, straw, etc., and may be thus briefly described:—A solvent is first prepared, which is an acid sulphite of an alkali or earth, that is, a solution of such sulphite with an excess of sulphurous acid. As the cheapest and most accessible base the inventor prefers lime. It has long been known that a solution of sulphite of lime, combined with free sulphurous acid, would, at a high temperature, dissolve the intercellular portions of vegetable fibres, leaving the fibres in a suitable condition for paper manufacture; but Mr. Francke claims to have determined the conditions under which this can be effected with rapidity, and in such a way as to preserve the strength of the fibres, and to have obtained a practical method of preparing pulp by his process. For his purpose he employs a moderately strong solution of the solvent at a high temperature, with gentle but constant agitation. The acid sulphite is produced by this process at small cost and at a temperature nearly high enough for use in the following way:—A tower or column is charged with fragments of limestone, which are kept wetted with a shower of water; fumes of sulphurous acid, produced by burning sulphur, or by roasting pyrites, etc., are then passed through the tower. The liquid which collects at the bottom of the tower is the desired solvent, which should have a strength of 4° to 5° B. It is not essential that the limestone should be pure, as magnesian limestone, etc., will answer equally well. The soluble alkalies, as soda and potassa, may also be used when their greater cost is not an objection. But for these alkalies the treatment is modified, as follows:—The tower is charged with inert porous material, such as coke, bricks, etc., and these are kept wetted by a shower of caustic alkali at 1° to 2° B., while the sulphurous acid fumes are passed through the tower. In like manner carbonate of soda or potassa may be used, but in this case the solution showered on the porous material should be stronger than that of the caustic alkali, so that it may contain approximately the same amount of real alkali. Whichever alkaline base be employed, the liquid collected at the bottom of the tower should have a strength of 4° to 5° B.; this being the acid sulphite of the base is used as the solvent employed for the manufacture of pulp. When wood is to be treated, it is freed as much as possible from resinous knots by boring and cutting them out, and is then cut—by preference obliquely—into chips of a ¼ to ¾ of an inch thick. Esparto, straw, and analogous fibres are cut into fragments. The fibrous material and solvent are charged into a digester heated by steam at a pressure of four or five atmospheres, and consequently capable of raising the temperature of the contents to about 300° F. As agitation greatly promotes the pulping of the materials, Mr. Francke employs a revolving cylindrical boiler, which is allowed to revolve while the charge is under treatment.

Ekman's Process.—In this process, which in some respects bears a resemblance to the preceding, native carbonate of magnesia (magnesite) is first calcined to convert it into magnesia; it is then placed in towers lined with lead, and sulphurous acid gas, obtained by the burning of sulphur in suitable furnaces, is passed through the mass, a stream of water being allowed to trickle down from the top of the towers. The supply of gas is so regulated that a continual formation of a solution of bisulphite of magnesium, of an uniform strength, is obtained; great care, however, is necessary to avoid excess and consequent loss of sulphurous acid by its conversion into sulphuric acid. In boiling, the fragments of wood, previously crushed by heavy rollers, are placed in a jacketed, lead-lined, cylindrical boiler, suspended on trunnions, so that it can be inverted to remove the charge. The pressure in the outer jacket is 70 lbs. per square inch, and that within the boiler is 90 lbs. per square inch. The boiling occupies twelve hours. This process has been extensively worked by the Bergvik and Ala Company, of Sweden, for many years with great success, and we understand that the company has been turned over to an English company—the Bergvik Company, Limited. The Ilford Mill and Northfleet Works have been largely supplied with sulphite pulp from the Swedish works.

One great drawback to the bisulphite processes is that the boiling cannot be effected in iron boilers unless these be lined with some material which will protect the iron from the destructive action of the bisulphite, which, being an acid salt, would exert more action upon the iron than upon the fibre itself, and the solution of iron thus formed would inevitably prove injurious to the colour of the fibre. In several of the systems adopted iron boilers lined with lead have been used, but the heavy cost of this material and its liability to expand unequally with the iron, especially at the high temperatures which the solvent necessarily attains under pressure, causes the lead to separate from the iron, while it is apt to bulge out in places, and thus becomes liable to crack and allow the acid liquor to find its way to the interior of the iron boiler which it was destined to protect. To overcome this objection to the simple lead lining, Dr. Mitscherlich patented a process which has been extensively adopted in Germany, and is now being carried out by several companies in different parts of America. This process is briefly described below.

Dr. Mitscherlich's Process.—The digester employed in this process is lined with thin sheet lead, which is cemented to the inner surface of the boiler by a cement composed of common tar and pitch, and the lead lining is then faced with glazed porcelain bricks. In this process a weaker bisulphite of lime is used than in Francke's, and the time of boiling is consequently considerably prolonged.

Ritter and Kellner have proposed to unite the inner surface of the boiler to its lead lining by interposing a soft metal alloy, fusible at a temperature lower than that of either metal, and it is claimed that the iron and lead are thus securely united, while the alloy being fusible under the normal working temperature of the digester, the lead lining can slide freely on a boiler shell.

Partington's Process.—This process, which has been for some time at work at Barrow, and for the further development of which a private company, entitled the Hull Chemical Wood Pulp Company, Limited, has been formed, consists in the employment of sulphite of lime as the disintegrating agent. The process consists in passing gaseous sulphurous acid—formed by burning sulphur in a retort, into which is forced a current of air at a pressure of 5 lbs. to the square inch—through a series of three vessels, connected by pipes, the vessels being charged with milk of lime. The first two of these vessels are closed air-tight, and the gas is then introduced, while the third vessel remains open; from this latter a continuous stream of nitrogen escapes, due to the removal of the oxygen by the burning sulphur from the air passed into the retort. This process is said to be a very economical one, so far as relates to the cost of materials used.

Blitz's Process.—This process consists of employing a mixture composed of bisulphite of soda 2 parts, caustic soda 1 part; and vanadate of ammonia 1 gramme, in hydrochloric acid 4 grammes to every 6 kilogrammes of the bisulphite. The wood, after being cut up in the ordinary way, is submitted to the action of the above mixture, under a pressure of three or four atmospheres, for from four to eight hours, and the pulp is then ground; it is said to possess some of the qualities of rag pulp and to look much like it.

McDougall's Boiler for Acid Processes.—This invention is intended to obviate the difficulties which arise in using lead-lined boilers, owing to the unequal expansion and contraction of the lead and the iron on their being alternately heated by steam and cooled, on the discharge of each successive batch of pulp. This invention consists in constructing the boilers with an intermediate packing of felt, or other compressible and elastic material, so that when the interior leaden vessel is heated, and thereby enlarged and pressed outwards by the steam, the compressible and elastic packing yields to the pressure and expansion. Also in the cooling of the vessels the packing responds to the contraction, and approximates to its original bulk and pressure between the two vessels, and so prevents the rupture or tearing of the lead and consequent leakage and other inconveniences. Another part of this invention consists in the construction of the outer iron or steel vessel in flanged sections, which are fitted to incase the interior leaden vessel with a space between the two vessels, into which the compressible and elastic materials are packed. In the construction of these vessels the iron or steel flanged sections are placed on to the leaden vessel and packed with the compressible and elastic lining in succession. As each section is packed it is screwed close up to the adjoining section by the screw bolts, fitted into corresponding holes in the flanges of the contiguous section until completed. This method of construction secures economy by the retention of the heat, which is effected by the packing between the two vessels. The materials used for the packing are caoutchouc, felt, flocks, asbestos, etc., and a space of about two inches between the vessels is preferred, into which the packing is filled.

Graham's Process.—This process consists in boiling fibrous substances in a solution of sulphurous acid, or a sulphite or bisulphite of soda, potash, magnesia, or lime, or other suitable base and water. The boiling is preferable conducted in a closed boiler, lined with lead, to protect it from the action of the chemical substances used, and is fitted with a valve which can be opened to allow the gases and volatile hydrocarbons contained in and around the fibres to escape. The method of carrying out the process has been thus described:—"In carrying out the process there is a constant loss of sulphurous acid gas going on, and consequently a continual weakening of the solution employed, to avoid which it is preferable to employ monosulphite of potash, soda, magnesia, lime, or other suitable base, and water. Either of these substances, or a suitable combination of them, and water are placed in the boiler with the fibrous substances to be treated, and the temperature raised to the boiling point. After the hydrocarbons, air, and gases natural to the fibrous substances have been driven out by the heat and allowed to escape, sulphurous acid, in its gaseous or liquid state, or in combination with either of the bases referred to, is pumped or injected into the boiler. There is thus forming in the closed boiler a solution containing an excess of sulphurous acid above that required to form, in combination with the base, a monosulphite. The operation of injecting sulphurous acids, or the sulphites, may be repeated from time to time during the boiling, so as to fully maintain, and if necessary increase, the strength and efficiency of the chemical solution. It is said that by this process a saving of the chemicals employed is effected, as little or no sulphurous acid gas is lost during the time the gaseous hydrocarbons, air, and other gaseous matters are being expelled from the fibrous materials. If an open vessel is used instead of a closed boiler, it will be necessary to keep the solution at a fairly uniform strength, and if necessary to increase the strength, but the result will be substantially the same; but as it is evident that, when using an open boiler, the excess of sulphurous acid supplied during the boiling will be constantly driven off as gas, it must be replaced by further injections, while the acid fumes may be conveyed away and condensed, so as to be available for further use. When the fibrous substances are boiled as above, with the addition of potash, soda, etc., during the boiling, the result will be equally beneficial. The inventor prefers to inject the sulphurous acid or its combinations into the boiler at the bottom, and to cause it to come in contact with the solution therein before reaching the fibrous materials. For this purpose there is formed a kind of chamber beneath the boiler, but separated from it by a perforated disc or diaphragm of lead or other suitable material not acted upon by the solution, so as to allow the latter to fill the chamber, to which is connected a pipe, through which the sulphurous acid or solutions of the sulphites is forced by any suitable apparatus.

Objections to the Acid or Bisulphite Processes.—While the various methods of boiling wood in caustic soda at high temperatures are well known to be open to serious objections, the acid treatment of wood also presents many disadvantages, which it is to be hoped may be yet overcome. In reference to this, Davis makes the following observations: —"In the acid treatment of wood for the purpose of converting the fibres into pulp for use in paper manufacture, the general practice has been to use alkaline solutions of soda, combined in various proportions with certain acids, such, for instance, as sulphurous acid, hydrochloric acid, etc. These solutions have been heated in digesting vessels, and the high temperature resulting from this process of heating developing a pressure of from six to seven atmospheres, the wood being disintegrated by the action of the boiling solutions, the gum, resinous constituents, and other incrustating or cementing substances that bind the fibres together are decomposed, destroyed, or dissolved, while pure cellulose, which constitutes the essential element of the ligneous fibres, is separated therefrom. To this end high temperatures had to be employed, otherwise the disintegration was found to be only partial, the wood remaining in a condition unfit for further treatment. The high temperature not unfrequently converts a large proportion of the resinous and gummy constituents of the wood into tar and pitch—that is to say, carbonaceous bodies that penetrate into the fibre and render its bleaching difficult, laborious, and costly, while the frequent washing and lixiviation necessary to bleach such products seriously affect the strength of the fibre and its whiteness, and also materially reduce the percentage of the product, in some instances to the extent of 18 per cent. These difficulties and detrimental results materially enhance the cost of production, while the fibre itself suffers considerably in strength from the repeated action of the chloride of lime.... The difficulties are chiefly due to the carbonisation of certain constituent parts of the fibres under temperatures exceeding 212° F., such carbonised matters being insoluble and incapable of being bleached, and as they permeate the fibre, cannot be entirely removed.

"To overcome these difficulties, the wood should be chemically treated at a temperature sufficiently low to ensure that the decomposition of the connecting substances of the fibres will remain chemically combined with the other elements, such as hydrogen, oxygen, and nitrogen, in order to obtain an increased product of superior quality and render the process more economical."

Sulphite Fibre and Resin.—A German manufacturer sent the following communication to the *Papier Zeitung*, which may be interesting to the users of sulphite pulp:—"In making [disintegrating] cellulose by the soda or sulphite process, the object in boiling is to loosen the incrusting particles in the wood, resin included, and to liberate the fibres. The resin is dissolved both in the soda and sulphite processes, but in the former it is at the same time saponified, and is consequently very easily washed out. In the case of sulphite fibre, however, the resin attaches itself by its own adhesiveness to the fibres, but can also be removed by as hot washing as possible, and adding a little hydrochloric acid, which produces a very great effect. At the same time, however, sulphite fibre loses in whiteness by thorough washing, and assumes a reddish-grey shade. As the paper manufacturer insists upon white fibre, the manufacturer of sulphite fibre not only often omits washing, but adds some sulphite solution (bisulphite of lime). This not only enables him to give his customers white fibre, but he also sells a quantity of the incrusting particles and sulphite residuum as cellulose.

"So long as the manufacturer looks more to white than to well-washed cellulose, or does not wash it well before working up the fibre, these annoyances cannot be avoided. Not only this, but other disadvantages will be added in the course of time, as the action of the sulphurous acid in the pulp will have very injurious consequences on metals—[and on the fibre itself?] especially iron—coming in contact with it. This should be the more avoided, as the whiteness of the unwashed cellulose is of very short duration. The paper made from it soon turns yellow and becomes brittle. Well-washed sulphite fibre, on the other hand—provided no mistakes have been made in the boiling process—makes a strong, grippy paper, which can withstand both air and sunlight. I have made no special studies as to resin, but believe that pine and fir act differently, especially with solvents."

Adamson's Process.—Mr. W. Adamson, of Philadelphia, obtained a patent in 1871 for the use of hydrocarbons in the treatment of wood. His process consisted in treating the wood with benzine in closed vessels, under a pressure of 5 to 10 lbs., according to the nature of the wood. His digester consisted of an upright cylinder, in which the wood-shavings were placed between two perforated diaphragms. The mass was heated beneath the lower diaphragm by a coil through which steam was passed. The vapours which were given off were allowed to escape through a pipe on the top of the digester, to which was connected a coil immersed in a vessel of cold water, and the condensed liquid then returned to the lower part of the digester. The remaining portion of the benzine in the digester, which was still liquid but saturated with the extracted matters, was drawn off through a faucet at the bottom. Benzine being a very cheap article in America, a similar process was recommended in another patent by the same author for extraction of pitch and tar from rags [tarpaulin, ropes, &c.?], and for removing oil from rags and cotton waste.

Sulphide Processes.—Many attempts were made about thirty years ago, and in subsequent years, to employ the soluble sulphides as a substitute for caustic soda in boiling wood and other fibres, but these processes do not appear to have been very successful. Later improvements in the construction of boilers or digesters, however, seem to have induced further experiments in this direction, and we understand that several sulphide processes are being worked on the Continent, the processes of MM. Dahl and Blitz being amongst them. One of the supposed advantages of these sulphides over caustic soda is that by evaporation and calcination of the liquors, or leys, by which the organic matters become destroyed, the original product would be recovered, which merely requires to be dissolved out for further use. There are, however, several important objections to the use of sulphides in this way, amongst which may be mentioned the deleterious vapours which they emit; and this alone would doubtless prevent their employment—at all events in this country.

II. Mechanical Processes.—Besides the various chemical methods of separating cellulose from woody fibres, before described, certain processes have been devised for reducing wood to the condition of pulp directly by mechanical means without the aid of any chemical substance whatsoever. In this direction Heinrich Voelter, of Wurtemburg, appears to have been the first to introduce a really practical process for the conversion of wood into pulp for paper-making, although, as far back as 1756, Dr. Schaeffer, of Bavaria, proposed to make paper from sawdust and shavings mechanically formed into pulp: the process was not successful, however, with the machinery then at his command.

Voelter's Process for Preparing Mechanical Wood Pulp.— In 1860-65 and 1873 Voelter obtained patents in this country for his methods of treating wood mechanically, and the process may be thus briefly described:—Blocks of wood, after the knots have been cut out by suitable tools, are pressed against a revolving grindstone, which reduces the material to a more or less fine condition, but not in a powdery form, and the disintegrated fibre is caused to press against a wire screen, which allows the finer particles to pass through, retaining the coarser particles for further treatment.

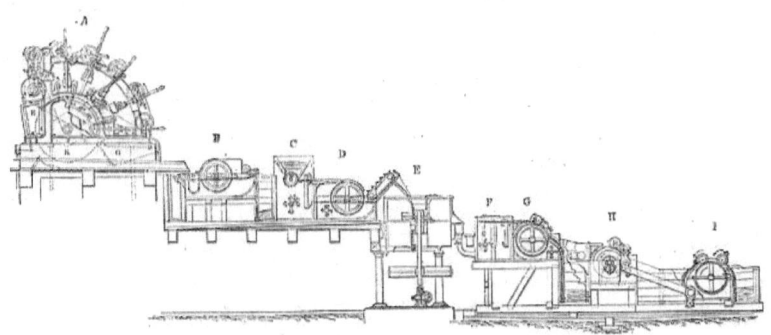

Fig. 17A.—Voelter's Wood-pulping Machine.

[*To face page 78.*

The apparatus employed, which is shown in Fig. 17A, consists of a pulping apparatus A, with vat K, in which the revolving stone S is placed; the blocks of wood are held against the stone at *p p*, and water is introduced at G, and the revolving stone carries the pulp against the screen E, which admits the passage of the finer particles of the wood, while the coarser particles are led by the trough F to the first refining cylinder B, after passing through an oscillating basket, which retains the coarser particles. From thence it is led through a distributing apparatus and hopper C, to be uniformly supplied to the refining cylinder D, these cylinders being of the ordinary construction, and, as usual, covered with fine gauze wire sieves. The ground material which fails to pass through the sieves is transferred by an elevator to the millstones E, which are of ordinary construction, and after leaving these unites with the finer fibres which pass through E, the whole now entering a mixing reservoir F, whence it is thrown on to the cylinder G, and the pulp which passes into this is distributed on to a similar cylinder H, the contents of which then passes through the last cylinder I, which is differently constructed to the others, inasmuch as its lower part is surrounded by an impervious leather jacket, so that the pulp ascends in order to enter it. The disintegrated fibres that are retained by the wires of the cylinders pass into the refiners, which consist of a pair of horizontal cylinders of sandstone, one of which (the upper one) only revolves, and by the action of these the coarser fibres become further reduced, the finer particles, as before, passing through the wire gauze of the cylinders, the operation being repeated in the same order until the whole of the fibres have passed through the sieves.

Thune's Process.—Mr. A. L. Thune, of Christiana, U.S.A., has recently patented an apparatus for disintegrating wood, which consists of a grinding apparatus connected to a turbine. In this arrangement the grindstone, fixed on a shaft, is worked by a turbine, and the wood, which is used in small blocks, is pressed against the stone by means of a series of hydraulic presses. The fine pulp is afterwards made into thick sheets by means of a board-machine, the pulp, mixed with water, passing down a shoot into a vat beneath, in which is a revolving cylinder covered with wire-cloth, which in its revolution carries with it a certain quantity of pulp in a continuous sheet; this is taken on to an endless travelling belt by means of a small couch-roll, and passes on to a pair of rolls, round the upper one of which the sheet becomes wound, and is removed when sufficiently thick.

CHAPTER VIII.

TREATMENT OF VARIOUS FIBRES.

Treatment of Straw.—Bentley and Jackson's Boiler.—Boiling the Straw.—Bertrams' Edge-runner.—M. A. C. Mellier's Process.—Manilla, Jute, etc.—Waste Paper.—Boiling Waste Paper.—Ryan's Process for Treating Waste Paper.

Treatment of Straw.—As a paper-making material, the employment of straw is of very early date, a patent for producing paper from straw having been taken out by Matthias Koops as far back as 1801. The material, however, was used in its unbleached state, and formed a very ugly paper. White paper was not obtained from straw until 1841, but no really practical method of treating this material was devised until about ten years later, in France, when MM. Coupier and Mellier introduced a process which, with subsequent modifications, has been extensively adopted. A great advance in the manufacture of paper from straw has since been effected by the introduction of various boilers, specially constructed for boiling the material at high pressures, and for keeping the alkaline liquors freely circulated amongst the fibre during the progress of the boiling. These boilers are of different forms—being either cylindrical or spherical—and are preferably of the revolving type, which causes the caustic ley employed in the boiling to become uniformly mixed with the fibre. Sometimes the vomiting boilers described elsewhere are used by paper-makers in preference to those referred to.

Bentley and Jackson's Boiler.—This boiler, a representation of which is shown in Fig. 18, is 7 feet in diameter, 18 feet long on the cylindrical surface, with hemispherical ends of Martin-Siemens steel plate 7/16 inch thick in the shell, and ½ inch thick in the ends. It is double riveted in the longitudinal seams, has two manholes 3 × 2, forged out of solid steel plate. Inside are two perforated lifting plates or shelves, each 1 foot wide, ¼ inch thick, the full length of the shell, and secured to the ends by strong angle-irons; it is supported on two turned cast-iron trunnions. These boilers are tested by hydraulic pressure to 120 lbs. per square inch.

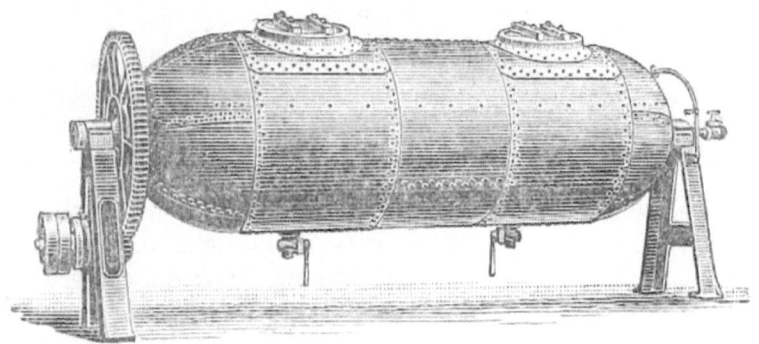

Fig. 18.

The varieties of straw generally used for paper-making in this country are wheat and oats, though rye and barley straws are also used, but in a lesser degree. The treatment of straw differs greatly at different mills, some makers using strong liquors and boiling at a lower pressure, while others prefer to use less caustic soda and boil at a higher pressure. There can be little doubt, however, that the high temperatures resulting from boiling at very high steam pressure must deteriorate the fibre considerably, causing subsequent loss of fibre in the processes of washing and bleaching.

Boiling the Straw.—The straw is first cut into short lengths of one or two inches by means of a chaff-cutter, or by a machine similar to a rag-cutter, and the cut material is then driven by an air-blast through a wooden tube into a chamber having coarse wire-gauze sides: a second chamber surrounds this, in which the dust from the straw collects as it passes through the wire gauze. The winnowed straw, freed from dust and dirt, is then conveyed in sacks to the boilers. In charging the boilers, a certain quantity of ley is first introduced, and steam also, and the cut straw then added, which soon becomes softened, and sinks to the bottom of the boiler, when further quantities of the material are added, until the full charge has been given. The requisite proportion of ley and water is then run in and the head of the boiler secured in its place. Steam is now turned on, until a pressure of 20 to 40 lbs., or even more, has been reached, when the boiling is kept up for 3½ to 8 hours, according to the pressure used and the strength of the alkaline liquor, which varies from 9° to 16° Tw. From 10 to 20 lbs. of caustic soda per cwt. of straw are generally required to boil the material thoroughly. The boiling being complete, steam is turned off, and when the boiler has somewhat cooled, the material, which is in the form of a pulp, is discharged by the pipes beneath into a large tank or strainer, the bottom of which is fitted with a series of plates having long narrow openings or slits, through which the liquor drains. The pulp is then washed with water, and again allowed to drain thoroughly, after which it is dug out and transferred to the potcher to be again washed and bleached. At some mills the straw is boiled whole and not subjected to any preliminary cutting In such cases the boiled straw, not being so fully pulped as when cut into short lengths, is emptied from the boiler through the manholes used for charging the material into the boiler.

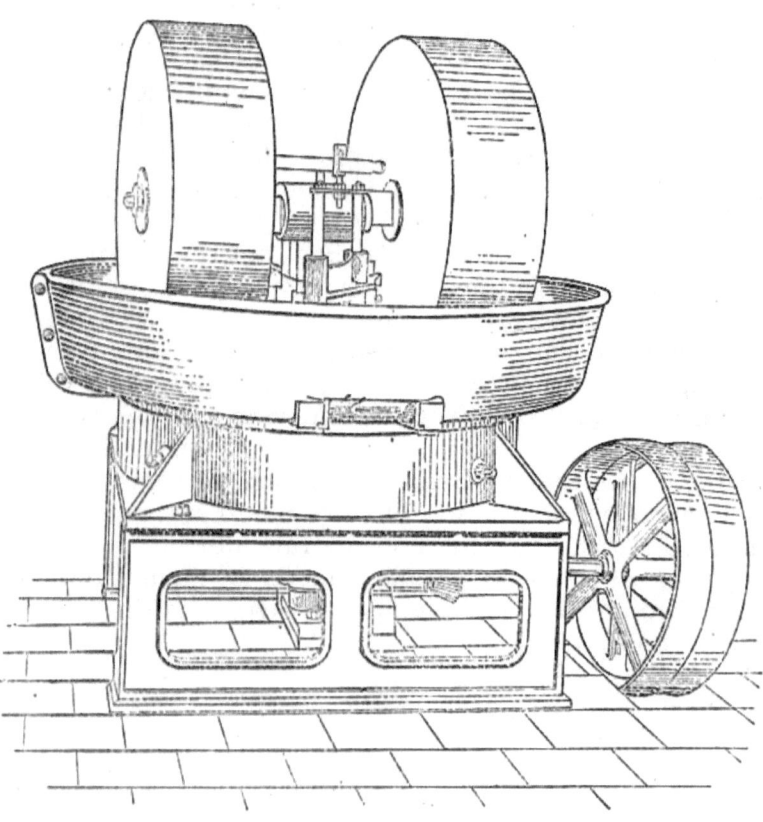

Fig. 19.

Bertrams' Edge-runner.—For the purpose of crushing the knots of the straw, and other hard particles derived from weeds, etc., a machine termed the "koller-gang" or "edge-runner" is sometimes employed. This machine, which is manufactured by Bertrams, Limited, and of which an illustration is given in Fig. 19, consists of two large millstones, made from hard red granite, the surfaces of which are sometimes grooved with V-shaped equidistant grooves. These stones are worked by a horizontal spindle, and are caused to revolve very rapidly in an iron basin, in which the washed pulp is placed, and by this means the knots and harder portions of the fibre not fully acted upon by the caustic alkali, become so reduced as to be more readily accessible to the action of the bleach, and thus a very superior straw pulp is produced. In using this machine in the way indicated, the washed pulp is mixed in a chest provided with agitators, with water, is then pumped into a second chest above it, from whence it flows into the basin shown in the engraving, while the stones are revolving.

M. A. C. Mellier's Process.—By this method the straw is first cut into small lengths as usual; it is then steeped for a few hours in hot water, and afterwards placed by preference in a jacketed boiler, the object being to heat the materials without weakening the ley by the direct introduction of steam into the body of the material. The boiler is to be heated to a pressure of 70 lbs. to the square inch, or to a temperature of about 310° F., by which means, it is said, a considerable saving of alkali is effected, as also time and fuel, as compared with the ordinary practice of boiling. The alkaline ley which M. Mellier prefers to use is from 2° to 3° B., or of the specific gravity of from 1·013 to 1·020, and in the proportion of about 70 gallons of such solution to each cwt. of straw. The boiler should revolve very slowly, making about 1 or 2 revolutions per minute. The boiling occupies about 3 hours, at the pressure named, when the steam is turned off and cold water passed through the jacket of the boiler, which assists in cooling the pulp, the water thus used being afterwards employed in washing the pulp. The pulp is then thoroughly washed until the last water runs off quite clear, when it is next steeped for about an hour in hot water acidulated with sulphuric acid, in the proportion of about 2 per cent. of the weight of the fibre. The pulp is then washed with cold water, when it is ready for bleaching in the usual way.

Manilla, Jute, etc.—Previous to boiling these fibres it is usual to cut them into short pieces by a machine such as is used for cutting straw, after which they are cleaned in a willowing and dusting machine. The boiling is then conducted in the same way as for esparto. Manilla fibre is not so much used in this country as in the United States, where its employment forms an important feature in the manufacture of certain kinds of paper. Some idea of the extent to which it is used by the paper-makers of America may be gleaned from the following statement of Mr. Wyatt:

—"Another large and important branch of the American paper trade are the mills running on news and Manilla paper. Many of these mills turn out a vast quantity of paper, running up to two hundred tons per week, besides making their own ground wood pulp. The American news is composed mainly of ground wood pulp, with an admixture of about 15 to 25 per cent. of sulphite wood or jute fibre, and not much loading, and the machines are run at high speed. What is termed Manilla paper is very largely used in the States, and much more so than with us for common writings, envelopes, and wrapping papers. The paper is composed of Manilla, jute fibre, old papers, etc., and is highly finished at the machine. I was told of one mill belonging to a large company running altogether six mills on news and Manilla, turning out, with one 96-inch machine and beater capacity of 1,800 lbs., and one Jordan, 10 to 12 tons of 2,000 lbs., of Manilla paper per day at an average speed of 200 feet per minute."

Jute is seldom reduced to the condition of a fine white pulp since the treatment necessary to obtain that condition would result in a weak fibre; it is usual, therefore, to only partially reduce the material, when a strong fibre is obtained, which, lacking in whiteness, is used for coarse papers. This also applies to Adamsonia, or Baobab, another description of bast obtained from the West Coast of Africa. These fibres are chiefly used for papers which require strength rather than whiteness of colour, such as wrapping papers, &c.

"Broke" paper is a term applied to paper which has been imperfectly formed on the paper machine or damaged while passing over the drying cylinders. Imperfect sheets when they are not sold as *retree*, and clean waste paper, also come under this designation and are re-converted into pulp after undergoing the treatment described below.

Waste Paper.—In treating waste paper for conversion into pulp for paper-making, it is doubtless advisable to separate, as far as can be done economically, papers which have been written upon with common ink, as old letters, documents, &c., from printed papers, since the latter require a more severe treatment than the former. While simple boiling in water containing a little soda-ash will discharge ordinary writing ink, printer's ink can only be extracted by using rather strong solutions of soda-ash or caustic soda; and even with this treatment it can only be rendered serviceable for an inferior paper, owing to the grey colour of the resulting pulp, due to the carbon of the printer's ink, upon which the alkali has no solvent effect.

Boiling Waste Paper.—This is sometimes effected in iron vats, about 8 feet deep and 8 feet in diameter at the bottom, and about 6 inches wider at the top. At the bottom of each vat is a false bottom, closely perforated with small holes. Steam is introduced by a pipe below the false bottom, which passes through the perforations and thus becomes uniformly distributed to all parts of the vat. To facilitate the emptying of the vats, the false bottoms have connected to them three or four iron rods, to the tops of which iron chains are hooked, and by this means the false bottom, carrying the mass of boiled paper can be raised by a steam hoisting engine or crane and deposited where desired. When the boiling is commenced, the vat should first be about one-fourth filled with a solution of soda-ash, and the steam then turned on. When the liquor boils, the papers having been previously dusted, are introduced gradually, and well distributed through the liquor; if they are thrown into the vat in large quantities at a time, and especially if they are in a compact state, the portions in contact may not be reached by the liquor, and an imperfect boiling will be the result. To ensure a uniform distribution of the boiling liquor over the surface of the material, an iron pipe extends from the centre of the false bottom to nearly the top of the vat, and this pipe is covered with a hood, which causes the soda liquor to be evenly spread over the whole mass. The vats are either cased with wood or coated with asbestos to prevent the escape of heat, and the vessel is covered with a flat iron cover, which is generally in two halves. The steam enters the tubs at the side, below the false bottom, and the exhausted liquor is drawn off through a valve connected to the bottom of the vat. In some mills the liquor is not drawn off after each boiling, but the boiled paper is hoisted from the vat as before described, and the liquor strengthened by the addition of from 10 to 20 lbs. of soda-ash for each 100 lbs. of the paper to be next boiled. Paper that is thickly coated with printing ink requires an extra dose of soda-ash. The boiling is continued for twelve to twenty-four hours according to the nature and condition of the waste paper under treatment.

Waste papers are frequently boiled, after dusting, in revolving boilers, in a solution of soda-ash or caustic soda, but it not unfrequently happens that some portions of the material become so agglomerated or half pulped during the boiling that the alkali fails to reach all the ink, and as this cannot be removed by the after processes of washing and breaking, it remains in the body of the pulp and necessarily forms a constituent part of the paper to be produced from it. The mass, when discharged from the boiler and drained is then conveyed to the washing-engine, in which it becomes broken and freed from alkali and so much of the ink as may have been dissolved or loosened, and it is afterwards treated in the beater and mixed with varying portions of other paper stock, according to the quality of paper to be produced. In some mills the boiled waste paper is disintegrated after boiling, by means of the edge-runner (Fig. 19).

Ryan's Process for Treating Waste Paper.—The following process for treating waste paper so as to produce a "first-class clean paper" therefrom, was patented by Mr. J. T. Ryan, of Ohio. The waste paper is first passed through a duster in the usual way, all thick old books being previously torn apart to separate the leaves. The papers are then boiled in a hot alkaline liquor without pulping them, whereby the alkali acts on the surfaces of the papers, and dissolves off, carrying away all the ink into the liquor. The papers, which are still in sheet form, are then drained as free as convenient from the alkaline liquor, and are next washed in the washing-engine, which leaves the material perfectly clean. It is then pulped in the beating-engine; and it is claimed that it can be formed into first-class paper without the addition of any new or expensive paper stock. The details of the process are thus given by the patentee: "Into a bucking-keir put a soda-ash solution having a density of 5° B., at 160° F., put in the stock, and shower for eight hours at a temperature of 160° F., without pulping the paper, then lift and drain, and cleanse well in the washing-engine; then pulp and form into paper. As the draining will always be imperfect, each charge removed will carry away some of the soda-ash solution, and leave the remainder of impaired strength. After each drainage add water to make up for loss in quantity of the solution, and add enough soda-ash solution at a density of 13° B., to bring all the liquor up to 5° B. at 160° F. In about eighteen working days the liquor will have accumulated considerable ink and other matter. Then blow one half of the liquor, and restore the quantity for proper working. None of the soda-ash solution is wasted, except such as falls to drain and what is blown out as last mentioned." In carrying out this process every care must be taken to guard against pulping before the alkali is washed out.

CHAPTER IX.

BLEACHING.

Bleaching Operation.—Sour Bleaching.—Bleaching with Chloride of Lime.—Donkin's Bleach Mixer.—Bleaching with Chlorine Gas (Glaser's Process).—Electrolytic Bleaching (C. Watt's Process).—Hermite's Process.—Andreoli's Process.—Thompson's Process.—Lunge's Process.—Zinc Bleach Liquor.—Alum Bleach Liquor.—New Method of Bleaching.

Bleaching Operation.—The half-stuff treated in the breaking-engine is run into the potcher, and the water it contains is lifted out as far as practicable by the washer; the spent liquor from the presses or drainers is then run in in lieu of water, and as much fresh bleaching liquor as may be required is then measured in, and in from two to six hours the pulp becomes perfectly white. "However well managed a mill may be," says Mr. Arnot, "it is scarcely possible to avoid having a small residue of unused chlorine in the liquid which drains from the bleaching stuff." The rule, therefore, is to use this liquor in the way above indicated, by which the unexhausted chlorine, operating upon fresh half-stuff, becomes available, and is, therefore, not wasted. "That as little of this residual chlorine as possible may remain in the stuff," Mr. Arnot further observes, "when put into the beating-engine, powerful hydraulic presses are employed to compress the stuff and squeeze out the liquid. These presses should be large enough to contain easily the whole contents of a poaching-engine, and of unexceptional workmanship. The perforated lining especially should be carefully prepared and properly secured. I have seen much trouble from negligent workmanship in this respect. Recently I examined a number of samples of press drainings, and found the unexhausted chlorine to vary very much—from a few grains of bleaching powder per gallon to about one ounce."

Sometimes it is the practice to partly fill the potcher with water, and the engine being set in motion, the half-stuff is gradually introduced until the full charge has been given, and the stuff is then washed for some time, after which the drum-washer is raised, and the bleaching liquor then run in, care being taken that the necessary quantity is not exceeded, otherwise the fibre will suffer injury from the chemical action of the bleaching agent. When vitriol is employed to liberate the hypochlorous acid, the vitriol, previously diluted with water, should be placed in a small lead-lined tank in such a position that the acid liquor may slowly trickle into the engine at the rate of 1 lb. of sulphuric acid in twenty minutes. As soon as the bleaching is complete the stuff is emptied into large stone chests, each of which will hold the contents of two engines. On the bottom of these chests are perforated zinc drainers, while a similar drainer runs up the back of each chest. The bleached stuff is allowed to remain as long as may be convenient in these chests, after which it is removed to the beating or refining engines. In some mills the bleaching is effected in the breaking-engine, while at other mills the operation is performed in the beating-engine.

In bleaching it is considered to be more advantageous to employ moderately strong liquors rather than weaker ones, inasmuch as the object is effected in less time than when weaker liquors are employed. An extreme in the opposite direction, however, must be avoided, since a very strong bleach will inevitably cause injury to the fibre. Sometimes the potchers are fitted with steam-pipes, in order that the diluted bleaching liquor may be heated, if required, to facilitate the operation. If the temperature be raised too high, however, the effect upon the fibre will be at least as injurious as if too strong a bleach were employed. It must also be borne in mind that in either case, after the pulp has been bleached and the liquor allowed to run off, the mass has to remain some time—even if pressed to remove as much of the liquor as possible—in direct contact with the products resulting from the decomposition, and probably some undecomposed hypochlorite also, which will continue their chemical action upon the fibre until removed by washing, or neutralised by one or other of the agents employed for the purpose.

Sour Bleaching.—When the bleaching liquor, after acting upon the half-stuff for some time, has become partially exhausted, dilute sulphuric acid—about one part acid to fifteen parts of water—is added, which, by liberating hypochlorous acid, hastens the bleaching considerably, and when the chemical action resulting from this treatment is nearly complete, the spent liquor is allowed to drain away, and fresh bleaching liquor is introduced, the strength being regulated by the progress made in the first case, which will depend upon the character of the fibre treated. In the second application of the bleach no acid is used. When sulphuric acid is added to the bleaching liquor, as above, the process is termed *sour bleaching*. Sometimes hydrochloric acid is used for this purpose, but in either case it is necessary to avoid employing the acid in too concentrated a state, or in too great a quantity, otherwise free chlorine will be liberated, which, besides being injurious to the health of the workmen and the surrounding machinery, also involves loss, while the colour and strength of the fibre itself will also be impaired. In some mills the bleaching is effected in the beating-engine, the bleaching liquor being pumped in while the machine is in motion.

Respecting the time which the bleaching operation should occupy, there appears to be some difference of opinion, or, at all events, the practice seems to vary in different mills, but there is, no doubt, an advantage, so far as ultimate yield is concerned, in moderately slow bleaching at a moderate temperature, inasmuch as there is less risk of chemical action upon the cellulose itself than when strong liquors are used, at a higher temperature, with a view to hasten the operation and economise the bleaching powder.

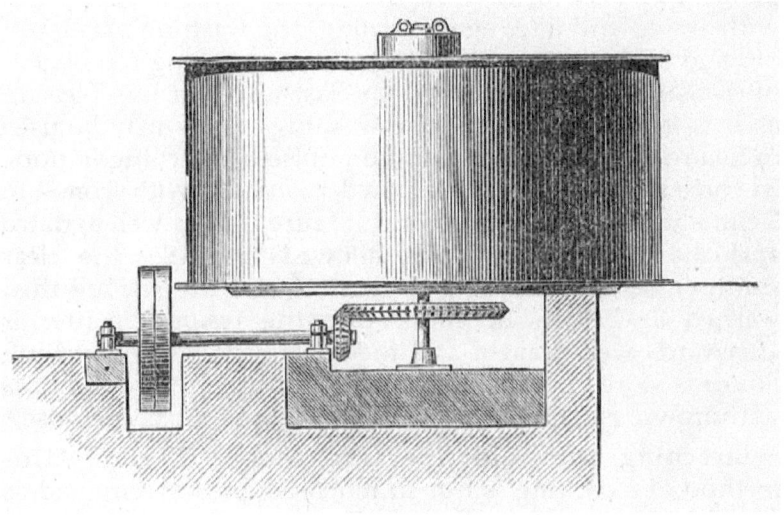

Fig. 20.

Bleaching with Chloride of Lime (*Preparation of the Bleaching Liquor*).—Chloride of lime, or hypochlorite of lime, commonly called bleaching powder, when well prepared, contains from 32 to 35 per cent. of active chlorine. Being readily decomposed by the air, and also by heat, this substance should always be stored in a cool and dry place until required for use. A solution of bleaching powder is generally prepared in large tanks lined with lead, which are provided with agitators or stirrers, so that the powder, when added to the water, may be freely diffused, and its active material dissolved in the liquid. A machine, or "bleach-mixer," manufactured by Messrs. Bryan Donkin and Co., of Bermondsey, is shown in Fig. 20, which is so constructed that the strong bleach liquor does not destroy it. The device for agitating the contents of the tank explains the principle of the machine. To prepare the bleaching liquor, about ½ lb. of chloride of lime to each gallon of water is used, which yields a liquor at about 6° T. When the required quantity of bleaching powder and water have been introduced into the mixer and sufficiently agitated, the vessel is allowed to rest until the residue, which chiefly consists of free lime and its carbonate, has subsided, when the clear liquor may be run off for use. When all the clear liquor has been drawn off the residue should be washed with water, and after again settling, the washing water run off, and fresh water added, these washings being repeated as often as necessary to remove the last traces of the "bleach," as it is technically called. The washing waters may be used in lieu of water in the preparation of fresh bleaching liquors. In some mills the bleaching powder is mixed with from 2 to 3 times its weight of water; the mixture is then well agitated and the residue afterwards allowed to settle, the clear solution being afterwards drawn off and the residue then washed as before. In either case the residual matter is afterwards well drained and then cast aside. The bleaching liquor is stored in large tanks ready for use, from which it is withdrawn as required by means of a syphon or otherwise.

Bleaching with Chlorine Gas (*Glaser's Process*).—This method of bleaching is not so much adopted in England as formerly, but has found much favour in Germany; indeed, within the past few years, namely, in March 3rd, 1880, a process was introduced by Mr. F. Carl Glaser for treating straw, in which, after boiling with caustic soda as usual, the pulp is bleached by the action of chlorine gas. The straw, after being separated from weeds by a slight or superficial picking, is cut into pieces of from $\frac{1}{3}$ to $\frac{2}{3}$ of an inch in length. The cut straw is then placed in a rotary boiler for about four hours, at a pressure of about 4 to 4½ atmospheres, in a solution composed of 29 lbs. of caustic soda at 71°, and 48 lbs. of calcined soda at 90°, rendered caustic, for every 220 lbs. of straw. After boiling, the dirty ley is drawn off, and the boiled straw subjected to two washings with water. It is then conveyed to the washing-engine, where it is washed for an hour; the drum of the machine should have a sieve or sifter, the apertures of which are about 60 to the square inch. The washed straw is next dried by centrifugal force in a hydro-extractor, until it contains about 70 per cent. of water, which is necessary for the action of the chlorine gas. To effect this, so as to obtain not very solid or close cakes of straw, the holes of the wire of the hydro-extractor should not be more than 50 to the square inch. The cakes of straw thus formed are then exposed to the action of chlorine in leaden chambers of the ordinary kind, in which they are placed in layers upon hurdles, or upon shelves. If the chlorine is produced by hydrochloric acid, for every 220 lbs. of unboiled straw, 51½ lbs. of the acid at 20° B., and a corresponding quantity of 70 per cent. peroxide of manganese are used. After the bleaching operation, the acid formed is removed by washing in a washing-engine. If a complete reduction of the fibres has not been effected by the bleaching, this may be completed by the aid of well-known machines, and either before or subsequent to the after-bleaching there is used for 220 lbs. of straw about 4½ lbs. of chloride of lime, at 35° [per cent.?] The patentee then gives the following explanation:—"As pine wood or fir is chemically freed from its colouring principle and transformed into fibres as well as cellulose, the object of the intense action of the chlorine is to destroy the mucilage of the straw, as well as the incrusting matters which have not been destroyed by the boiling with caustic soda, and consequently to strip or expose and open the fibres." It will be readily seen that this process bears a close resemblance to Mr. C. Watt's wood-pulp process.

Electrolytic Bleaching (*C. Watt, jun.'s, Process*).—At the present time, when the means of obtaining the electric current for practical purposes in the arts have so far exceeded that which would have been deemed probable some forty years since, we find that many ingenious processes, which were found to be unpractical at that time from the want of cheap electrical power, have since reappeared in the form of patented inventions, which would seem to possess every merit—but originality.

So long ago as September 25th, 1851, the author's brother, Mr. Charles Watt, obtained a patent for, amongst other claims, decomposing chlorides of sodium and potassium, and of the metals of the alkaline earths into hypochlorites by electricity. It may be well to make a few extracts here from his specification in order that some of the subsequent patents, to which we shall refer, may be traced to what may, perhaps, be considered their true origin. In the specification in question, the inventor says:—"The third part of my invention consists of a mode of converting chlorides of potassium and sodium, and of the metals of the alkaline earths, into hypochlorites and chlorates, by means of a succession of decompositions in the solution of the salt operated upon, when induced by the agency of electricity.... Electricity first decomposes the chloride, the chlorine being eliminated at one of the electrodes, and the alkaline or earthy metallic base at the other electrode.... The liberated chlorine will, when it is set free, combine with a portion of alkali or alkaline earth in the solution, and a hypochlorite will be formed. The hypochlorite thus formed will, by the continued action of heat, be resolved partly into a chlorate of the alkali or alkaline earth, and partly into a chloride of the metallic base, and the chloride will again be subjected to decomposition, and a hypochlorite formed.... If I desire to produce a hypochlorite of the alkali or earth, I merely keep the vessel warm ... and continue the process until as much of the saline matter has been converted into a hypochlorite as may be required for the purpose to which the solution is to be applied. This mode of forming a hypochlorite of the alkalies and alkaline earths may be used for preparing a bath for the purpose of bleaching various kinds of goods, and the bath may be strengthened [recuperated] from time to time by the action of the electric current."

Thus it will be seen that this specification clearly described a process by which the chlorides of sodium and potassium, and of the metals of the alkaline earths (chloride of magnesium, for example), may be converted into hypochlorites by electrolysis, and the hypochlorite solution obtained used for the purposes of bleaching. It would appear difficult to conceive how any subsequent patent for accomplishing the same thing, and using essentially the same means, can claim originality in the face of such "prior publication" as was effected by the usual "Blue-book," which any person can buy for eightpence.

Hermite's Process.—The following description of this process has been furnished by the engineers engaged in connection with the process to the *Paper Trade Review*: —"Briefly described, the Hermite process consists in manufacturing a solution of high bleaching power by electrolysing an aqueous solution of magnesium chloride. The salt is decomposed by the current at the same time as the water. The nascent chlorine, liberated from the magnesium chloride, and the nascent oxygen, liberated from the water, unite at the positive pole, and produce an unstable oxygen compound of chlorine of very high bleaching power. The hydrogen and magnesium go to the negative pole; this last decomposes the water and forms magnesium oxide, whilst the hydrogen is disengaged. If in this liquid coloured vegetable fibre is introduced, the oxygen compound acts on the colouring matter, oxidising it. Chlorine combines with the hydrogen to form hydrochloric acid, which finding itself in the presence of magnesium in the liquid combines with it, and forms the initial chloride of magnesium."

Andreoli's Process.—This process consists, avowedly, in bleaching pulps "by means of hypochlorite of sodium, produced by electrolytical decomposition of a solution of chloride of sodium." In carrying out his process, M. Andreoli uses as an electrolyte "concentrated or non-concentrated sea-water, or a solution of chloride of sodium, the specific gravity of which varies according to the quality and nature of the materials to be treated. Generally the solution to be electrolysed works better with a density of 8° to 12° B., but although salt is cheap, and the solution when exhausted may be regenerated by passing an electric current, I always endeavour to have when possible (*sic*) a weak solution, and with some kinds of pulp an electrolyte having the density of sea-water (3° B.) is sufficiently strong to bleach."

The foregoing are the only electrolytic processes for bleaching fibres that need recording, and we fancy there will be little difficulty in tracing the resemblance between the two latter and the process of Mr. C. Watt.

Thompson's Process.—This process, for which a patent was obtained on February 3rd, 1883, may be thus briefly described:—In bleaching linen fabrics the material is boiled for about three hours in a solution of cyanide of potassium or sodium—about half an ounce of the salt to each gallon of water—to remove the resinous matter from the fibre, so that the cellulose may be exposed to the action of the bleach. The fabric is then washed, and again boiled for three hours more in a similar solution, and after being again washed is ready for bleaching. With cotton the preliminary boiling is not necessary, unless the material is greasy, in which case a solution of half the strength and two hours' boiling is sufficient. In ordinary cases cotton is not boiled at all, but is simply washed in cold water and squeezed. In bleaching, all vegetable fibres are treated in the same way, the only difference being in point of time. The cotton or linen, after being treated as described, is then piled somewhat loosely in an air-tight vessel, 9 lbs. of cloth to the cubic foot of space being considered sufficient. The vessel is then filled with a weak solution of bleaching liquor, consisting of about one ounce of dry bleaching powder to each gallon of water. "After the vessel has been filled, the liquor is immediately run out, and is replaced by an atmosphere of carbonic acid, which quickly liberates the chlorine on the fibre, and thus decomposes the water, uniting with the hydrogen and liberating the oxygen, the result of which, is to bleach the fibre or fabric. In about an hour the whole of the bleaching liquor in the fibre will have been thus decomposed, and this operation must be repeated until the material is of the proper whiteness to be withdrawn from the action of the chlorine. The material is then washed and squeezed. Chlorine, however, always leaves these materials of a yellowish white." To remove this tint, the material is passed through a solution of oxalic acid—about 2 oz. to the gallon—squeezed as it passes out of this solution, and then passed through another solution made by dissolving ¼ grain of triethyl rose aniline to the gallon of water, or 20 grains of indigo, as may be preferred. To this solution oxalic acid is added until it becomes of an opaque but bright turquoise blue. The material, after washing, is then white.

The patent describes and illustrates the apparatus to be used in conjunction with certain parts of existing apparatus used in bleaching.

Lunge's Process.—In this process acetic acid is used in place of hydrochloric or sulphuric acids, etc., to set free the chlorine or hypochlorous acid, in the ordinary method of bleaching with hypochlorite of lime, or bleaching powder, which, the inventor says, "combines all the advantages of the materials formerly employed, without any of their drawbacks.... The price is no impediment, for a minimal quantity is sufficient, the same being regenerated over and over again. At first acetic acid and chloride of lime decompose into calcium acetate and free hypochlorous acid.

In the bleaching process the latter yields its oxygen, hydrochloric acid being formed. The latter instantly acts upon the calcium acetate; calcium chloride is formed and acetic acid is regenerated, which decomposes a fresh quantity of chloride of lime, and so forth. Consequently the smallest quantity of acetic acid suffices for splitting up any amount of chloride of lime.... The hydrochloric acid formed is never present in the free state, as it instantly acts upon the calcium acetate. This is very important, since hydrochloric acid weakens the fibre by prolonged contact, whilst acetic acid is quite harmless. Since there are no insoluble calcium salts present, the operation of 'souring' after bleaching is quite unnecessary; this not merely saves the expense of acid, and of the subsequent washing of the fabrics, but it also avoids the danger, especially present in the case of stout fabrics, of leaving some of the acid in the stuff, which concentrates on drying and weakens the fibre; it may also prove injurious in subsequent dyeing operations. But in the new process no free acid is present except acetic acid, which has no action upon fibre, even in its concentrated state and at a high temperature."

The acetic acid may be employed in various ways, including the following:—A small quantity of the acid may be added from the first to the bleaching liquor; or the fabric, after being treated in the ordinary way with a solution of the bleaching powder, may be steeped, without previous washing, in water containing a little acetic acid; or the fabric may be steeped in water acidulated with acetic acid, and bleaching liquor afterwards run in slowly and gradually, with continuous agitation in the usual way. In the case of hard water, or of impure bleaching liquors, a good deal of the acetic acid would be consumed in neutralising the lime; in this case, some hydrochloric or sulphuric acid may be added, but only sufficient for the purpose, so that no acid but hypochlorous or acetic acid exists in the free state. The process is applicable to the bleaching of vegetable fibres, whether spun or in the unspun state, and for bleaching paper pulp made from rags, wood, straw, esparto, etc. Besides acetic acid, any other weak organic acid of an analogous nature may be used.

Zinc Bleach Liquor.—Strong acids are often objectionable for liberating chlorine from bleaching powder, and especially in bleaching some classes of paper pulp. If a solution of sulphate of zinc be added to one of bleaching powder, sulphate of lime is precipitated, and the zinc hypochlorite formed at once splits up into zinc oxide and a solution of free hypochlorous acid. Chloride of zinc acts similarly; for a saturated solution of zinc in hydrochloric acid decomposes as much bleaching powder as half its weight of concentrated oil of vitriol.—*Varrentrapp.* Consequently zinc salts can be employed in place of sulphuric acid, and thus bleach the paper pulp very quickly. When this mixture is employed in bleaching pulp, the precipitated sulphate of lime resulting from the reaction and also the oxide of zinc formed, remain in the pulp, and serve as loading materials.

Alum Bleach Liquor.—Orioli[19] recommended for use, in paper-mills especially, a bleach liquor made by decomposing equivalent quantities of a solution of chloride of lime and sulphate of alumina, formerly known as *Wilson's Bleach Liquor.* Sulphate of lime is precipitated, and hypochlorite of aluminium remains in solution; this being a very unstable salt can be applied for bleaching without the addition of an acid, splitting up into aluminium chloride and active oxygen. Consequently the liquid always remains neutral, and the difficulty caused by the obstinate retention of free acid in the fibre, by which it is strongly acted upon in drying, in this case does not exist. The aluminium chloride also acts as an antiseptic, so that the paper stock may be kept for many months without undergoing fermentation or other decomposition. The solution is allowed to act for about ten minutes in the engine.—*Lunge.*

New Method of Bleaching.—Young's Paraffin Oil Company have recently introduced what they term an "intermediate oil for paper-making," to be used with alkali in the boiling of rags and esparto, for the purpose of increasing the bleaching power of the powder, and producing a softer pulp, at the same time having no smell. Several well-known paper-makers have tried, and speak favourably of it. The quantity of oil to be added to the caustic varies for different stock, but may be said to average about 1½ gallon per ton.
[20]

CHAPTER X.

BEATING OR REFINING.

Beating.—Mr. Dunbar's Observations on Beating.—Mr. Arnot on Beating Engines.—Mr. Wyatt on American Refining-Engines.—The Beating Engine.—Forbes' Beating-Engine.—Umpherston's Beating Engine.—Operation of Beating.—Test for Chlorine.—Blending.

Beating.—One of the most important operations in the manufacture of first-class paper is that of *beating*, by which the half-stuff becomes reduced to a fine state of division, and the fibres which, in the condition of half-stuff, are more or less loosely held together in a clotted state, become separated, and are thus put into a condition in which they will intertwine with each other, or *felt*, as it is termed, when submitted to the vibratory motion of the wire-cloth of the paper machine. The beating-engine, or beater, as it is commonly called, much resembles in construction the washing- and breaking-engine, but since it is required to still further reduce the pulp to a condition suitable for paper-making, the knives of this engine are more numerous and are made to revolve more rapidly. In this engine the half-stuff is cleansed from bleach, hydrochloric or sulphuric acid—whichever acid may have been used in the bleaching—chloride of calcium, and the various products resulting from the decomposition of the chloride of lime. In this engine, also, the loading, sizing, and colouring materials are worked up with the pulp, and the stuff fully prepared for its final transfer direct to the paper-machine. Before describing the various forms of beating-engines which have been from time to time introduced, including some of the most recent types, to which special attention will be drawn, we purpose quoting some observations of well-known experts in paper manufacture which will be read with interest, since they fully explain the importance that attaches to the proper manipulation of the beating-engine for the production of paper of high quality.

Mr. Dunbar's Observations on Beating.—There is no

operation of the paper-mill that requires more careful attention and experienced judgment than that of beating, or refining, to bring the pulp to the finest possible condition for paper-making; in this department, Mr. Dunbar urges, "none but thoroughly efficient men should be employed, for it is here that the paper is really made—that is, the quality of the paper produced at the paper-machine will be in proportion to the treatment the material has received; and if the half-stuff sent to the beating-engines is not subjected to judicious manipulation and careful preparation for the special paper to be made, all future doctoring will prove unsatisfactory."

Mr. Arnot on Beating Engines.—On this subject Mr. Arnot says:—"Upon the management of the beating-engine the character of the paper produced largely depends. What is wanted is not a mincing or grinding of the fibre, but a drawing out or separation of the fibres one from another; in fact, the name of the machine indicates pretty accurately the nature of the action required—beating. Long, fine fibres can only be produced [obtained] by keeping the roll slightly up off the bed-plate, and giving it time to do the work. Sharp action between the roll and the bed-plate will, no doubt, make speedy work of the fibre, but the result will be short particles of fibre only, which will not interlace to make a strong felt. Indeed, the action I refer to will reduce the long, strong fibre of linen to little better than that of wood or straw. Practice and careful observation can alone make a good beater-man, and for the finer classes of paper none but careful, experienced men should be entrusted with the management of the beating-engine. Sometimes the operation is conducted in two successive engines, the first being called the intermediate beater, but I have hitherto failed to see wherein the advantage of this system lies. The time usually occupied in beating esparto for printing-paper is about four hours, while for rags the time may vary from four to twelve hours, or even more." This, however, depends upon the nature of the rags themselves, and the purposes to which they are to be applied.

Mr. Wyatt on American Refining-Engines.—Referring to the engines adopted in America, Mr. Wyatt says:—"There are various modifications of the original Jordan, the principal ones being the Marshall, Jeffers, and improved Jordan; but I gathered that experience proves the Jordan type to be the most practical and efficient in the end, and is one of the most generally used. One Jordan is required for each machine, refining all the stuff supplied to it. The roll, or plug, runs from 350 to 400 revolutions per minute, the horse-power consumed varying from 25 to 40 horse-power according to the work done, and an engine will do up to 1,000 lbs. of pulp per hour. The time saved in the beating-engine by the use of the Jordan is just about one-third of what would otherwise be necessary, that is to say, pulp requiring otherwise six hours beating only takes four hours if finished in the Jordan. The half-beaten pulp is emptied into a stuff-chest, and the Jordan is furnished with a small stuff-pump and service-box, just as at the paper-machine what the Jordan does not take flows back again into the chest: the pulp from the Jordan is run into the ordinary machine stuff-chests. The finished pulp can be taken from the Jordan at three different levels from the circumference of the roll, or plug. If the pulp is wanted 'free,' it is drawn from the bottom of the engine; if wanted 'wet,' or well greased, it is drawn from the top; and if medium from the centre."

The Beating-Engine.—The ordinary form of beater consists of a cast-iron trough 13 feet 6 inches long × 6 feet 6 inches wide, and the bottom is dish-shaped, so as to prevent the pulp from lodging, which would inevitably be the case if the bottom were flat, as the pulp would be apt to lodge in the angles formed by the junction of the bottom with the vertical walls of the trough. The iron trough is fitted with a cast-iron roll, 3 feet 6 inches × 3 feet 6 inches, which is provided with 69 "roll-bars," or knives, arranged in 23 groups of 3 bars each; this roll is suspended upon a malleable iron shaft 5 inches in diameter, resting upon side levers; suitable gearing is attached by which the roll can be lifted or lowered at will, the action being uniformly equal on both sides, by which the knives of the roll are kept uniform with those of the bed-plate beneath. The bed-plate, furnished with 20 steel knives, of the same length as the roll, is placed immediately beneath the roll. When the knives of the bed-plate are straight they are fitted into the plate-box at an angle, but in some cases they are bent at a slight angle, when they are termed *elbow plates*. There have been, however, many improvements in the beating-engine introduced of late years, some of which are of considerable importance, and to some of these we will now direct attention. Although our own manufacturers have introduced improvements in beaters which have been fully recognised by the trade, the American engineers have not been behindhand in devising modifications which appear to have some important advantages. The Jordan beater, which has been extensively adopted in the States, consists of a roll in the form of a truncated cone, furnished with knives in the usual way; this revolves in a box of a similar form, fitted with knives in the direction of its length, but at slightly different angles. In this engine the stuff enters at the narrow end through a box having an arrangement which regulates its flow, and the pulp is discharged by several openings in the cover at the wider end. In an engine invented by Mr. Kingsland there is a circular chamber furnished with knives covering its sides; between this is a circular plate, also fitted with knives, which revolves. The stuff enters through a pipe in the centre of one of the sides of the chamber, and flows out through an opening in the opposite side.

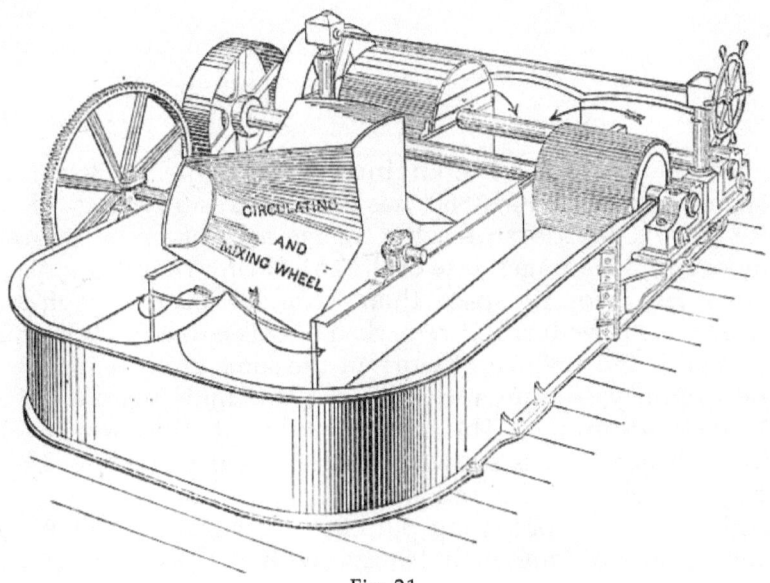

Fig. 21.

Forbes' Beating Engine.—This engine, an illustration of which is given in Fig. 21, is manufactured by Bertrams, Limited, of St. Katherine's Works, Edinburgh. The engine has three chambers, two rolls, and a mixing wheel; the rolls, only one of which is uncovered in the engraving, are fixed in the outer channels, and the mixing wheel is placed in the middle channel. By this arrangement the pulp flows alternately into the two outer channels, and after passing through the rolls again it enters the centre channel at the opposite end.

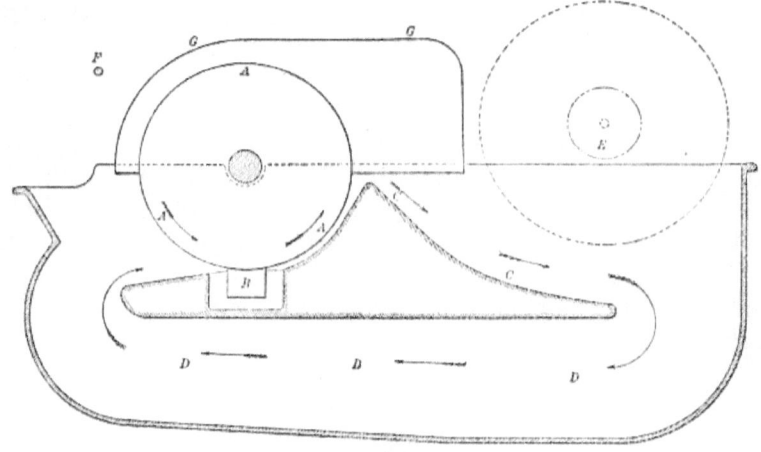

Fig. 22.

Umpherston's Beating Engine.—This engine, for which a patent was granted in 1880, has been successfully adopted at the *Daily Chronicle* and other mills, and presents several important advantages, one of the chief being that it occupies much less ground space than ordinary beating-engines. Indeed, we have heard it remarked of this engine that it will do double the amount of work in the same ground space as the ordinary engine, and this, in some mills, would be a decided advantage. The construction of this beater, a drawing of which is shown in Fig. 22, is thus described by the patentee:—"In the common and almost universal form of engines used for preparing pulp for paper-making, the pulp travels horizontally in a trough with semi-circular ends, and straight sides, partly divided longitudinally by a partition called the midfeather, around which the pulp flows from the back of the roll to its front, where it passes under the roll and over the bottom working-plate, and is again delivered over the back fall to pass again round the midfeather to the front of the roll. In the course of these repeated revolutions part of the pulp near the circumference of the tub has much farther to travel than the part near the midfeather, and consequently is not so often operated upon, and the pulp is thus unequally treated. As an improvement upon this form of tub, I make it so that the pulp passes from the back of the roll to its front through a longitudinal passage under the back fall, the pulp thus moving as through an inverted syphon, the superincumbent weight of the semi-fluid pulp, as delivered over the back fall of the roll, pressing it along this passage and upwards, to enter again in front of the roll. The roll A, bottom plate B, and the form of the back fall C, are similar to those of ordinary engines, but the trough is formed with the passage D under the bottom plate B, so that the semi-fluid contents of the engine, in travelling from the back fall C to the front of the roll A, pass by means of the passage D under the bottom plate B in the direction indicated by the arrows, the superincumbent weight of the semi-fluid pulp, as it is delivered over the back fall C at the back of the roll A, pressing it along the under passage D and upwards to the front of the roll A. The position of a drum-washer is shown at E, and at F is seen a section of the cross shaft for raising or lowering both ends of the roll A simultaneously; G is the roll cover, which may be of any usual form. By this invention the semi-fluid pulp is acted upon in a more effective manner, and its particles are also more equally treated than has hitherto been the case."

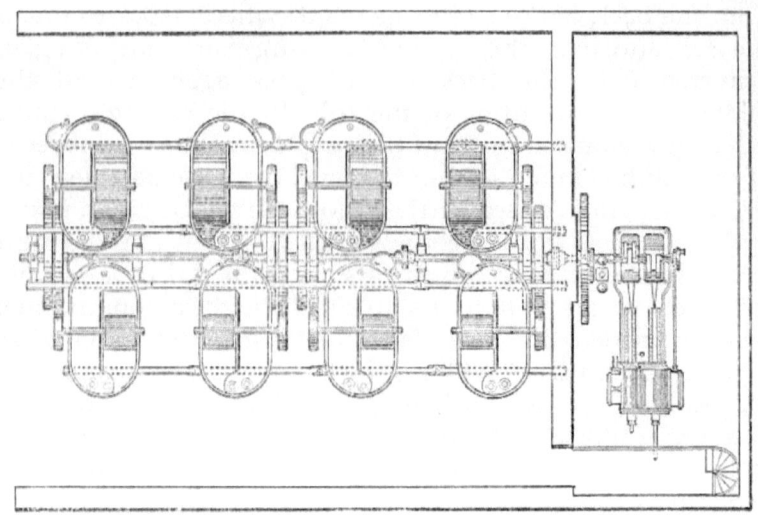

Fig. 23.

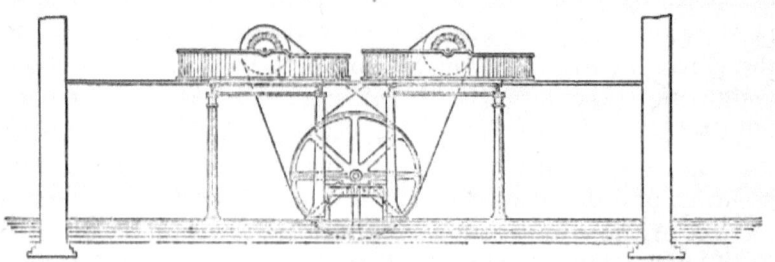

Fig. 24.

The beating-engines are usually driven from a separate engine, but Messrs. Bertrams have introduced a system of direct driving for these engines by which, it is said, there is a considerable saving in power. The accompanying engravings, Figs. 23 and 24, show a series of eight beaters, each carrying 300 lbs. of pulp, driven by one of their compound direct-driving steam-engines, and now being worked at the Forth Paper Mills.

Operation of Beating.—Having referred to some of the more important improvements connected with the beating-engines, we will proceed to explain the operation of beating as briefly as possible. The bleached half-stuff is removed from the tray of the press in caked masses, and in this condition is conveyed in trucks or boxes to the beating-engine. The first thing to be attended to is the removal of the last traces of chlorine from the pulp, which, if not effectually done, would cause injury to the size, and also corrode the strainer plates and wire-gauze of the paper-machine. It is possible to wash out the chlorine by an abundant application of pure water, but this method of removing the chlorine is very tedious and occupies a long time, while it also involves the use of enormous quantities of water—a serious consideration in some mills; to this may be added the still more important fact that by the method of washing out the chlorine a considerable loss of fibre takes place. The plan most usually adopted is to neutralise the chlorine left in the pulp by the application of suitable chemical agents, whereby the chlorine is rendered inert. These agents, technically termed "antichlors," are sometimes objected to, however, although they are in themselves practically harmless so far as their action upon cellulose is concerned. Mr. Arnot, who has considered this subject very thoroughly, says:—"I do not think there is much in this objection, as those agents that are soluble pass through the wire of the machine almost completely, while those that are insoluble are in the finest possible state of division and pearly white. The chemical agent most largely used is hyposulphite of soda, but hyposulphite of lime is also employed, and those agents, known by the name of 'antichlor,' are put into the engine in such a quantity as will ensure the neutralisation of the whole of the chlorine. The products of the reaction, when the soda salts are used, are chloride of sodium (common salt) and sulphate of soda (Glauber's salt), and, when the lime salt is used, chloride of calcium and sulphate of lime, the latter identical with the pearl hardening so well known as a loading agent." From this it will be seen that little or no harm can possibly occur either to the fibre or the metal work of the machine by the employment of the neutralising agents named, and when it is borne in mind that the simple washing of the pulp would occupy the beating-engine for a lengthened period and exhaust a considerable quantity of water—which, as we have said, would in some mills be a serious matter—the adoption of the neutralising method would undoubtedly have the preference.

The engine, being partly filled with water, is set in motion, and the bleached half-stuff introduced in small quantities at a time, each portion being allowed to become thoroughly mixed with the water before the next batch is added. The charging of the beater with half-stuff is kept up until the mass becomes so thick that it will only just move in the trough under the action of the revolving roll. If the beater is of the older type, portions of the pulp are liable to lodge in corners, to remove which the "beater-man" uses a wooden paddle, with which tool he also pushes the slowly moving pulp in the direction of the roll, especially when the stiff mass appears to move too slowly. At this stage the neutralisation of the chlorine in the pulp is effected, which is done by adding a solution of hyposulphite of soda, a little at a time, until the liquor ceases to redden blue litmus paper, strips of which should be dipped into the pulp every few minutes until the paper persistently retains its blue colour. This operation should be conducted with great care, so as to exactly neutralise the traces of chlorine without adding an excess of the hyposulphite of soda. Besides this salt, other substances are used as "antichlors," as, for example, hyposulphite of lime, which is prepared by boiling milk of lime (slacked lime made into a thin mixture with water) and flour of sulphur in an iron vessel until the latter is dissolved, when, after cooling and settling, the resulting solution, which is of an orange-yellow colour, is ready for use. One great objection to the use of hyposulphite of lime, however, is that when decomposed by the chloride of lime remaining in the pulp sulphur is set free, which, mingling with pulp, will impart to it a yellow tint; besides this, in passing over the drying cylinders of the machine the sulphur present in the paper may attract oxygen from the air, converting it into sulphuric acid, which must inevitably prove injurious to the manufactured paper. Sulphite of soda has also been used as an antichlor, and is said to be preferable to hyposulphite of soda,[21] inasmuch as the latter salt is liable to decompose with the liberation of free acid, which is not the case with the sulphite of soda.

Test for Chlorine.—Instead of relying solely upon the litmus paper test when applying the antichlor, the following test for chlorine may also be used with advantage:—Take 2 drachms (120 grains) of white starch, and make it into a paste with a little cold water; then pour over it about half a pint of boiling water, stirring briskly; to this add 1 drachm of iodide of potassium, and stir until dissolved and well incorporated with the starch solution. The mixture is then to be allowed to cool, when it is ready for use. A few drops of this mixture dropped upon a small sample of the pulp will indicate if any chlorine be present by the spot assuming a blue colour; if such be not the case, the pulp may be considered free from chlorine.

During the beating, the roll, which should make not less than 220 revolutions per minute, is lowered, a little at a time, so that the cutting edges of the bars and plate may be brought together gradually and equally until the pulp is reduced to the desired condition. The pulp is made long or short according to the quality of paper to be produced; news papers, which require strength, are made of long-fibred pulp, while writing paper, or paper of fine texture, is made of shorter pulp. The stuff should be what is called "mellowed" in the engine, which is effected by a judicious working of the roll, not lowering it suddenly but gradually, and not much at one time, on the plate, until the pulp attains the fineness required. This is generally arrived at in about three and a half to four hours, though sometimes the beating of pulp from rags is continued for more than double that time. It should be added that if the cutting edges of the roll and plate are brought together suddenly and too closely, the fibre will be cut, and as a consequence the paper produced will be tender.

Esparto, which, in the process of boiling becomes reduced to such a soft condition that the fibres may be readily separated by the fingers, does not require such excessive beating as rags; indeed, the perfect disintegration of the fibres of esparto is practically accomplished in about half the time occupied by rags, and often much less, but this of course depends upon the nature of the esparto itself and upon the thoroughness of the boiling. Wood pulps also require but moderate beating, since the process of disintegration is generally pretty effectually accomplished by the processes to which the raw material is subjected in the course of manufacture into half-stuff, which is the condition in which this paper material is furnished to the manufacturer.

Blending.—To produce papers of the different qualities required by the trade, a system of blending is adopted, which may be effected—(1) by mixing the materials in the raw state, or the rags, previous to boiling; and (2) blending the half-stuff in the beating-engine. The latter method, however, is generally preferred. Sometimes, also, pulps of different character are beaten separately and then mixed in the stuff-chests, where they are mixed as thoroughly as possible before passing on to the machine, but this method would be less likely to ensure a perfect mixture of the respective pulps than would be effected with proper care in the beater. The proportions of the several materials to be blended is also a matter of important consideration. In blending esparto with rag stuff, if the former be in excess it becomes reduced to the proper condition before the latter is sufficiently fine, which causes the rag fibre to appear in "knots and threads" in the manufactured paper. But if the rag stuff be allowed to predominate, the beating is conducted as though no esparto were present, by which, while the rag stuff becomes reduced to the proper length of fibre, the esparto, which is still further reduced, in mingling with the longer fibre of the rags forms what is called a "close" paper. Mr. Dunbar, in his useful little work, "The Practical Paper-maker," furnishes a series of receipts for blending for high-class papers, as also the proportions of colouring matter to be used, which the reader will do well to consult. For news papers, esparto and straw pulps are generally used, in varying proportions according to the nature and quality of the esparto; these proportions have to be regulated according to the judgment of the paper-maker, and vary greatly at different mills. A large quantity of sulphite and other wood pulps are also used, those coming from Scandinavia and Germany being especially suited to the requirements of the English manufacturer. Mechanical wood pulp is also used in a moderate degree—sometimes up to 15 per cent., in some English mills, but it is said that in

Germany this paper stock is sometimes used to the extent of 90 per cent.

CHAPTER XI.

LOADING.—SIZING.—COLOURING.

Loading.—Sizing.—French Method of Preparing Engine Size.—Zinc Soaps in Sizing.—Colouring.—Animal or Tub-Sizing.—Preparation of Animal Size.—American Method of Sizing.—Machine-Sizing.—Double-sized Paper.—Mr. Wyatt's Remarks on Sizing.

Loading.—The very finest qualities of paper are usually made without the addition of any *loading*, as it is called, but for most other papers more or less loading material is added, according to the quality of paper to be produced. The loading material used for ordinary qualities is kaolin, or china clay, and for the better qualities sulphate of lime or *pearl hardening*, as it is termed in the trade. China clay, as it occurs in commerce, is in the form of soft lumps and powder, is nearly white, and when rubbed between the finger and thumb should present no hard particles of gritty matter. To prepare it for mixing with the pulp it is first worked up into a thin cream with water, which is usually done in a vessel furnished with an agitating arrangement by which the clay becomes intimately mixed with the water. The cream is then strained through a fine sieve to separate any impurities present, and is then allowed to flow into the beating-engine containing the stuff while in motion, by which it soon becomes mingled with the pulp. The proportion of china clay or other loading material which is to be introduced into the pulp depends upon the quality of the fibre and the requirements of the manufacturer, some makers using less of the material than others. From 3 per cent. to 10 or 15 per cent. appears to be about the extreme range for employing the material as a necessary ingredient, in the production of various classes of paper, above which figures the addition of loading material may be considered as an adulteration. Sometimes nearly twice the largest amount named is employed, no doubt to meet the exigences of keen competition—from foreign sources especially.

One effect of the loading, whether it be china clay or

sulphate of lime, is to close the pores of the paper, whereby a smoother surface is obtained, while at the same time, if the material has been used in proportions suited to the quality of the fibre, and not in immoderate excess, a stronger paper is produced. A species of asbestos termed *agalite* has been introduced as a loading material, and since it has a fibrous texture, it blends with the fibres of the pulp, forming, as it were, a vegeto-mineral paper. It is stated that as much as 90 per cent. of the agalite used in the beating-engine enters into the manufactured paper, while not much more than half the china clay used is held by the pulp.

Sizing.—"Engine sizing," as it is termed, consists in adding certain ingredients to the pulp while in the beating-engine. The materials generally used are alum and resin soap, in proportions suitable to the paper to be produced. Resin soap is formed by boiling ordinary resin in a jacketed pan such as is used by soapmakers for preparing small quantities of fancy or other soaps, with a solution of soda crystals in the following proportions: Resin, 16 lbs.; soda crystals dissolved in water, 8 lbs.; and the boiling is kept up for about two hours, or until a soap is produced which is perfectly soluble in water. The method of preparing this soap as conducted at the soapworks has been described in the author's work on soap-making,[22] p. 64, from which the following abstract is taken: "Put into a pan capable of holding about 12 gallons, 2¼ gallons of fresh caustic soda ley at 30° B. Apply gentle heat, and when the ley begins to boil throw in, every few minutes, in small quantities at a time, finely powdered and sifted resin until 37 lbs. have been introduced. The mixture must be well stirred the whole time to prevent the resin from 'clogging' and adhering to the pan. It is important to moderate the heat, as the resin soap has a great tendency to expand and an excess of heat would cause it to boil over. The heat, however, must be kept to near the boiling point, otherwise the mass will become thick and of a very dark colour. When kept at near the boiling point it is always clear and its colour of a reddish yellow. If, during the boiling, the resin soap rises and threatens to overflow, the heat must be checked by throwing in a little cold water, only using sufficient to effect this object. It is absolutely necessary to stir the mass continually, otherwise the resin will agglomerate in masses and thus prevent the alkali from acting freely upon it. The boiling takes about two hours, when the soap is run into an iron frame and allowed to cool. It is very important that the resin used is freed from particles of wood, straw, etc., for which purpose it should be passed through a tolerably fine sieve."

Respecting the preparation of resin soap, Davis says:—"The proportion of resin used to each pound of soda ash varies in different mills, 3, 4, or even 5 lbs. of resin being used to each pound of soda ash. The proportion of resin, soda ash, and water, can be best determined by practical experience, as no prescription could be devised which would be suitable to every case." M. d'Arcet, who modified the proportions recommended by M. Bracconot, recommends for the preparation of resin soap—

Powdered resin	4·80 parts.
Soda crystals at 80° (French, alkalimeter)	2·22 "
Water	100 "

Theoretically speaking, only 2·45 parts of alum would be required to precipitate the resin; but the waters, which are almost always calcareous, neutralise part of the alum. Crystals of soda are much more expensive than soda ash, but on account of their greater purity they are sometimes preferred to the latter. At the present day the resin soap is preferably made by dissolving ordinary resin with a solution of carbonate of soda under boiling heat in a steam-jacketed boiler, the class of paper to be made governing the quantity of resin to be employed. The boiling usually requires from two to eight hours, according to the relative proportions of soda ash and resin used—the greater the proportion of soda used the less time is required for boiling—the process being completed when a sample of the soap formed is completely soluble in water.... About 3 lbs. of resin to 1 lb. of soda is the usual proportion. The resin soap is cooled after boiling by running it into iron tanks, where it is allowed to settle, the soap forming a dense syrup-like mass, and the colouring matters and other admixtures of the resin rising to the top are easily removed. It is important to run off the mother liquor (ley) containing the excess of alkali, for when the soap is used it consumes the alum to neutralise it."

When the impurities and ley have been removed the soap is dissolved in water, and if, from imperfect boiling, a portion of the resin is found not to have been saponified, a small quantity of a strong solution of soda crystals is added to the water used for dissolving the soap.

Where starch is used for stiffening purposes, the soap is mixed with a quantity of starch paste in the proportion of 1½ part of starch to 1 part of resin soap. Some manufacturers, Mr. Davis states, mix the starch paste with the kaolin in lieu of mixing it with the resin soap. In either case the materials should be thoroughly strained before being added to the pulp. From 3 to 4 lbs. of the mixture of resin soap and starch paste to each 100 lbs. of dry pulp are about the proportions in which the size is generally used, but the quantity added to the pulp in the beater depends upon whether the paper is to be soft-sized or hard-sized.

Sizing is chiefly applied to papers which are to be written upon with ordinary inks, and also, with a few exceptions, to printing papers, the object being to close the pores of the paper and render it non-absorbent, by which the spreading or running of the ink is effectually prevented. While the finest lines may be written upon a well-sized paper (as ordinary writing paper, for example) without spreading in the least degree, a similar stroke of the pen upon blotting paper, tissue, or unsized printing paper would spread in all directions, owing to the highly absorptive property of the cellulose.

The sizing of the pulp is conducted as follows:—After the loading material has been introduced and well mixed, the resin soap, previously dissolved in water, a little carbonate of soda being sometimes added, is mixed with a paste of starch prepared by dissolving starch in boiling water, and the mixture of soap and starch is then passed through a fine sieve to keep back any particles or lumps that may be present. The proportion of the materials used in sizing vary at the different mills, each manufacturer having formulæ of his own; about 1 part of resin size to 3 of starch paste, and, say, from 9 to 12 lbs. of the mixture, may be used for 300 lbs. of pulp; and, if preferred, the respective ingredients may be put into the engine separately, a method adopted at some mills. Some manufacturers of the finest papers, instead of dissolving the starch in hot water, make it into a thin paste with cold water, in which condition it is introduced into the pulp, the object being to impart to the paper a particular feeling to the touch which is not obtainable by other means.

The mixture of resin size and starch paste, with or without the addition of water, is added to the pulp in the beater, in which the pulp is circulating, and the engine allowed to run until the materials are well incorporated in the pulp. At this stage a solution of alum (about 28 to 30 lbs. for 300 lbs. of pulp), or of sulphate of alumina,[23] is introduced, which causes the resin soap to become "separated," the sulphuric acid of the alum uniting with the alkali of the soap and setting the resin and alumina free in the form of minute particles; the resin in the subsequent drying on the calenders becomes fused, as it were, and thus cements the fibres and alumina together, at the same time rendering them non-absorbent and improved in whiteness by the precipitated alumina. Sometimes ordinary soap is added to the resin soap, which is said to impart a higher finish to the paper in the operation of calendering.

The so-called "concentrated alum," which contains a higher percentage of sulphate of alumina than the crystallised alum, is considered the most economical in use, being proportionately cheaper, and the variety known as "pearl alum" is specially recommended. "Aluminous cake" is another preparation which has found favour in many mills, but since it sometimes contains a large excess of free sulphuric acid it requires to be used with caution, since this acid, although it will brighten the colour of some aniline dyes, will discharge the colour from others, while at the same time it may injuriously affect the brass-wire cloths of the paper machine. The alum solution should be prepared in a lead-lined tank, fitted with a steam pipe for heating the contents when required.

The proportions of the materials used in sizing differ

considerably in different mills, but the following may be taken as an average for common writing and printing papers:—

Per 100 parts of dried pulp 10 to 12 parts of resin.
" " " 20 " 30 " starch.
" " " 10 " 12 " alum.

To the sizing solution is generally added from 30 to 50 parts of kaolin. When a colour is present on which alum would have a prejudicial effect this is usually replaced by about one-third of its weight of sulphate of zinc. Many mineral substances have from time to time been added to paper stock, principally to increase its weight, and in 1858 Sholl took out a patent for adding carbonate of lime, a substance which, however, had long been fraudulently used in order to increase the weight, but he found it to have the property of fixing the ink in the pores of the paper, thus rendering it immovable. The only useful addition is kaolin, or some similar aluminous compound, as it attaches itself to the fibre, and, while giving the required opacity and a good surface, takes both printing and writing ink well, and has the advantage, from a manufacturer's point of view, of increasing the weight. It has been proposed that small quantities of glycerine be added to the pulp, in order to give the paper greater flexibility, and especially to give copying-paper the quality of taking up colour readily.[24]

French Method of Preparing Engine Size.—Thirteen pails of water are boiled in a copper-jacketed pan capable of holding about 150 gallons; 90 lbs. of soda crystals are then introduced and allowed to dissolve, when 200 lbs. of finely-powdered resin are gradually introduced, with constant stirring, and the boiling is sustained for about two hours after the last portion of resin has been added. A further addition of water is now made by putting in five pails of cold water, and the water is then boiled for an hour and a half longer. The resin soap is then transferred to stock-chests, in which it is allowed to remain for ten days or longer, fresh batches being prepared in rotation, to meet the requirements of the mill.

To determine whether an excess of resin soap or of alum has been added to the pulp, red and blue litmus papers should be employed, the former turning blue if an excess of resin soap be present, and the latter red when alum or sulphate of alumina is in excess. For uncoloured papers the aluminous material should be added until the pulp becomes faintly acid, which will be indicated by the blue litmus paper turning slightly red when immersed in the pulp.

Besides resin soap, various substances have been proposed as sizing materials, including wax dissolved in a strong solution of caustic soda and precipitated with alum, but the cost would be an objection to the use of this material except for the highest classes of paper. It is stated that 12 lbs. of gum tragacanth to each 500 lbs. of resin has been used in preparing some kinds of engine-sized papers, and is said to impart to them an appearance equal to that of tub-sized papers.

Zinc Soaps in Sizing.—According to a paragraph in the *Papermakers' Monthly Journal*, a somewhat novel method of sizing is employed in Germany, which consists in the precipitation in the stock of zinc soaps. Cottonseed oil soap or Castille soap is worked up in the engine with the stuff, and after it has become well mixed with the pulp a solution of sulphate of zinc is added, which results in the formation of a white and heavy zinc soap, which is insoluble, and adheres well to the fibres. The weight and whiteness of the zinc soap are the main points in favour of this method, which is said to yield good results.

Colouring.—The pulp, after passing through the various processes described, although apparently white, invariably presents a yellow tinge when converted into paper. To obviate this it is usual to "kill" the yellow tint by adding to the pulp small quantities of blue and pink colouring matters. The blue colours generally used are ultramarine, smalts, and various aniline blues, and the pinks are usually prepared from cochineal, either in a liquid form or as "lakes" (compounds of cochineal and alumina) or aniline dyes, the former being preferable, as it is not injuriously affected by the alum used in sizing. The ultramarine should be of good quality, otherwise it will become decomposed, and its colouring property destroyed by the action of the alum, but more especially so if the alum contains an excess of free acid. Smalts blue, which is a kind of coloured glass, is not affected by acids. In preparing the colouring matters for mixing with the pulp they must first be mixed with water, and the liquid should then be strained, to keep back any solid particles that may be present in the material. Aniline blues should be dissolved in hot water, or alcohol, and then diluted. Samples of the pulp are examined from time to time until the desired effect is produced, which the practised eye of the beater-man can readily determine.

Animal or Tub-sizing.—Another process of sizing, termed "animal-sizing," "tub-sizing," or "surface-sizing," is also adopted in the manufacture of certain classes of paper, and is either accomplished by hand or on the machine. The former method having been elsewhere described (p. 132) we will now describe the operation of sizing on the machine, to which the term tub-sizing is also applied. The size employed, which is prepared from what are called "glue pieces," or clippings of "limed" and unhaired skins of animals, requires to be as colourless as possible, in order that the colour of the paper may not be injuriously affected by it.

Preparation of Animal Size.—This operation is generally conducted at the mill, the materials from which the size is produced being the cuttings or parings of animal skins and hides, or *pelts*, which have undergone the processes of "liming" and unhairing preparatory to being tanned. The cuttings, or *pates*, commonly called "glue pieces," are first soaked in a mixture of lime and water, placed in large tubs for several days, after which they are put into a wooden cylinder, or drum, five or six feet in diameter, and about ten feet in length, which revolves upon a horizontal shaft, which, being hollow, admits the passage of water to the interior of the drum. The drum is perforated, and revolves in a large tank, while a continuous stream of water is allowed to pass through it, and the dirty water escapes through the perforations in the drum. When the cuttings are sufficiently cleansed in this way, they are transferred to an iron copper, furnished with a false bottom and steam-pipe, or a jacketed pan. The cuttings are next covered with water; steam is then turned on, and the liquid brought to a temperature below boiling point, or say, about 180° to 190° F., it being very important that the liquid should not actually boil. This operation is carefully kept up for twelve to sixteen hours, according to the nature of the cuttings, by which time all the material excepting any membranous or fatty matters that may be present, will have become dissolved and a solution of gelatine obtained. The liquor is then allowed to settle for a short time to allow fatty matters to rise to the surface and membranous substances to deposit, and the fatty matters must afterwards be carefully removed by skimming. The liquor should next be strained to separate any floating particles of a membranous character. Sometimes the gelatine solution is clarified by adding a small quantity of powdered lime, which is thoroughly mixed by stirring, after which it is allowed to rest. When it is found that the impurities and lime deposit too slowly, a little weak sulphuric acid is added, which, forming an insoluble sulphate of lime, the solid matters quickly subside, leaving the liquor quite clear. The solution is next filtered through felt, and is afterwards treated with a solution of alum, which at first causes the liquid to thicken and become nearly solid, but it becomes fluid again, however, on the addition of more alum solution. When this condition is finally attained, the liquid is ready for use in the process of sizing. The addition of the alum (which should not contain any free acid) to the gelatine greatly improves its sizing property, besides preserving it from decomposition. The treatment of the glue pieces for the purpose of obtaining gelatine solutions is fully described in the author's work on "Leather Manufacture," p. 401.[25]

American Method of Sizing.—Another method of preparing size, and which is adopted in America, is the following:—In large paper mills the size is generally prepared in a room devoted to the purpose, and is commonly situated near the machine. The finest grades of light hide and skin clippings are used for No. 1 letter papers, but less costly stock is employed for the lower grades of animal-sized papers. To preserve the glue pieces the tanners and tawers macerate the clippings in milk of lime and afterwards dry them. As the clippings require to be freed from the lime, the first treatment they receive at the paper-mill is to put them in large wooden tubs partly filled with water, in which they are allowed to soak for several days. They are afterwards more perfectly cleansed by means of a drum-washer, such as we have before described. Fresh hide and skin clippings, that is, those which have not been limed and dried at the tanneries, and which are occasionally purchased by the paper manufacturers, require to be used as soon as possible after they arrive at the mill as they readily decompose, and are placed in tubs partly filled with water, in which 2 per cent. by weight of caustic lime has been dissolved. The pieces, if from calfskins, are allowed to remain in the lime bath for ten to fifteen days, clippings of sheepskins fifteen to twenty days, and trimmings from heavy hides, as ox, etc., twenty-five to thirty days, the milk of lime being renewed once or twice a week, and the material well stirred from time to time. The glue-stock, as it is sometimes termed, is afterwards thoroughly washed in the drum-washer, and when this operation is complete the material is spread out in the yard to drain, and when sufficiently dried is ready for boiling, or may be stored until required for use.

To prepare size from the material treated as described, it is placed in a boiler of cast or wrought-iron or copper, furnished with a perforated false bottom, and capable of holding from 100 to 400 lbs. of the raw material, according to the requirements of the mill. Several such boilers may be placed close to each other. At the bottom of the boiler is a stop-cock for drawing off the gelatine solution when required. When the requisite charge of glue-stock has been introduced into the boiler, water is poured over it and steam turned on, which passes through a pipe fixed beneath the false bottom, and care is taken that the temperature of the contents of the boiler should not exceed 200° F., which heat is kept up for ten to eighteen hours, according to the nature of the materials treated. The gelatine solution is drawn off from the boiler as it is formed, into wooden tubs, and at the same time carefully strained to remove membranous matters and suchlike impurities. Several boilings are made from the same batch of glue-stock, and all the solutions are afterwards mixed together in the receiving tubs, and a solution of alum is added in such proportions as to be recognised by tasting the liquor. One object in adding the alum being to prevent the gelatine from decomposing, more of this substance should be added in warm than in cold weather.

When the solutions are cool they are ready for use, and the gelatine is removed from the receiving tubs and dissolved in a separate tub as required for use, the dissolving tub being provided with a steam-pipe. The proportion of water—which should only be lukewarm—used in dissolving the gelatine varies from a quarter to half the bulk of the latter, the nature of the fibre and thickness of the paper regulating the proportion of water to gelatine, the strength of the size liquors being greater for thin papers and weak fibres than for thick papers and strong fibres.

The operation of sizing is considered one of the most difficult and uncertain with which the paper-maker has to deal, since the material (gelatine) is greatly influenced by the conditions of the atmosphere, both as regards its temperature and humidity, while the temperature of the liquid size itself has also an important influence on the success of the operation. The condition of the paper, again, also affects the result, for if it be highly porous it will probably be weak, and consequently there may be considerable waste during the process of sizing from the necessary handling it is subjected to; moreover, should the paper have been blued with ultramarine, a strongly offensive odour is often imparted to it; this, however, may be obviated by employing fresh size and drying the paper as completely as possible. There are two systems of animal-sizing employed at the mill, namely, hand-sizing and machine-sizing, which is also called tub-sizing, the former being applied to papers of the finest quality. Papers that have been made by the machine, after being cut into sheets, are hand-sized, as described in the next chapter.

Machine-Sizing.—The lower-priced papers, to be machine-sized, are first partly dried over a few cylinders, after which the paper passes through a tank containing liquid size, from whence it passes between two rollers, which squeeze out the superfluous size; it is then wound on to a reel on which it remains some time to enable the size to thoroughly permeate the paper, after which it is wound on to another reel, and from thence it passes over a series of wooden drums or cylinders, each of which is furnished with a revolving fan; by this means the paper becomes dried slowly, whereby a more perfect sizing of the material is effected.

Double-Sized Paper.—This term is applied to paper which, after being sized in the engine in the usual way, is afterwards "surface sized," as it is called, with animal size in the manner described.

Respecting the drying of paper after it has been tub-sized there seems to be some difference of opinion as to whether it is best to hang it in a loft to dry or to dry it over the cylinders of a drying machine. Upon this point the New York *Paper Trade Journal* makes the following remarks:—"When the paper is passed through the size-tub, it is again wet; the fibres expand, and their hold on each other is relaxed. Now it must make a difference to the subsequent strength and quality of this paper whether it be hung up in a loft to dry or run over a drying machine. If it is hung in the loft no strain is put upon it and the fibres are at liberty to shrink, or slowly contract, in all directions; whereas if it is run over a drying machine, consisting of from 50 to 100 reels, the longitudinal strain prevents the fibres from shrinking and reassuming their normal position in that direction. Attempts have been made to obviate this defect by regulating the speed of each section of the machine in such a manner as to allow for the shrinking, but this only remedies the evil by preventing the paper from breaking as it travels over the machine. Everything else being equal, it would seem that loft-dried paper must be superior to that dried over the drying machine. Our home manufacturers endorse this view, inasmuch as they continue to prefer the system of loft-drying to the less expensive machine methods."

Mr. Wyatt's Remarks on Sizing.—Mr. James W. Wyatt, in a paper on the "Art of Paper-making,"[26] makes the following observations on engine-sizing and animal-sizing which will be read with interest:—"Engine-sizing renders the paper fully as non-absorbent as animal size. The latter penetrates the sheets only slightly and forms a coating or skin on each surface, whereas the engine size surrounds each fibre and impregnates the whole mass. Surface-sizing, however, produces a stronger, firmer sheet, and is smoother for the pen to travel over; the manufacturer also gets the benefit in the price of the paper of the additional weight of the size, amounting to 7 per cent. on the average. On the other hand, as the animal size is mostly a skin on the surface, if the coating be broken anywhere by the use of a knife in scratching, the paper will only imperfectly resist ink in that place, a great disadvantage for account and office-books and ledgers. Engine-sized paper is much cheaper to produce than animal sized, and is therefore used principally for the lower qualities of writings and for almost all kinds of printings where firmness and smoothness is not so much a desideratum. Most tub-sized papers have a certain portion of engine size mixed with the pulp. This not only ensures the thorough sizing of the sheet, but also is a measure of economy in reducing the absorbing power of the paper for the animal size. Papers for ledgers and office-work are best given an extra proportion of engine size to ensure their ink-resisting properties, and they are also sized by hand in animal size and loft dried." The following rough estimate of the comparative cost in materials and wages of engine-sizing and animal-sizing paper may be of interest:

Engine-sizing, per 20,000 lbs.:—

	£	s.	d.	
Materials	5	2	0	
Wages	0	12	6	
				d.
				Cost
Total	£5	14	6	per 0·068
				lb. =

Animal-sizing, per 20,000 lbs.:—

	£	s.	d.
Materials	36	0	0

Wages		4	10	0
	Total	£40	10	0

CHAPTER XII.

MAKING PAPER BY HAND.

The Vat and Mould.—Making the Paper.—Sizing and Finishing.

Under the old system of making paper by hand, the rags were reduced to a fine state of division by a process of *retting*, or slow putrefaction. The rags were first washed in water, and then piled in heaps, in which condition they were allowed to remain until they became tender, that is, readily pulled asunder by the fingers. During the decomposition the rags not unfrequently became rotten in some portions of the heaps, thus involving considerable loss of fibre. The rags were next placed in a strong chest, in which iron-shod stamping rods were fitted, and these by their continued action gradually reduced them to a pulp. The stampers were eventually superseded by the beating-engine, the invention of a Dutchman, which received and still retains the name of the "Hollander." Other machines, as the duster, washing and breaking engines, and the beating engine, have entirely taken the place of the older system, which required the work of forty pairs of stamps for twenty-four hours to produce one hundredweight of paper.

The Vat and Mould.—The pulp being prepared, is conveyed from the beaters to the working vat, where it is diluted with water. The vat is a wooden or stone vessel about 5 feet square and 4 feet deep, being somewhat wider at the top than at the bottom. A steam-pipe is supplied to the vat, so that the pulp and water may be heated to a convenient temperature for working, and an agitator is also furnished to keep the pulp and water uniformly mixed. The mould in which the pulp is raised from the vat to form a sheet of paper, consists of a wooden frame, neatly joined at the corners, with wooden bars running across, about 1½ inch apart, and flush with the top edge of the frame. Across these again, in the length of the frame, wires are laid, about fifteen or twenty in an inch, which are placed parallel to each other. A series of stronger wires are laid along the cross-bars, to which the other wires are fastened; these give to what is termed "laid" paper, the ribbed or "water-marked" lines noticeable in hand-made paper. Upon the mould is fitted a movable frame, called the *deckle* or *deckel*, which must fit very neatly or the edges of the paper will be rough. The mould and deckle form together a kind of shallow tray of wire. Sometimes the mould is divided by narrow ribs of wood, so that two or four sheets of paper may be made in one operation. Connected with the vat is a slanting board, called the *bridge*, with copper fillets attached lengthwise upon it, so that the mould may slide easily along the bridge.

Making the Paper.—When preparing for work, the vat-man stands on one side of the vat, and has on his left hand a smaller board, one end of which is fastened to the bridge, while the other rests on the side of the vat. An assistant, called the *coucher*, is at hand, whose duty it is to handle the frames or moulds containing the pulp after they have passed through the hands of the vat-man or maker. The latter now takes in his hand a mould, and lays it upon the deckle; he then dips the mould, with its deckle in its proper place, into the vat of agitated pulp, and lifts up as much of the pulp as will form a sheet of paper. This, as will be readily seen, requires the greatest dexterity, since the workman has nothing but his sense of feeling to guide him. It is said, however, that practice gives him such a nicety of feeling in this respect that he can make sheet after sheet of the largest-sized drawing papers with a difference in weight of not more than one or two grains in any two of them. Great skill is also required to hold the mould in a perfectly horizontal position, otherwise during the felting and settling of the pulp the sheet of paper would be thicker on one part than another. The mould being held lengthwise, that is, with the long parallel wires running from right to left hand, he gives the mould a gentle shake from his chest forward and back again, which is called the *fore-right shake*; this shake takes place across the wires, not in the direction of their length. He next gives a shake from right to left, and back again, the respective movements thus propelling the pulp in four directions. The vat-man now pushes the mould along the small board on his left, and removes the deckle, which he connects to another mould and proceeds to form another sheet of paper, and so on. The coucher, taking the first mould in hand, turns it upside down upon a piece of woollen felt-cloth, then removing the mould, he takes another piece of felt and lays it over the sheet and returns the mould by pushing it along the bridge to the vat-man, when he receives in return a second mould to be treated as before.

In the above way felts and paper are laid alternately until a pile of six or eight quires is produced, which is afterwards submitted to pressure in a very powerful press. When sufficiently compressed, the machine is relaxed, and the felts are then drawn out, on the opposite side, by an operative, called a *layer*, who places the felts one by one upon a board, and the sheets of paper upon another board. The coucher then uses the felts again for further operations. Two men and a boy only are employed in this part of the work. In the evening all the paper made during the day is put into another press, and subjected to moderate pressure to obliterate the felt marks and expel a further portion of the water. On the following day the paper is all separated, which is called *parting*, again pressed, and is then transferred to the drying-loft. The drying is effected by suspending the sheets of paper upon a series of ropes, attached to wooden supports; ropes of cow-hair are used for the purpose, as this material does not stain the paper.

Sizing and Finishing.—When the paper is dry, it is taken down and laid carefully in heaps ready for sizing, which is the next operation to which the paper is subjected. The preparation of the size from animal skins, etc., is described in Chapter XI. When preparing to size the paper, the workman takes several quires of the paper, and carefully spreads the sheets out in the liquid size, which is placed in a large tub, taking care that each sheet is uniformly moistened before introducing the next. The superfluous size is afterwards pressed out, and the paper then "parted" into separate sheets, which are again subjected to pressure, and finally transferred to the drying-room, where they are allowed to dry slowly. When dry, the paper is conveyed to the finishing-house, to be again pressed and looked over by women, who, being furnished with small knives, pick out knots and other imperfections and separate the perfect from the imperfect sheets. The paper is now again pressed, and then handed to the finisher, to be counted into reams and packed, the reams being afterwards pressed and finally tied up and conveyed to the warehouse for sale. When the paper is required to be hot-pressed, this is done by placing each sheet of paper alternately between two smoothed sheets of pasteboard, and between each group of fifty pasteboards is placed a hot plate of iron, and the pile then submitted to heavy pressure, whereby the surface of writing paper acquires a fine, smooth surface.

CHAPTER XIII.

MAKING PAPER BY MACHINERY.

The Fourdrinier Machine.—Bertrams' Large Paper Machine.—Stuff Chests.—Strainers.—Revolving Strainer and Knotter.—Self-cleansing Strainer.—Roeckner's Pulp Strainers.—The Machine Wire and its Accessories.—Conical Pulp Saver.—The Dandy Roll.—Water Marking.—De la Rue's Improvements in Water-marks.—Suction Boxes.—Couch Rolls.—Press Rolls.—Drying Cylinders.—Smoothing Rolls.—Single Cylinder Machine.

The Fourdrinier Machine.—It is just ninety years since Louis Robert, a Frenchman, devised a machine for making a continuous web of paper on an endless wire-cloth, to which rotary motion was applied, thus producing a sheet of paper of indefinite length. The idea was subsequently improved upon by Messrs. Fourdrinier, who adopted and improved upon M. Robert's machine, and with the valuable aid of Mr. Bryan Donkin, a young and gifted machinist, in the employ of Mr. Hall, engineer, of Dartford, constructed a self-acting machine, or working model, in 1803, which, from its effectiveness and general excellency of workmanship, created at the time a profound sensation. This machine was erected at Frogmore, Hertfordshire; and in 1804 a second machine was made and put up at Two-Waters, Herts, which was completely successful, and the manufacture of continuous paper became one of the most useful and important inventions of the age. From that period the "Fourdrinier," with some important improvements introduced by Mr. Donkin, gradually, but surely, became established as an absolutely indispensable machine in every paper-mill all over the world. Although the machine has been still further improved from time to time, those of recent construction differ but little in principle from the original machine. An illustration of the machine is shown in Fig. 25, the detailed parts of which are expressed on the engraving.

Bertrams' Large Paper Machine.—The principal aim in the construction of the paper-making machine has been to imitate, and in some particulars to improve, the operations involved in the art of making paper by hand, but apart from the greater width and length of paper which can be produced by the machine, the increased rapidity of its powers of production are so great that one machine can turn out as much paper in three minutes as could be accomplished by the older system in as many weeks. The drawing represents the modern paper-machine as manufactured by Bertrams, Limited, who supplied one of these machines to Mr. Edward Lloyd, for the *Daily Chronicle* Mill, at Sittingbourne, which runs a wire 40 feet long by 126 inches wide, this being, we believe, the largest and widest paper-machine in the world. It is provided with 20 cylinders, chilled calenders, double-drum reeling motion, with slitting appliance for preparing webs to go direct to the printer's office without the assistance of a re-reeling machine, and is driven by a pair of coupled condensing steam-engines. On our recent visit to Mr. Lloyd's mill we were much struck with the excellent working of this splendid machine.

In the illustration, as will be seen, there are two sets of drying cylinders, while small cylinders, or felt drying-rolls, from 16 to 24 inches in diameter, are introduced to the felts of the cylinders, before the smoothing-rolls, which discharge the moisture with which the felts are impregnated from the damp paper, whereby a considerable saving in felts is effected. Messrs. Bertram state that the highest speed yet attained has been by their own machinery, and is 270 feet of paper per minute.

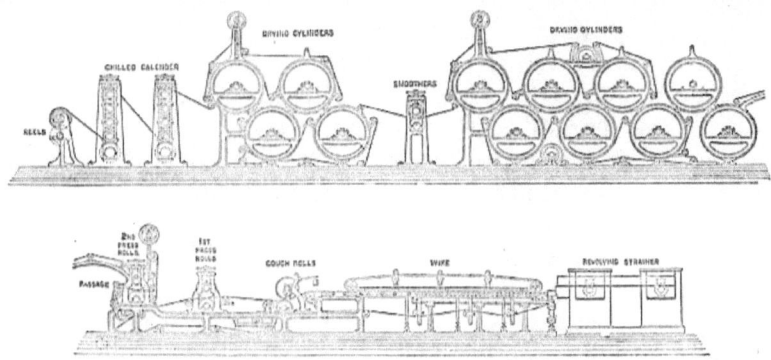

Fig. 25.

The progress of the pulp after it leaves the beating-engines for conversion into paper may be described as follows:—The valve at the bottom of the beating-engine is opened, when the pulp flows through a pipe into the stuff-chests, which are generally situated below the level of the engines. The beaters are then rinsed with clean water to remove any pulp that may still cling to them, the rinsing water passing also into the stuff-chests.

Stuff-chests.—These are large vessels of a cylindrical form, so that the pulp may have no corners to lodge in, and are generally made of wood, though sometimes they are made of cast-iron plates bolted together. The chests are of various dimensions, according to the requirements of the mill, being usually about 12 feet in diameter and 6 feet deep, having a capacity for 1,000 to 1,200 lbs. of stuff. To keep the pulp well mixed in the stuff-chest, of which two are usually employed for each machine, a vertical shaft, carrying two horizontal arms, each extending nearly across the interior of the chest, are provided, which are only allowed to revolve at a moderate speed, that is, about two or three revolutions per minute, otherwise the pulp would be liable to work up into knots, and thus form a defective paper. Motion being given to the shaft, the rotating arms keep the pulp and water uniformly mixed, at the same time preventing the pulp from sinking to the bottom of the stuff-chest.

The pulp is next transferred to a regulating box, or "supply box," by means of a pump called the *stuff-pump*. The regulating-box, which has the effect of keeping a regular supply of pulp in the machine, is provided with two overflow pipes, which carry back to the stuff-chests any superfluous pulp that may have entered them, by which the stuff in the regulating-box is kept at a uniform level, while the machine is supplied with a regular and uniform quantity of the diluted pulp. The stuff-pump conveys the pulp through a valve in the bottom of the regulating-box in a greater quantity than is actually required, the superfluity returning to the stuff-chests by the overflow pipes; thus the supply-box, being always kept full, furnishes a regular and uniform supply of pulp to the sand-tables, or sand-traps as they are sometimes called. *Sand-tables* are large wooden troughs, varying in size at different mills, but Mr. Dunbar gives the following proportions for a first-class sand-trap; namely, 14 feet long by 8 feet wide, and 8 inches deep. The bottom of the trap is covered with felt, sometimes old first-press felt being used, and is divided into several compartments by thin bars of lead or iron, or strips of wood, which keep the felt in position, and also retain any particles of sand or other heavy solid matter that may be accidentally present in the pulp. For the purpose of diluting the pulp for the machine, there is, attached to the inlet of the sand-traps, a box with two supply-taps, one for the delivery of pulp, and the other for water; and these being turned on, the pulp and water flow over the sand-traps, and the diluted pulp then falls into the strainers, which, while allowing the fine pulp to pass freely, keep back all lumps of twisted fibre, and particles of unboiled fibre, which latter, if not removed, would appear as specks on the surface of the finished paper.

The Strainers are formed of brass or bronze plates, in which are cut a very large number of narrow slits, which gradually widen downward, so as to prevent the pulp from lodging. Each plate has about 510 slits, and several plates, connected together by bolts, constitutes the complete strainer. When in use, the strainer receives a jogging motion, which is communicated to it by means of small ratchet wheels keyed on shafts passing beneath the machine; this causes the fibres to pass more freely through the slits. There are many different forms of strainers, which have been the subject of numerous patents. It will be sufficient, however, to give one or two examples of improved strainers which have been more recently adopted by manufacturers.

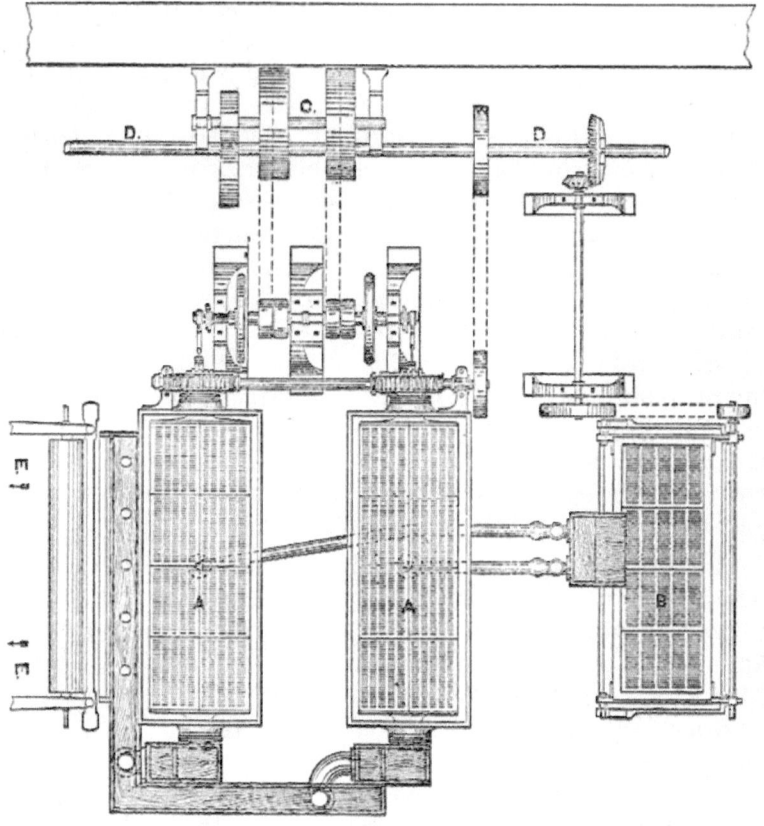

Fig. 26.

Revolving Strainer and Knotter.—The revolving strainer, which was invented by the late senior partner in the firm of Messrs. G. and W. Bertram (now Bertrams, Limited), has since been extensively adopted, and the present firm have introduced a patent knotter in conjunction with the apparatus, the complete arrangement of which is shown in Fig. 26. The standard size for these revolving strainers is 7 feet long by 18⅜ inches wide on each side of the four surfaces. The vats are of cast iron, and the apparatus is supplied with driving gear, bellows, regulating boxes and spouts, as necessary. The firm also supply these strainers with White's patent discs, and Annandale and Watson's arrangement. A A are two revolving strainers, as applied to the paper-machine, showing gearing for strainers and bellows. B is the patent knotter as used for two strainers. C is the counter-shaft overhead. D D is the back shaft of the machine, and E E the wire of the paper-machine.

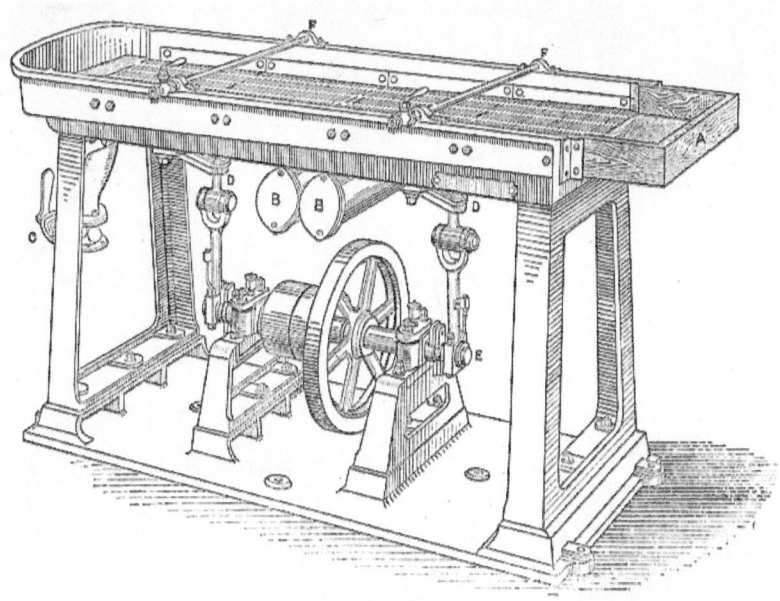

Fig. 27.

Self-cleansing Strainer.—The same firm also introduced this form of strainer, an illustration of which is given in Fig. 27. The action of the strainer is described as follows:—

The pulp flows on to the strainer at A, and passes away through the pipes B B. At C is a valve for the discharge of waste pulp. The strainer plates have an inclination of about 1 inch in the direction of their length, and in those which are nearest to A, where the pulp enters, the slits are wider, the knots being pushed forward by the energy of the flow. The vacuum pumps, D D, are worked from the shaft E. The tubes F F are for supplying water to the plates, by which the coarser particles of the pulp are pushed forward, and the slits are thus kept clean. The strainer will pass from 18 to 20 tons of the finest paper per week.

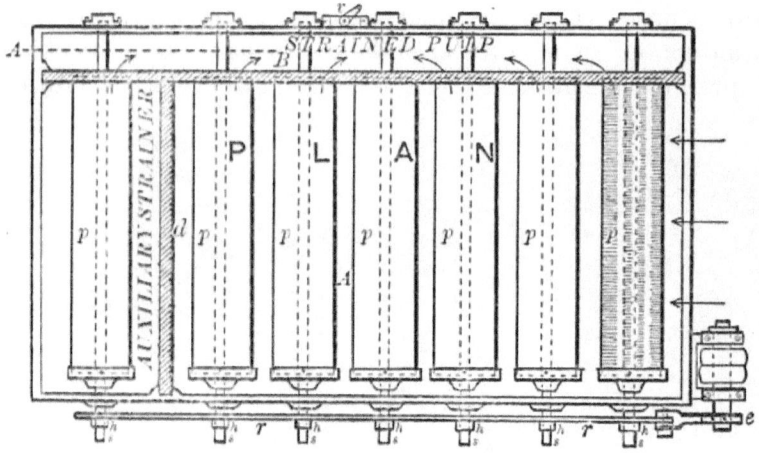

Fig. 28.

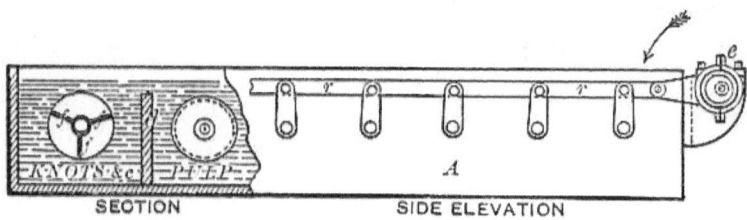

Fig. 29.

Roeckner's Pulp Strainers.—This invention consists in constructing boxes, with one or both ends open, forming the strainers, fixed, or to slide in or out, so as to be readily cleaned. One or more fans are fitted in these boxes, and are put in motion from the outside, so as to cause what is called "suction" through the strainers. One or a number of such boxes are fixed into a vat, the open ends discharging the pulp which has passed through the strainers to the paper-machine, and can be so arranged that all the fans are worked on one shaft. The vat may be divided into compartments, so that the stuff flows from one to the other. Instead of boxes, the strainers may be formed of tubes, in which suitable slits or perforations have been provided. The tubes will be perfectly closed at one end, and the strained pulp, after passing through them, will be delivered to the paper-machine from their open ends, which may fit into a ring, so that when cleaning is required they may be easily lifted out or in. The suction is provided inside these tubes by the fans, which are oscillated by suitable gear from the outside of the vat. The strainers may, instead of being stationary, be attached to the fans and oscillate with them, in which case the open ends would have to be attached to the vat by an indiarubber or cloth ring, or the strainers may oscillate whilst the fans are stationary. Any number of these strainers may be fixed into vats, disposed vertically or otherwise. In the vat A, Fig. 28, which receives the pulp to be strained, are several tubes, $p\ p\ p$, with one end open, having slits in them similar to strainer plates. Inside of these are two, three, or more plates, $f\ f\ f$, Fig. 29, running the full length of the tube fixed to the shafts, $s\ s\ s$, and to the sides of the tubes, which serve as fans, besides giving strength to the tubes. The shafts $s\ s\ s$ are carried in bearings at each end, and have each one end projecting through, upon which are keyed levers, $h\ h\ h$, which, being connected to a rod r, worked by an eccentric, e, at the end, gives an oscillating motion to the tubes and fans. Any number of tubes may be in the vat, and may either work separately or divided. With several tubes it is preferable to have them arranged as shown in the drawing by division plate d, so that the accumulated "knots," &c., may flow finally into the end compartment (which will form an auxiliary strainer), and may be mixed with more water, so that the fine pulp still contained in the stuff can flow away through the slits and the knots, &c., be taken out when necessary. The tubes should be placed so far apart that a workman can get his hand between. The closed ends work free in the stuff, while the open ends run through indiarubber sheet or other material, fitted so well to the tube that the fibre can only get through the slits of the tube to flow on to the paper-machine through the channel at side by the sluice v. The arrows indicate the direction of the flow of pulp.

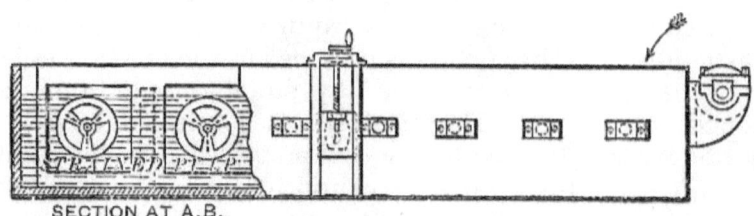

SECTION AT A.B.

Fig. 30.

Mr. Dunbar says, "the straining power necessary to pass and clean pulp in an efficient manner for 25 tons of finished paper per week is two revolving strainers, consisting of four rows of plates, or 7 feet by 18 inches of straining surface on each of the four sides, the plates being cut No. 2½ Watson's gauge."

After passing through the strainers the pulp should be absolutely free from knots or objectionable particles of any kind, and in a proper condition for conversion into paper.

The Machine Wire and its Accessories.—On leaving the strainers the pulp passes into a vat, in which is a horizontal agitator, which causes the pulp and water to become well mixed, and ready to flow on to the endless wire-cloth of the machine. The wire-cloth is made of exceedingly fine wire, the meshes ranging from 60 threads and upwards to the inch, there being sometimes as many as 1,900 holes per square inch, but the meshes usually employed run from 2,000 to 6,000 per square inch. The ends of the cloth are united by being sewn with very fine wire. The width of the wire-cloth varies considerably, the greatest width being, we believe, that supplied for the large machine at Mr. Edward Lloyd's mill at Sittingbourne, which is 126 inches. The length of the wire-cloth is generally from 35 to 40 feet, the latter being considered preferable. Beneath the wire is placed a shallow box called the "save-all," which receives the water as it flows through the wire cloth from the pulp. In order to effect a further saving of pulp which escapes through the meshes of the wire-cloth, a machine called a "pulp-saver" is used at some mills, through which the backwater, as it leaves the box or save-all referred to, is passed.

The wire-cloth is supported by a series of brass tube rolls, which are so placed as to render the layer of pulp on the wire absolutely uniform, by which a regular thickness of the finished paper is ensured. The wire is attached to a malleable iron frame, having a sole-plate of cast iron, and carries a brass or copper breast-roll, 18 inches in diameter, a guide-roll 7 inches in diameter, and four brass or copper rolls 5 inches in diameter under the wire, with shafts extending through the rolls, and furnished with brass bushes and brackets, and a self-acting guide upon the 7-inch guide-roll. The tube-rolls or "carrying tubes" are carried upon brass bearings. Attached to the sole-plate of the wire framing are three cast-iron stands on each side for supporting the save-all beneath the wire. To regulate the width of the paper there is on the top of the wire a set of brass "deckles," carried on a brass frame passing over the first suction box, of which there are two, and supported on the wire frame by iron studs fixed in the frame. At each end of the deckle-frame is a pulley for carrying the deckle-strap, with three similar pulleys for expanding it. The deckle-frame is furnished with two endless straps of india-rubber, these straps keeping the pulp to the width required for forming ledges at the sides of the web.

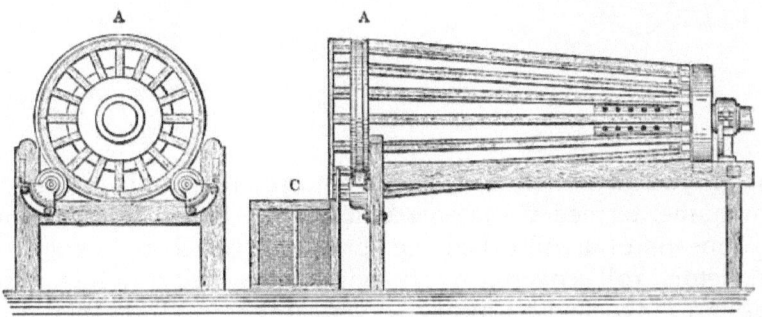

Fig. 31.

The Conical Pulp-saver, which is shown in Fig. 31, was invented by the late Mr. George Bertram and Mr. Paisley, and is manufactured by Bertrams, Limited. Its use is to extract fibres from the washing water before going into the river or otherwise. For the water from the drum-washer, washing and beating engines, and for the water from the paper-making machine, save-all, &c., it has proved itself of great utility. It is simple in construction, small in cost, takes up little room, and is easily repaired. When placed to receive the washings from the beaters or paper-machine, the pulp saved, if kept clean, can always be re-used. A is a conical drum which is covered with wire-cloth, and it is made to revolve slowly by suitable gearing. The water enters by the pipe B, which is perforated, as shown, and passes through the meshes of the gauze, while the pulp gradually finds its way to the wider end of the drum, where it escapes into the box C, and can be conveyed again to the beating-engines.

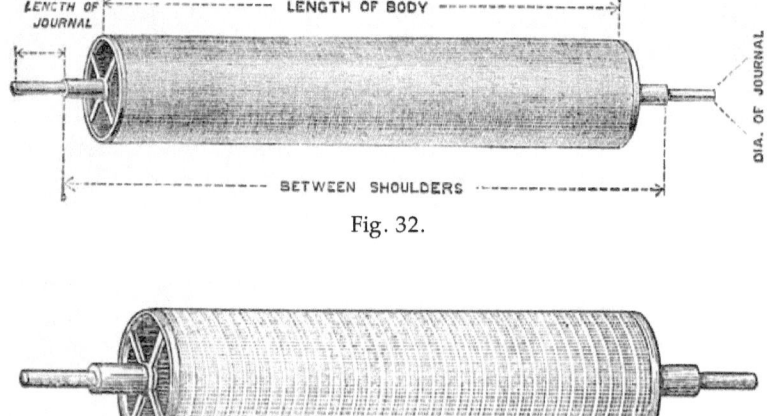

Fig. 32.

Fig. 33.

The Dandy-roll.—When it is required to produce a design or name, termed a *water-mark*, upon the paper, this is done by means of a roll called the *dandy-roll*, which consists of a skeleton roll covered with wire-cloth, upon which the design is worked by means of very fine wire. If the paper is required to be alike on both sides, without any specific pattern or name upon it, the roll is simply covered with wire-cloth, the impressions from which upon the moist pulp correspond with those of the machine-wire on the under surface. By this means paper known as "wove" paper is produced. A dandy-roll of this character is shown in Fig. 32. "Laid" paper, as it is termed, is distinguished by a dandy-roll having a series of equidistant transverse wires on the upper surface of the wire cylinder, as shown in Fig. 33, the effect of which is to produce parallel lines on the paper, caused by the pulp being thinner where the moist paper is impressed by the raised wires, which renders the lines more transparent than the rest of the paper. The dandy-roll, which is usually about 7 inches in diameter, corresponds in length to the width of wire on which it rests, and is placed over the wire-cloth between the suction-boxes. The journals of the roll turn in slits in two vertical stands, one behind the machine frame and the other in front of it. The roll, however, rests with its whole weight on the wire, and revolves by the progressive motion of the wire. The stands which support the roll prevent it from being influenced by the lateral motion of the wire. By thus running over the surface of the pulp when the wire is in motion, this roll presses out a considerable quantity of water, at the same time rendering the paper closer and finer in texture. Dandy-rolls of various lengths, and bearing different designs or patterns, are kept at the paper-mills, and great care is exercised to preserve them from injury.

Water-Marking.—Dr. Ure describes the following processes for producing a design for a line water-mark:—1. The design is engraved on some yielding surface in the same way as on a copper-plate, and afterwards, by immersing the plate in a solution of copper sulphate, and producing an electrotype in the usual way, by which all the interstices become so filled up as to give a casting of pure copper. This casting, on being removed from the sulphate bath, is ready for attaching to the wire gauze of the dandy-roll. 2. The design is first engraved on a steel die, the parts required to give the greatest effect being cut deepest; the die, after being hardened, is forced by a steam hammer into some yielding material, such as copper, and all of this metal which remains above the plain surface of the steel is subsequently removed by suitable means; the portion representing the design being left untouched would then be attached to the wire-gauze as before. Light and shade can be communicated to the mark by a modification of the above process, for which purpose an electrotype of the raised surface of a design is first taken, and afterwards a second electrotype from this latter, which consequently will be identical with the original surface.

These two are then mounted on lead or gutta-percha, and employed as dies to give impression to fine copper-wire gauze, which is then employed as a mould. Thus absolute uniformity, such as could not be attained by the old system of stitching wires together, is now attained in bank-notes by the adoption of the above method. It may be mentioned that when the moulds were formed by stitching the fine wires together to form a design, no less than 1,056 wires, with 67,584 twists, and involving some hundreds of thousands of stitches, were required to form a pair of £5 note moulds, and it was obviously impossible that the designs should remain absolutely identical.

Sometimes water-marks are produced by depressing the surface of the dandy-roll in the form of a design, which causes the paper to be thicker where the design is than in the rest of the sheet of paper. This modification was invented by Dr. De la Rue.

De La Rue's Improvements in Water-marks.—By one method, patented in 1869, dandy-rolls, having a surface of embossed wire-gauze, are used; the indentations in the gauze are inwards, causing a thickening of the paper where they are brought in contact with it. These thickenings correspond in form to the configuration of the design or water-mark. The inventor has also affixed wire to the surface of such dandy-rolls so as to form projections, in order to thin the paper where the projections come in contact with it, by which means light lines are obtained in the water-mark, strengthening the effect of the thickened opaque design.

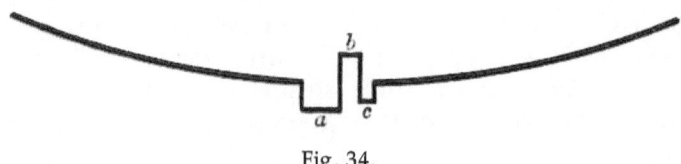

Fig. 34.

By another patent, dated May, 1884, No. 8348, the inventor forms the surface of the dandy-roll of wire-gauze embossed in such a manner that parts of the surface of the gauze, corresponding to the configuration of the design of the water-mark, are raised, and project out from the general surface, and other parts corresponding to the line shading of the design are depressed below the level of the general surface. The accompanying drawing, Fig. 34, shows diagrammatically, and greatly enlarged, a section of a portion of the surface of a dandy-roll made in accordance with this invention. *a* represents the section of a ridge or projection raised on the surface of the gauze; *b* represents the section of a groove or depression in the wire-gauze, which, with other similar grooves, serves to produce an opaque shading to the design. *c* is an auxiliary ridge or projection, serving to define the shading line, and to intensify it by driving the pulp into the groove or depression *b*. Further effects may be obtained by attaching wires to the dandy-roll, either in the usual way, where the surface is unembossed, or upon the raised parts *a*, which give the configuration to the water-mark. In place of forming the ridges or projections *a*, which produce the configuration of the water-mark, by raising portions of the wire-gauze above the general surface, they may be formed by sewing on suitably shaped slips of wire-gauze, or of sheet metal perforated all over with fine holes, on to the surface of the gauze which is embossed with the grooves *b*, but it is much to be preferred that both the ridges *a* and the grooves *b* should be produced by embossing the gauze. Water-marks may also be produced by placing sheets of finished paper in contact with plates of copper or zinc, bearing a design in relief, and submitting them to heavy pressure.

Suction-Boxes.—These boxes, which are fitted under the wire, are made of wood, and are open at the top, the edges being lined with vulcanite. The ends of the boxes are movable, so that they may be adjusted to suit the width of the paper required; they are also provided with air-cocks for regulating the vacuum, which is obtained by means of two sets of vacuum pumps, having three 6-inch barrels to each set: a vacuum pump of this form is shown in Fig. 35. As the wire travels over these boxes, the action of the pumps draws the wire upon them with sufficient pressure to render them air-tight; by this means a large portion of the water which the pulp still retains at this point becomes extracted, thereby giving to it such a degree of consistency that it can stand the pressure of the couch-rolls without injury. The backwater extracted by the suction-boxes, as also that collected in the save-all, is added to a fresh supply of pulp before it flows on to the sand-tables.

Couch-Rolls.—At the extreme end of the wire-cloth from the breast-roll, and inside the wire, is the under couch-roll, from which the wire receives its motion. This roll, which is of brass, is usually about 14 inches in diameter, is carried upon a cast-iron framing with brass bearings, and is ground to a working joint with the top roll, which is also of brass, and 20 inches in diameter. Both these rolls are covered with a seamless coating of woollen felt. The upper roll rests upon the lower one, and the wire-cloth, and the web of paper upon it, pass between the rolls, receiving gentle pressure, by which the paper becomes deprived of

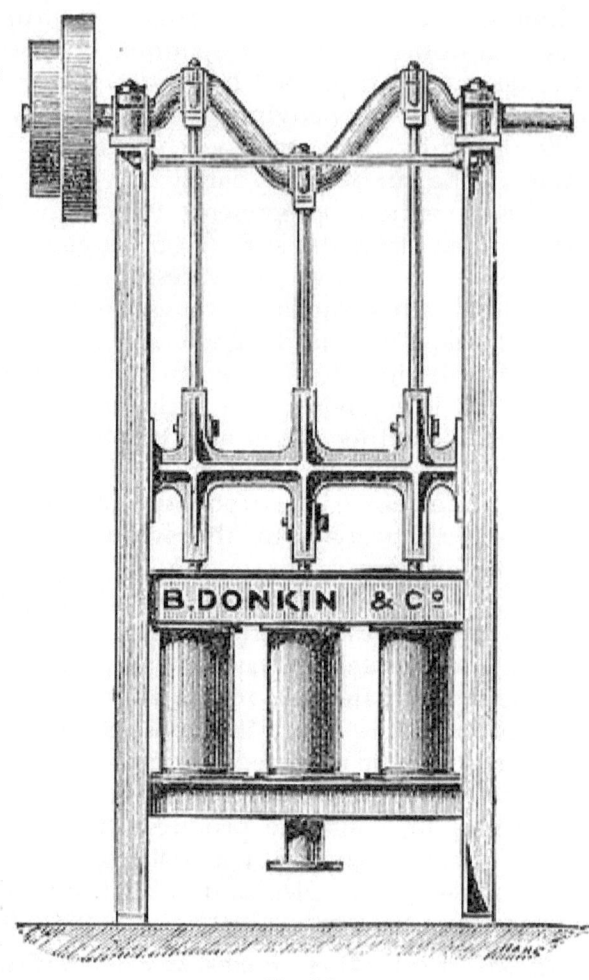

Fig. 35.

more water, rendering it still more compact. It is at this stage that the web of paper leaves the wire-cloth, and passes on to a continuously revolving and endless web of woollen felt, termed the "wet felt," from the moist condition of the paper. This felt, which is carried on wooden rollers, is about 20 feet long, and is manufactured with considerable care.

The Press-Rolls.—The paper now passes on to the *first press-rolls*, which deprive it of a still further quantity of water, and put it in a condition to bear gentle handling without injury. The upper roll is fitted with a contrivance termed the "doctor," which keeps the roll clean by removing fragments of paper that may have become attached to it. The doctor is furnished with a knife which passes along the entire length of the roll, pressing against it from end to end. These rolls are generally of iron, jacketed with brass, the under one being 14 inches in diameter, and the top roll 16 inches. Sometimes this roll is made of fine-grained cast-iron. When the roll is of iron the doctor blade is steel; but when this roll is brass the knife is of the same material. The under surface of the paper, which has been in contact with the felt, and necessarily being in a moist condition, receives more or less an impression from the felt over which it travelled, while the upper surface, on the other hand, will have been rendered smooth by the pressure of the top roll of the first press. To modify this, and to render both surfaces of the paper as nearly uniform as possible, the paper passes through another set of rolls, termed the *second press-rolls*, in which the paper becomes reversed, which is effected by causing it to enter at the back of the rolls, which rotate in a reverse direction to those of the first press, by which the under or wire side of the paper comes in contact with the top roll of the press. By this arrangement the underside of the paper is rendered equally smooth with the upper surface. The second set of press-rolls is provided with an endless felt of its own, which is usually both stronger and thicker than that used in connection with the first press-rolls. In some mills each set of press-rolls is provided with a doctor, to prevent the web of paper from adhering to the metal. Sometimes the doctor knives are made from vulcanite, a material which would seem specially suited for a purpose of this kind. From this point the paper passes to the first set of drying cylinders.

The Drying Cylinders.—The invention of the steam drying cylinder is due to Mr. T. B. Crompton, who, in the year 1821, obtained a patent for this useful addition to the paper-machine. Since that period, however, the system of drying the paper by steam-heat has been brought to a high state of perfection; not only this, but the number of cylinders has gradually increased, while the heat to which they are raised has proportionately decreased, and as a consequence the size, which is injuriously affected by rapid drying, is gradually deprived of its moisture, and thus renders the paper closer and stronger, while at the same time a very rapid speed can be maintained. The drying cylinders in the machine shown in the engraving are 4 feet in diameter and 12 in number, being arranged in two groups of 8 and 4 cylinders respectively, and in the aggregate present a very large drying surface, it being very important that the operation should be effected gradually, more especially at its earlier stages. There is a passage between the second press-roll and the cylinders, through which the machine-men can pass from one side of the machine to the other. The first two or three of the first section of cylinders are only moderately heated, and having no felt on them, allow the moisture from the paper to escape freely. The next five cylinders, however, are provided with felts, which press the paper against the heated surfaces, by which it becomes smooth and flattened, thus putting it into a proper condition for passing between the *smoothing-rolls*. The cylinders are heated by steam, and are generally of decreasing diameter, to allow for the shrinking of the paper during the drying.

Smoothing-Rolls.—These consist of highly polished cast-iron rolls, heated by steam. The paper being in a somewhat moist condition when it passes through these rolls, they have the effect of producing a fine smooth surface.

The paper next passes over the last four drying cylinders, all being provided with felts, to keep the paper closely pressed against their heating surfaces, by which the remaining moisture becomes expelled and the paper rendered perfectly dry. The paper now passes through the calender rolls, and is then wound on to reels at the extreme end of the machinery. The operation of calendering will be treated in the next chapter.

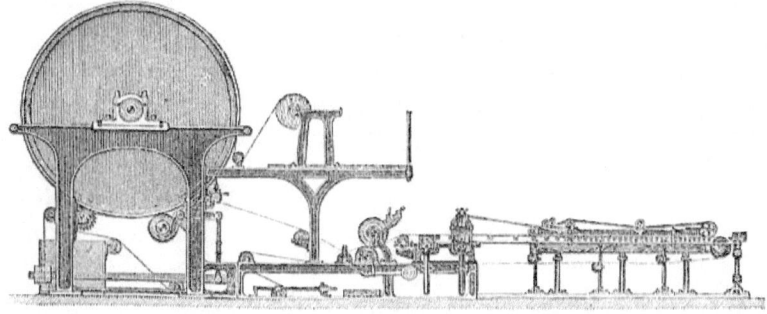

Fig. 36.

Single Cylinder Machine.—For the manufacture of thin papers, as also for papers which are required to be glazed on one side only, a single cylinder machine, called the Yankee machine, has been introduced, a representation of which is shown in Fig. 36. It is constructed on the same principle as the larger Fourdrinier machine up to the couching-rolls, when the paper leaves the wire-cloth and passes on to an endless felt running round the top couch-roll, and passes from thence to a large drying cylinder, which is about 10 feet in diameter and heated by steam, the surface of which is highly polished, giving to the surface of the paper in contact with it a high gloss. There is attached to the machine an arrangement for washing the felt for the purpose of cooling and opening it out after passing through a cold press-roll and the hot drying cylinder. This machine, as manufactured by Messrs. Bentley and Jackson, for cap, skip, and thin papers, consists of a rocking frame, and wrought-iron side bars, fitted with brass bearings, the necessary brass and copper tube-rolls, couch-rolls, with driving shaft, stands and pulley; self-acting wire guide, brass deckle sides and pulleys, brass slice, vacuum boxes, pipes and cocks; wet felt frame, with the necessary water pipes and cocks, and carriages to carry the couch-rolls and felt-rolls; the necessary wet felt-rolls and a felt washing apparatus; one bottom press-roll carried by brass steps, and fitted with compound levers and weight; one large cast-iron drying cylinder about 10 feet in diameter, and fitted with a central shaft, steam admission and water delivery nozzles, two water lifters and pipes, a manhole and vacuum valve, a large spur driving wheel, spur pinion, driving shaft and pulley; massive cast-iron framework, with pedestals to carry the cylinder; traversing steel doctor and frames; copper leading roll and carriages, a pair of reeling stands fitted with brass steps, friction pulleys and plates, regulating screws, etc.; a wooden platform and iron guard rail, all carried by strong cast-iron framing; the necessary pulp and backwater pumps, shake, knotter, stuff chests, service cistern, pipes and valves, shafting, pedestals, change wheels, pulleys, &c. These machines can be obtained of any desired width.

CHAPTER XIV.

CALENDERING, CUTTING, AND FINISHING.

Web-glazing.—Glazing Calender.—Damping-Rolls.—Finishing.—Plate Glazing.—Donkin's Glazing Press.—Mr. Wyatt on American Super-calendering.—Mr. Arnot on Finishing.—Cutting.—Revolving Knife Cutter.—Bertrams' Single-sheet Cutter.—Packing the Finished Paper.—Sizes of Paper.

To impart a higher gloss, or, as it is technically termed "glaze," to paper after it leaves the machine, it has to be subjected to further calendering, which is accomplished either in the web, or in sheets, according to the quality of the paper.

Web-Glazing.
—*Glazing Calender.*—
When paper has to be glazed in the web, it is passed between a series of rolls, which are constructed upon several different systems. In one form of this machine the rolls are alternately of finely polished iron, and compressed paper, or cotton, the iron rolls being

Fig. 37.

bored hollow to admit of their being connected to steam pipes, for heating them when necessary. In this machine there are eight rolls, the centre pair being both paper rolls, which have an effect equivalent to reversing the paper, by which both sides are made alike. Another form of glazing calender, of American origin, but which has been improved upon by our own engineers, consists of a stack of rolls made from chilled iron, the surfaces of which are ground and finished with exquisite precision upon a system adopted in America. A representation of this calender as manufactured by Messrs. Bentley and Jackson is given in Fig. 37. Such rolls as require heating are bored through, and their ends fitted with brass junctions and cocks, to regulate the admission of steam. The standards are of cast iron, planed and fitted with phosphor bronze bearings; the bearings to carry the top roll of the stack are furnished with wrought-iron screws and hand wheels, and wrought-iron lifting links can be attached to raise one or more of the rolls, according to the finish required on the paper. Compound levers are also supplied, to regulate and adjust the pressure on the ends of the rolls.

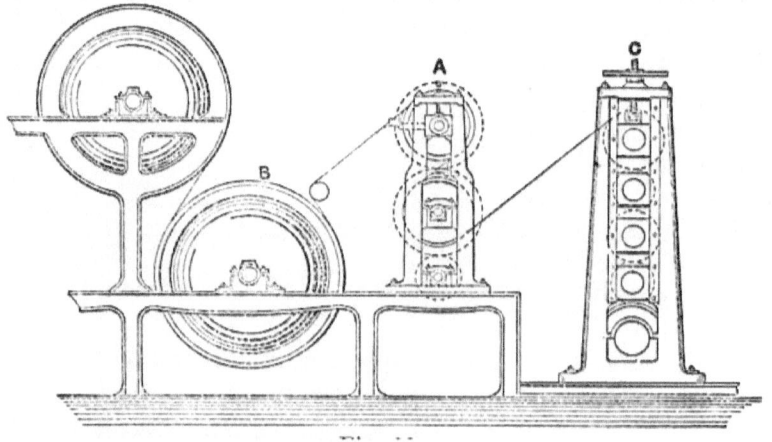

Fig. 38.

Damping Rolls.—An important improvement in connection with the calendering of paper was introduced by Messrs. G. and W. Bertram a few years since, by which a higher finish is given to the paper than had previously been attainable. This consists of a damping apparatus A (Fig. 38) which is placed between the last drying cylinders B of the machine and the glazing calenders C. The damping-rolls consist of two brass or copper rolls, about 14 inches in diameter, through which a constant stream of cold water is passed, while a line of steam jets, issued from finely-perforated pipes, plays over the face of the rolls. The cold water within the rolls condenses the steam, thereby imparting a uniform moisture to the under surface of the paper, which enables it to take a better surface when passing through the glazing rolls. The steam-pipes can be regulated so as to give any amount of dampness required by adjusting the steam cocks accordingly. By reference to the engraving, it will be observed from the disposition of the rolls that the web of paper is reversed, thus equalising the moisture on both sides, by which the paper-maker is enabled to produce an evenly-finished paper.

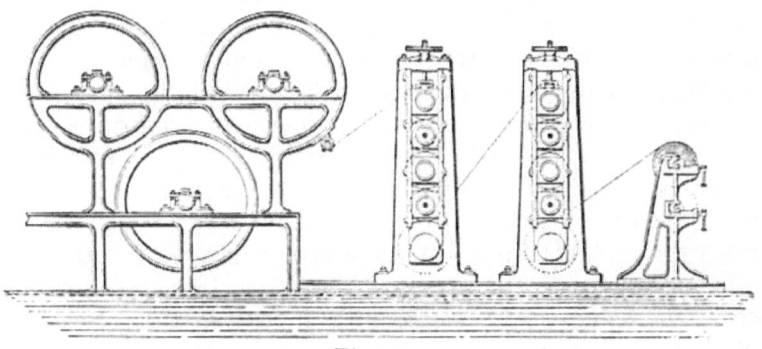

Fig. 39.

The chilled-iron glazing-rolls, as originally introduced, were fitted up in stacks of seven, and sometimes as many as nine rolls, but it was found in practice that so large a number of rolls gave unsatisfactory results; the heavy pressure, acting on the paper immediately after leaving the drying cylinders, had the effect of "crushing" the paper, giving it a thin feel. It is now considered preferable to use calenders having not more than four, or at most five rolls. An arrangement of this description, manufactured by Bertrams, is represented in Fig. 39. The system recommended by Mr. Dunbar is to employ three sets of rolls, disposed as follows:—"First, a set of three rolls; second, a set to consist of four rolls, and a stack of five to give the finishing or dry surface. With this arrangement of calenders, and the assistance of the damping apparatus, any desired surface can be got by varying and regulating the drying of the paper, which any careful machine-man can do with ordinary attention."

Finishing.—To give a still higher finish to the paper, it is subjected to what is termed "friction-glazing," which consists in passing it through a stack of rolls, formed alternately of small iron rolls and larger paper ones, the iron rolls revolving at a much higher speed than the paper-rolls. The effect of this final glazing operation gives the paper a very fine surface.

Plate-Glazing.—*Donkin's Glazing Press.*—This term, which is also called "super-calendering," is applied to a method of glazing hand-made paper, and is also adopted for the better qualities of machine-made paper. It consists in placing sheets of paper between highly polished plates of either copper or zinc, the latter being more generally used. The metal plates, with the sheets of paper placed alternately between them, are made up into packs or "handfuls" (the operation being usually performed by women), and these are passed between two powerful rolls, giving a pressure of from twenty to thirty tons, and each pack, consisting of about forty plates and as many sheets, is passed through the rolls several times, the pressure being regulated by means of screws or levers and weights acting on the ends of the top roll. A machine for glazing paper in packs, manufactured by Messrs. Bryan Donkin and Co., is shown in Fig. 40. Some descriptions of paper, as "antique" and "old style," for example, are surfaced with good cardboard instead of copper or zinc plates. As soon as the handful has passed through the rollers, the motion of the machine is reversed, by which means the pack is made to pass forwards and backwards repeatedly, according to the extent of gloss or smoothness required.

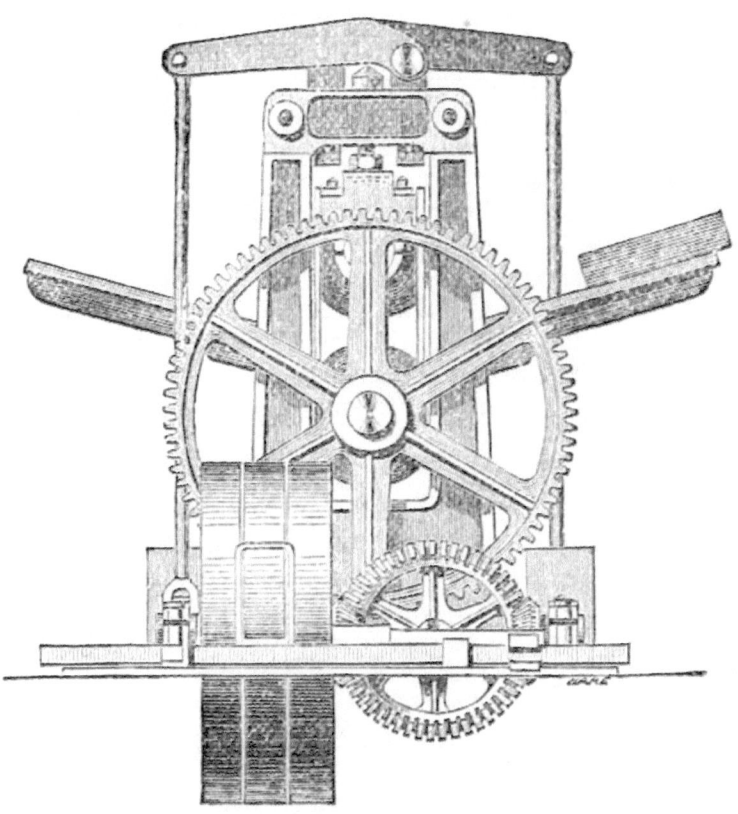

Fig. 40.

Mr. Wyatt on American Super-calendering.—Mr. Wyatt, on a recent visit to America, had many opportunities of witnessing the systems of manufacture adopted there, and subsequently delivered an interesting address to the members of the Paper-Makers' Club,[27] in which he acknowledged the superiority of the high-class printing papers for book-work, which has so often been the subject of recognition in this country. Indeed, if we compare the surface of the paper used even for ordinary technical journals in America and that generally adopted for our own periodicals of a similar class, we are constrained to admit that the difference is in favour of our transatlantic competitors. "In the manufacture of high-class super-calendered printing papers," Mr. Wyatt observes, "for fine book-work, or as they call them book papers, the Americans certainly excel. Whether this be due to the kind of raw material used, to the almost universal use of the refining-engine, which renders the pulp very soft and mellow, or to the state of perfection to which they have brought the art of super-calendering, or perhaps due to all three, I could not exactly determine. The material generally used for this class of paper is poplar chemical fibre and waste paper to the extent of 50 per cent., and even up to 75 and 80 per cent. of the total fibre, the balance being rags, or, in cheaper qualities, sulphite wood pulp; the stuff is all mixed together in large beaters, holding from 800 lbs. up to 1,500 lbs. of pulp, where it is about half beaten, and then finished in one or other form of refining-engine.

"The Americans have, I think, more thoroughly studied the question of super-calendering paper than we, and in this respect get better results and better work. The paper is mostly slit and trimmed on the paper-machine, and reeled up in from two to four widths by an ingenious contrivance called the *Manning-winder*, which automatically keeps the tension constant on each of the reels, whatever the diameter, and is super-calendered in narrow widths on small calenders. These calenders are from 36 inches to 42 inches wide, and consist of a stack of 9 to 11 rolls, alternately chilled iron, and cotton or paper; the paper is passed through the rolls two or three times, never less than twice, under great pressure applied by hand-screws. The power required is very high, being from 40 to 50 h.p. for each calender, and the speed from 450 feet up to 600 feet per minute. The paper is not usually damped before calendering, but is left rather under-dried from the machine; neither is steam heat used in the rolls, which get very warm, owing to the high speed at which they run. The rolls are driven entirely by straps, the arrangements for the fast and slow speed and for reeling on and off the paper being well designed and worked out; the main strap, running at high speed, runs on a loose pulley on the shaft of the bottom roll, by means of a powerful friction clutch; this pulley can be made a tight one. On this same bottom shaft is keyed a multiple V-shaped grooved friction pulley. Another, and independent shaft, driven from the main shaft by a crossed belt, has a small grooved pulley keyed on it, which can be thrown in and out of gear with the large grooved pulley. Strap-driving is thus secured throughout, and the speed can be increased gradually without jerks, from the starting up to the fastest speed by working the levers, gearing the friction clutch and pulleys slowly."

In reference to the high finish of American papers, we are disposed to attribute this mainly to the nature of the chief raw material used—wood fibre. In the year 1854, when specimens of Mr. Charles Watt's wood-fibre paper were first printed upon, the remarkable gloss of the wood paper attracted much attention, and it was noticed that the impression of the ink appeared to be well *on the surface of the paper*, and not, as was often the case with ordinary printing papers of the time, partially absorbed by the paper itself. Mr. Wyatt states that poplar chemical fibre and waste paper to the extent of 50 per cent., and even up to 75 and 80 per cent., are used, the balance being rags; now since the waste paper in all probability would be composed largely of wood fibre, and as, in the cheaper qualities, sulphite wood pulp is used in lieu of rags, it will be fair to assume that the chief basis of the highly-finished papers for which the Americans are justly famous is wood fibre, and we believe that there is no other variety of cellulose which is so susceptible of producing a naturally glossy paper as that which is obtained from wood by the soda process.

Mr. Arnot on Finishing.—Mr. Arnot makes the following observations respecting the finishing of paper:—"The paper may be slit into widths, suitable for wet calenders, or may be cut up into sheets, and glazed by the plate or board calenders. The former method of surfacing or finishing has come extensively into use in recent times, the labour involved being much less than in the older method of finishing in sheets. Still, however, the plate calenders are kept at work upon the higher classes of goods, it being possible to give almost any degree of surface to good paper by that means. There is little doubt, too, that the paper glazed by the plate rolls retains its original softness to a greater degree than that passed through web calenders. In the latter it is exposed in one thickness to great pressure, and is thinned in consequence; whereas, when the sheets are made up into piles, along with copper or zinc plates, there is a certain amount of spring or elasticity in the treatment which largely counteracts the crushing action of the rolls. The web calenders consist of a series of rollers erected in a vertical frame, and between these the paper winds, beginning at the top and coming downwards, so that the pressure gradually increases as the paper moves on its journey. It will be observed that the under rolls have to bear the weight of the upper ones, and that consequently the pressure on the paper will be greater the lower down it descends. Many of the rollers themselves are now made of paper, and as these possess a slight degree of elasticity, and take a high polish, they are alternated with iron rollers with good effect. The paper-rolls are made by sliding an immense number of circular sheets, perforated in the centre, on to an iron core or shaft, pressing these close together by hydraulic action, and trimming them off on the lathe. The plate or broad calenders consist only of two rollers, the upper one heavily weighted, preferably by compound levers. Between these rollers the sheets of paper, alternated with plates of copper or zinc, and made up into bundles about an inch in thickness, are passed backwards and forwards, the reciprocating action being produced by the movement of a lever in the hand of an attendant. The metal and paper sheets of different bundles may be interchanged, and the process repeated with the effect of increasing the beauty and equality of the finish."

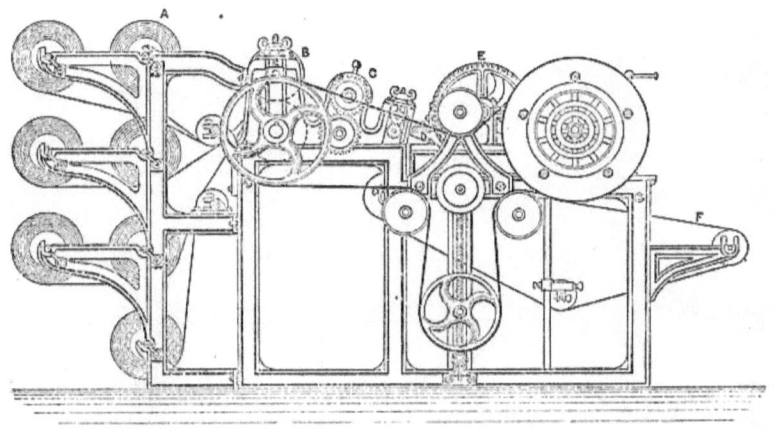

Fig. 41.

Cutting.—*Revolving Knife.*—When paper is to be used in a continuous printing-machine, or, as is often the case, has to be exported in the web, it is supplied in rolls; otherwise it is cut into sheets before leaving the mill. The form of cutter generally used is what is termed the *revolving knife-cutter*, an illustration of which, as manufactured by Bertrams, Limited, is shown in Fig. 41. At A is shown a series of webs, the paper from which is drawn forward by the rolls, B, and is then slit into suitable widths, and the margin at the same time pared by circular knives, one of which is shown at C. It then passes through a pair of leading-rolls, after which it comes in contact with a knife, D, attached to a revolving drum, E, pressing against a dead knife not shown in the engraving. The sheets, as they are thus cut, drop upon a travelling felt or apron, F, from which they are lifted and placed in piles, by boys or girls standing on each side of the felt. These machines will cut eight webs at one time.

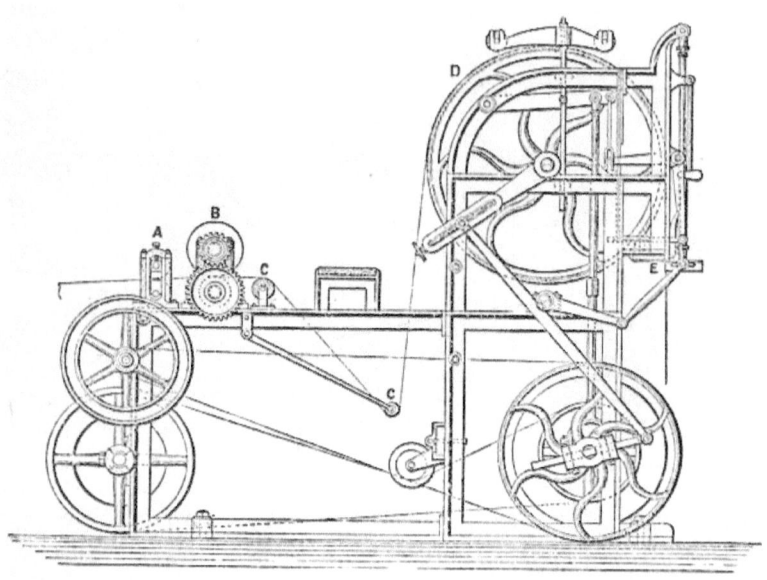

Fig. 42.

Bertrams' Single-sheet Cutter.—In cases where it is necessary that the sheets should be cut with great uniformity, as in the case of paper bearing a water-mark, in which it is requisite that the design should appear exactly in the centre of the sheet, the ordinary cutter is not found to be sufficiently reliable; a machine termed a "single-sheet cutter" is therefore used for this purpose, of which an illustration is shown in Fig. 42. The paper is led direct from the paper-machine, or from a reel frame, to the drawing-in rolls, A; after which it passes through the circular slitting-knives, B; from here it is led by the roller C to a large wood-covered drum, D, and at the front of this drum the sheets are cut by the cross-cutting knives, E. There are two cast-iron tapered cones, with belt guide for adjusting the speed; a fly-wheel to promote steadiness in working; a series of wrought-iron levers, cranks, eccentrics, shafts, etc., for accurately regulating the travel of paper and the cut of the horizontal knives; a small pasting table is also fitted across the machine for mending broken sheets.

Packing the Finished Paper.—The paper, after it leaves

the cutting-machine, is conveyed to the *finishing-house*, where it is carefully examined by women, who cast aside all defective or damaged sheets, which, under the trade names of "imperfections" or "retree," are sometimes disposed of, at a lower rate, to the customer for whom the order is executed. In the warehouse these imperfections are marked with a capital R on the wrapper, or two crosses, thus **X X**. If the paper is broken, it is sometimes marked B **X X**; it is not generally the custom, however, to sell imperfections, but to return them to the beater-man, to be re-converted into pulp. The perfect sheets are then counted, and packed up in reams consisting of 480 to 516 sheets.

Sizes of Paper.—The various sizes of paper are known in the stationery trade under different designations, as demy, crown, double crown, royal, imperial, etc. As paper is generally purchased according to weight, the various weights per ream are also distinguished with the size of the paper, as 16 lb. demy, 22 lb. double crown, and so on. The following table shows the sizes of some of the writing and printing papers in common use:—

Name.	Writing Papers.	Printing Papers.
	Inches.	Inches.
Foolscap	17 × 13¼	17 × 13¼
Small post (or post)	18¾ × 15¼	18¾ × 15¼
Crown		20 × 15
Double crown		30 × 20
Demy		22½ × 17¾
Royal		25 × 20
Imperial		30 × 22
Double demy		35½ × 22½
Double royal		40 × 25

CHAPTER XV.

COLOURED PAPERS.

Coloured Papers.—Colouring Matters used in Paper-Making.—American Combinations for Colouring.—Mixing Colouring Materials with Pulp.—Colouring Paper for Artificial Flowers.—Stains for Glazed Papers.—Stains for Morocco Papers.—Stains for Satin Papers.

Coloured Papers.—There are several methods by which any desired shade of colour may be imparted to paper, which are as follows:—

1. By blending with the pulp in the beating-engine some insoluble substance, such as smalts blue—a kind of glass coloured by oxide of cobalt—ultramarine, yellow ochre, etc.

2. By adding a coloured liquid, which simply dyes or stains the fibre.

3. By using rags which are already coloured, in proportions to give the required shade, in which case of course the process of bleaching must be omitted.

4. By employing two substances, as yellow prussiate of potash (ferrocyanide of potassium) and a persalt of iron, for example, which, when combined, yield the requisite blue tint—Prussian blue.

By this latter method the buff shade given to what is termed *toned paper* is effected, by using a solution of copperas (sulphate of iron) and an alkaline solution, or by using a solution of pernitrate of iron. In experimenting in this direction we have found that a mixture of solutions of sulphate of iron and bichromate of potassa produce an agreeable and permanent buff tint. The solutions may be added to the pulp alternately, or may be first mixed and then at once put into the beater. From 2 to 3 ozs. of each salt for each gallon of water may be used if the solutions are to be mixed before using; but when applied separately the solutions may be used in a more concentrated condition.

Colouring Matters used in Paper-Making.—The following substances, used either alone or mixed in suitable proportions, are employed in colouring pulp for paper-making:—

Smalts blue.
Prussian blue.
Indigo blue.
Aniline blues.
Aniline reds, including eosine.
Cochineal, for pink, etc.
Brazil wood, which imparts either a fine red or orange-brown colour, according to the treatment it has undergone.
Logwood, for violet colours.
Chrome yellow and orange chrome.
Orange mineral.
Copperas, for mixing with other substances.
Venetian red.
Yellow ochre.
Quercitron, or oak-bark.
Nutgalls.
Lamp black.

Blue.—The coarser kind of paper used for packing is prepared from rags blued with indigo, which, when reduced to pulp, are not subjected to the process of bleaching. The finer kinds of paper are blued in various ways, but the chief material used is what is known as artificial ultramarine, of which there are many qualities in the market, to which reference is made in another chapter. Prussian blue is also used, but this is usually produced directly in the beating-engine by adding in solution, 95 parts of sulphate of iron and 100 parts of ferrocyanide of potassium (yellow prussiate of potash). Smalts blue, which was formerly much used before the introduction of artificial ultramarine, is still preferred for high-classed papers as the colour is more permanent. To obtain smalts in an exceedingly fine state of division the best plan is to grind the colour in a little water, and then to separate the finest particle by the process of *elutriation*, that is, by diffusing the reduced mass through a large volume of water, and after allowing the larger particles to subside, pouring off the liquor in which the finer particles are suspended, to a separate vessel, in which they are allowed to subside. If this operation is carefully conducted the smalts may be obtained in an exceedingly fine state of division, and we have found that in this state the colour blends well with the pulp, and has little or no disposition to sink through it, but produces a uniform colouring throughout.

American Combinations for Colouring.—Hofmann gives the following examples of the combination of colours which have been adopted by American manufacturers:—

Yellow Gold Envelope of fine quality is made of—

Bichromate of potash	10 lbs.
Nitrate of lead	18 "
Orange mineral	56 "
Porous alum	30 "

each substance being separately dissolved and added to 400 lbs. of pulp.

Orange-red Gold Envelope:—

Bichromate of potash	7 lbs.
Nitrate of lead	10½ "
Orange mineral	60 "
Porous alum	20 "

These substances are dissolved separately and added to 400 lbs. of pulp.

Buff Envelope of fine deep shade is made from—

Bichromate of potash	3 lbs.
Nitrate of lead	5 "
Orange mineral	10 "
American ochre	20 "
Porous alum	30 "

Some half-stuff of red jute bagging. For 400 lbs. of pulp.

Tea-Colour is made from a decoction of quercitron bark, the liquid being poured into the engine, and 2 lbs. of copperas in solution are added for every gallon of the bark extract. A little ultramarine may be used to brighten the colour.

Drab.—Venetian red, well washed, added to a pulp of tea-colour made as above will give a fine drab.

Brown is composed of several colours, or a very fine dark green tea-colour brown, containing tea, buff, drab, and ink-grey, may be made of—

Quercitron bark liquid	15 gals.

Bicarbonate of soda	2	lbs.
Venetian red	4	"
Extract of nutgalls	2½	"
Copperas	18	"
Porous alum	30	"

The above proportions are for 400 lbs. of pulp.

The large proportion of alum prescribed in all the above examples serves as a mordant, and also, with the addition of resin soap, for sizing. All the above mixtures should be passed through a No. 60 wire-cloth into the beating-engine.

Mixing Colouring Materials with Pulp.—It will be readily understood that when paper is sized in the pulp, as Mr. Hofmann points out, the resinous alumina surrounds the fibres and prevents the colouring materials from penetrating them. In such cases the colouring materials are only loosely held, and a portion must therefore be lost in the machine. If added to the pulp before it is sized they become thoroughly mixed with the fibres, and with them enveloped by the size. The pulp should always be coloured before it is sized, except in cases where the alum or resin soap would injure the colours, or be injured by them. While the pulp is being sized and coloured, the finishing touch is given by the engine-man, who examines it and empties it into the stuff-chest.

Colouring Paper for Artificial Flowers.—Davis gives the following recipes for colouring one ream of paper of medium weight and size, sap colours only being used, and principally those containing much colouring matter. The gum arabic given in the recipes is dissolved in the sap-liquor.

Blue (dark) 1.—Mix 1 gallon of tincture of Berlin blue with 2 ozs. each of wax soap and gum tragacanth. 2. Mix ¾ gallon of tincture of Berlin blue with 2 ozs. of wax soap, and 4¼ ozs. of gum tragacanth.

Crimson.—Mix 1 gallon of liquor of Brazil wood compounded with borax, 2 ozs. wax soap and 8¾ ozs. of gum arabic.

Green.—1. Take ½ gallon of liquor of sap-green[28], 4¼ ozs.

of indigo rubbed up fine, 1 oz. of wax soap, and 4½ ozs. of gum arabic. 2. ½ gallon of sap-green liquor, 4¼ ozs. of distilled verdigris, 1 oz. of wax soap, and 4½ ozs. of gum arabic.

Yellow (golden).—Mix 6½ ozs. of gamboge with 2 ozs. of wax soap.

Yellow (lemon).—1. Compound 1 gallon of juice of Persian berries with 2 ozs. of wax soap and 8¾ ozs. of gum arabic. 2. Add to 1 gallon of quercitron liquor, compounded with solution of tin, 2 ozs. of wax soap, and 8¾ ozs. of gum arabic.

Yellow (pale).—Mix 1 gallon of fustic, 2 ozs. of wax soap, and 8¾ ozs. gum arabic.

Yellow (green).—Compound 1 gallon of sap-green liquor with 2 ozs. each of distilled verdigris and wax soap, and 8¾ ozs. of gum arabic.

Red (dark).—1 gallon of Brazil-wood liquor, 2 ozs. of wax soap, and 8¾ ozs. of gum arabic.

Rose Colour.—Mix 1 gallon of cochineal liquor with 2 ozs. of wax soap, and 8¾ ozs. of gum arabic.

Scarlet.—1. Mix 1 gallon of Brazil wood liquor compounded with alum and a solution of copper, with 2 ozs. of wax soap, and 8¾ ozs. of gum arabic. 2. Mix 1 gallon of cochineal liquor compounded with citrate of tin, with 2 ozs. of wax soap, and 8¾ ozs. of gum arabic.

Stains for Glazed Papers.—Owing to the cheapness of these papers glue is used in lieu of the more expensive gums; 1 lb. of glue dissolved in 1¼ gallon of water; the proportions of colouring materials are given for 1 ream of paper of medium weight and size.

Black.—1. Dissolve 1 lb. of glue in 1¼ gallon of water; triturate this with lampblack (1 lb.) previously rubbed up in rye whiskey; Frankfort black, 2¾ lbs.; Paris blue, 2 ozs.; wax soap, 1 oz.; then add liquor of logwood, 1½ lb. 2. 1½ gallon of liquor of logwood compounded with sulphate of iron, 1 oz. of wax soap, and 4½ ozs. of gum arabic.

Blue (azure).—1¼ gallon of glue liquor, as before, mixed with 1½ lb. Berlin blue, 2¾ lbs. powdered chalk, 2¼ ozs. of light mineral blue, and 2 ozs. of wax soap.

Blue (dark).—Mix with 1¼ gallon of glue liquor, 4½ lbs. of powdered chalk, 4¼ ozs. of Paris blue, and 2 ozs. of wax soap.

Blue (pale).—1. Mix ½ gallon of tincture of Berlin blue and 1 oz. of wax soap with 3½ ozs. of solution of gum tragacanth. 2. Take 1¼ gallon of glue liquor and mix with 4 lbs. of powdered chalk and 2 ozs. each of Paris blue and wax soap.

Brown (dark).—1. 1¼ gallon of glue liquor, mixed with 6 lbs. each of colcothar (jewellers' rouge) and English pink, 1½ lb. of powdered chalk, and 2 ozs. of wax soap. 2. Dissolve 1 oz. of wax soap and 4½ ozs. of gum arabic in ½ gallon of good Brazil-wood liquor, and add a like quantity of tincture of gallnuts.

Green (copper).—Mix in 1¼ gallon of glue liquor 4 lbs. of English verdigris, 1½ lb. of powdered chalk, and 4 ozs. of wax soap.

Green (pale).—Mix with 1¼ gallon of glue liquor 1 lb. of Bremen blue, 8½ ozs. of whiting, 1 oz. of pale chrome yellow, and 2 ozs. of wax soap.

Lemon Colour.—Mix in 1¼ gallon of glue liquor 13 ozs. of lemon chrome, 2 lbs. of powdered chalk, and 2 ozs. of wax soap.

Orange-Yellow.—Mix in 1¼ gallon of glue liquor 2 lbs. of lemon chrome, 1 lb. of Turkish minium, 2 lbs. of white lead, and 2 ozs. of wax soap.

Red (cherry).—Mix in 1¼ gallon of glue liquor 8½ lbs. of Turkey red, previously mixed up with ¼ gallon of Brazil-wood liquor, and 2 ozs. of wax soap.

Red (dark).—Mix ¾ gallon of Brazil-wood liquor with wax soap 1 oz., and gum arabic 4½ ozs.

Red (pale).—To 1¼ gallon of glue liquor is to be added 8¼ lbs. of Turkey red previously rubbed up with 2 ozs. of wax soap.

Violet.—4½ ozs. of gum arabic, and 1 oz. of wax soap are to be mixed with ½ gallon of good logwood liquor. When the gum is dissolved, mix with it enough potash to form a mordant.

Stains for Morocco Papers.—For 1 ream of paper of medium size and weight the following recipes are recommended:—

Black.—8¾ ozs. of good parchment shavings are dissolved in 1½ gallon of water; into this liquid is to be stirred lampblack, 1 lb., Frankfort black, 3 lbs., and Paris blue, 1¾ oz.

Blue (dark).—Dissolve parchment shavings, as before, and mix in 8¼ lbs. of white lead and 4½ lbs. of Paris blue.

Blue (light).—Dissolve parchment shavings, as before, and mix in 8¾ lbs. of white lead and 2¼ ozs. of Paris blue.

Green (dark).—Dissolve 13 ozs. of parchment shavings in 2½ gallons of water, and mix in 10 lbs. of Schweinfurth green.

Green (pale).—Prepare solution of parchment as in the last, and mix with 8¾ lbs. of Schweinfurth green and 1 lb. of fine Paris blue.

Orange-Yellow.—8¾ ozs. of parchment shavings are to be dissolved in 1½ gallon of water, and then mixed with 1½ lb. of lemon chrome, 8¾ ozs. of orange chrome, and 1 lb. of white lead.

Red (dark).—To the same quantity of parchment liquor as the last is to be added 7¾ lbs. of fine cinnabar, and 1 lb. of Turkey red.

Red (pale).—To the same quantity of parchment liquor add 8¾ ozs. of Turkey red.

Violet (light).—To 1½ gallon of parchment liquor add 4¼ lbs. of white lead, 13 ozs. of light mineral blue, and 8¾ ozs. of scarlet lake.

Violet (dark).—To 1½ gallon of parchment liquor add 3¾ lbs. of white lead, 1 lb. of pale mineral blue, and 8¾ ozs. of scarlet lake.

Yellow (pale).—To 1½ gallon of parchment liquor add 2 lbs. of light chrome yellow and 8¾ ozs. of white lead.

Stains for Satin Papers.—For each ream of paper of medium weight and size the following recipes are given:—

Blue (azure).—13 ozs. of parchment are dissolved in 2½ gallons of water and mixed with 3 lbs. of Bremen blue, 1¾ lb. of English mineral blue, and 4½ ozs. of wax soap.

Blue (light).—8¾ ozs. of parchment are to be dissolved in 1½ gallon of water, and to be mixed with light chrome yellow, 13 ozs.; colcothar, 6½ ozs.; Frankfort black, 2 ozs.; powdered chalk 3 lbs., and wax soap, 3½ ozs.

Brown (reddish).—1½ gallon of parchment liquor as the last, to which is added yellow ochre, 1 lb.; light chrome yellow, 4½ ozs.; white lead, 1 lb.; red ochre, 1 oz., and wax soap, 3½ ozs.

Brown (light).—1½ gallon of parchment liquor, as before, to which is added 13 ozs. of light chrome yellow, 6½ ozs. of colcothar, 2 ozs. of Frankfort black, 3 lbs. of powdered chalk, and 3½ ozs. of wax soap.

Grey (light).—1½ gallon of parchment liquor is mixed with 4¼ lbs. of powdered chalk, 8¾ ozs. of Frankfort black, 1 oz. of Paris blue, and 3½ ozs. of wax soap.

Grey (bluish).—To the above quantity of parchment liquor add 4¼ lbs. of powdered chalk, 1 lb. of light mineral blue, 4¼ ozs. of English green, 1¾ oz. of Frankfort black, and 3½ ozs. of wax soap.

Green (brownish).—To the same quantity of parchment liquor add Schweinfurth green, 1 lb.; mineral green, 8¾ ozs.; burnt umber and English pink, of each 4¼ ozs.; whiting, 1 lb., and wax soap, 3½ ozs.

Green (light).—To the same quantity of parchment liquor add English green and powdered chalk, of each 2¾ lbs., and 3½ ozs. of wax soap.

Lemon Colour.—To the same quantity of parchment liquor add lemon chrome, 1½ lb.; white lead 1 lb., and wax soap, 3½ ozs.

Orange-Yellow.—Parchment liquor as before, 1½ gallon, to which is added lemon chrome, 4¼ lbs.; Turkey red, 8¾ ozs.; white lead, 1 lb., and wax soap, 3½ ozs.

Rose Colour.—1½ gallon of parchment liquor as before, to which is added ¾ gallon of rose colour prepared from Brazil wood and chalk, and 6½ lbs. of wax soap.

Violet (light).—1½ gallon of parchment liquor as above, mixed with light mineral blue and scarlet lake, of each 1½ lb.; white lead, 1 lb., and wax soap, 3½ ozs.

White.—To 1½ gallons of parchment liquor is added fine Kremnitz white, 8¾ lbs., Bremen blue, 4¼ ozs., and wax soap, 3½ ozs.

Silver White.—1½ gallon of parchment liquor mixed with Kremnitz white, 8¾ lbs., Frankfort black, 8¾ ozs., and wax soap, 3½ ozs.

Pale Yellow.—1½ gallon of parchment liquor, to which is added 4½ lbs. of light chrome yellow, 1 lb. of powdered chalk, and 3½ ozs. of wax soap.

CHAPTER XVI.

MISCELLANEOUS PAPERS.

Waterproof Paper.—Scoffern and Tidcombe's process.—Dr. Wright's process for preparing Cupro-Ammonium.—Jouglet's process.—Waterproof Composition for Paper.—Toughening Paper.—Morfit's process.—Transparent Paper.—Tracing Paper.—Varnished Paper.—Oiled Paper.—Lithographic Paper.—Cork Paper.—New Japanese Paper.—Blotting Paper.—Parchment Paper.—Test Papers.

Waterproof Paper.—*Scoffern and Tidcombe's Process.*—In this process, for which a patent was granted in 1875, the well-known solubility of cellulose in cupro-ammonium is taken advantage of, for the purpose of producing waterproof paper by destroying its absorptive properties. After the paper is made and dried in the usual way by the paper-making machine, it is led through a bath of cupro-ammonium, having a roll or rollers therein, or in connection therewith, either on reels on which the paper is reeled, or from the continuous web of paper itself directly from the machine, and from this bath it is led over a table of wire-cloth, or india-rubber, or over a series of rollers forming a table, under which steam-pipes are placed for the purpose of "setting," or partially drying, the web; it is then led over suitable reels in a hot-air chamber to season or finish the treated paper, which is then cut as the paper runs, by the ordinary cutting machine, into the required sheets. The chamber in which the paper is treated is ventilated as follows:—Over the bath and hot-air chamber is another chamber having openings leading into the hot-air chamber, and at these openings a steam-blast, or fan-blast, is applied, which ventilates the chamber in which the paper is heated, and drives the ammonia into contact with either sulphurous or hydrochloric acid, and by this means the ammonia is recovered in a solid form which would otherwise be wasted.

The inventors also incorporate hydrated oxide of copper with paper pulp, so that after it is made into paper it has only to be subjected to the action of ammonia, as ordinarily done, or to the action of gaseous ammonia mingled with steam. Brown papers are strengthened and glazed by passing them through a bath of pulp containing cupro-ammonium, either with or without pitch, tar, or other resinous matters. It is well known that by passing paper through a cupro-ammonium bath it is surface dissolved and glazed by its own material, and if it be desired to unite two or more sheets together this is the most economical way of conducting the operation; but if it be desired to strengthen and glaze a single thickness of paper or millboard, it is considered undesirable to make the glaze by dissolving a portion of the paper itself. In this case the inventors pass the web or sheet of paper through a bath, not of cupro-ammonium simply, but of cupro-ammonium in which ligneous material is already dissolved; and when the glazing of brown paper is to be effected, they prefer to fortify the bath with tar, pitch, marine glue, or other resinous materials. By this process, panels and tiles may be manufactured from millboard, or thick sheets of ligneous material made from pulp already incorporated with hydrated oxide of copper. The panels, etc., are passed, by means of an endless web, through a bath of ammoniacal solution, or the vapour of ammonia and steam, and the tiles or panels may be surface-glazed by exposing them while moist to the action of fluo-silicic acid gas, by which silica is deposited in the material and on its surface.

Dr. Wright's Process for preparing Cupro-ammonium.— This process, which has been adopted at the Willesden Paper Mills, may be thus briefly described:—In the first part of the process, metallic copper, in small lumps, solid metal, or clippings, etc., is covered with a solution of ammonia in water, or with a weak solution of cupro-ammonium hydrate, containing an amount of free ammonia in solution dependent upon the strength of the copper solution ultimately required; a current of air is then caused to pass through the whole by means of an air-pump, in such a manner that the bubbles of air pass over and amongst the fragments of metallic copper, which, if in small particles, may be advantageously kept in suspension by any convenient agitator. In a few hours the liquid becomes saturated with as much copper as it can dissolve, the rate of solution varying with the form of the vessel containing the materials, the strength of the ammoniacal fluid, and the rate of the passage of the stream of air. To carry this process into effect, metallic copper in fragments of convenient size is loosely piled inside a vertical tube or tower, and water is allowed to trickle from a pipe over the copper so as to keep its surface moist. At the base of the tower a current of air, mixed with ammonia gas, is caused to pass into the tower, so as to ascend upwards, meeting the descending water as it trickles over the copper. Under these conditions the copper becomes oxidised, and the water dissolves firstly the ammonia gas, and, secondly, the oxide of copper formed, so that the liquor which passes out at the base of the tower is a solution of cupro-ammonium hydrate, the strength of which depends on the proportions subsisting between the bulk of the mass of copper, the quantity of water trickling over it, and the amount of air and ammonia gas supplied in a given time. As an example of the method of carrying out the above process, the inventor proceeds as follows:—He constructs a vertical iron tower which may be ten inches in internal diameter and ten feet in height, and this is filled with scraps of sheet copper. On this water is allowed to trickle, whilst at the base of the tower a mixture of air and gaseous ammonia is allowed to pass upwards through the tower, by which a solution of cupro-ammonium is formed, which is allowed to trickle out at the base of the tower into a tank. It has been found advantageous to use a series of towers, allowing the air and ammonia gas that pass out at the top of the first tower to enter at the bottom of the second tower, and so on successively throughout the series. The weaker solutions produced in the later towers of the series are used instead of water in the earlier towers, so that practically all the ammonia gas originally used is obtained in the form of cupro-ammonium hydrate solution, issuing from the first tower of the series.

The cupro-ammonium process, as carried on at the Willesden Mills, is applied to ropes, netting, etc., by immersing them in a solution of cupro-ammonium, which, when they are subsequently dried, gives them a varnished appearance, while at the same time, the fibres having become cemented together by the action of the cupro-ammonium, their strength is increased. By the same process paper, canvas, and other manufactured articles are rendered waterproof. A concentrated solution of cupro-ammonium may also be used for securing envelopes, whereby the adhesion of the surfaces of the paper is rendered perfect, and the only means of opening the envelope is by cutting or tearing the paper.

Jouglet's Process.—This process, which with modifications has been adopted by others, is based on the solvent action on cellulose of a solution of oxide of copper in ammonia. A quantity of this solution is placed in a tank, and the paper rapidly passed over and in contact with the surface of the liquid, by means of suitable rollers in motion. The paper is afterwards pressed between a pair of rolls and dried by the ordinary drying cylinders. The brief contact of the paper with the liquid occasions just sufficient action on the cellulose to have the effect of an impermeable varnish.

Waterproof Composition for Paper.—The following composition for rendering paper waterproof for roofing and flooring purposes has been patented in America.[29] By preference good, hard manilla paper is selected, and a composition of the following ingredients is applied with a brush, or by means of rollers:—Glue, 2 lbs., is dissolved in 3 gallons of crude petroleum, of about the density of 33° B. at 60° F.; 35 gallons of resin oil, and about half a pint of oil of eucalyptus, which will have the effect of destroying the objectionable odour of the resin oil. To this mixture is further added about 4 gallons of any ordinary drier. The above ingredients are to be thoroughly mixed by agitation, and the composition brushed over the paper in a room heated to about 80° F., and allowed to dry. It is said that paper thus coated will exclude wind, cold, dampness, and dust.

Toughening Paper.—*Morfit's Process.*—The object of the following process is to produce a paper "toughened in a degree and quality distinctively from any other in the market," and is applicable to all kinds of paper, but more particularly to those made with inferior grades of pulp for printing newspapers, and for wrapping papers. The means employed are the seaweeds which form glutinous liquors with water, such as Carrageen, or Irish moss, Agar-agar, and the like. Any of such seaweeds may be employed, either separately or mixed with another of its kind, according to the judgment of the operator and the sort of paper to be manufactured, but some seaweeds are superior to others for this purpose. The raw seaweed is first washed, and then boiled with water until all the soluble matter has been extracted, and the resulting liquor is then strained. The hot strained liquor forms the bath in which sheets of paper or pulp are to be treated. If desired, resin soap and aluminous cake may be added to the glutinous liquor, but these "serve rather to size and make the paper rustle than increase its toughness." If the paper is to be treated in the form of sheets or web, it is to be passed, as it leaves the wire-cloth in which it is formed, through a hot solution of the seaweed alone, or mixed with resinous soap and aluminous cake, and dried by means of suitable machinery. To apply it to the pulp, the latter is to be diffused in the hot liquor, and the sheets or web made therefrom in the usual manner. The proper proportions of seaweed, resinous soap, and aluminous cake will vary with the kind of pulp and sheets under treatment, and must be adjusted as the judgment of the operator determines best for each operation.

Transparent Paper.—There are several methods of rendering paper transparent, amongst which the following has been recommended:—

Boiled and bleached linseed oil	120	parts.
Lead turnings	6	"
Oxide of zinc	30	"
Venice turpentine	3	"

The above ingredients are placed in an iron or other suitable vessel, in which they are thoroughly mixed, and the whole then boiled for about eight hours. The mixture is then allowed to cool, when it is again well stirred and the following substances added:—White copal, 30 parts; gum sandarac, 2 parts, these ingredients being well incorporated by stirring.

Tracing Paper.—Sheets of smooth unsized paper are laid

flat on a table, and then carefully coated on one side only with a varnish composed of Canada balsam and oil of turpentine. The brush used for this purpose must be a clean sash tool, and when the first sheet has been varnished in this way it is to be hung across a line to dry. The operation is then to be applied to fresh sheets in succession until the required quantity of paper has been treated. In the event of one coating of the varnish not rendering the paper sufficiently transparent, a second coating may be applied when the first coating has become quite dry.

Varnished Paper.—When it is desired to varnish the surface of paper, card-work, pasteboard, etc., it must first be rendered non-absorbent with two or three coatings of size, which will also prevent the varnish from acting upon any colour or design which may be impressed upon the paper. The size may be made by dissolving isinglass in boiling water, or by boiling clean parchment cuttings in water until a clear solution is formed, which, after straining, is ready for use. If necessary, for very delicate purposes, the size thus prepared may be clarified with a little white of egg. The size should be applied, as in the former case, with a clean sash tool, but the touch should be light, especially for the first coating, lest the inks or colours should run or become bleared. When dry, the varnish may be applied in the usual way.

Oiled Paper.—Sheets of paper are brushed over with boiled linseed oil, and then hung up to dry. Paper thus prepared is waterproof, and has been used as a substitute for bladder and gut skins for covering jam pots, etc., but the introduction of parchment paper has almost entirely superseded it.

Lithographic Paper.—This paper, which is written upon with lithographic ink, may be prepared by either of the following formulæ:—1. Take starch, 6 ozs.; gum arabic, 2 ozs.; alum, 1 oz. Make a strong solution of each separately in hot water, then mix the whole and strain the liquor through gauze. It must be applied to one side of the paper while still warm by means of a soft brush or sponge; a second or third coating may be given as the preceding one becomes dry. The paper is finally pressed to render it smooth. 2. The paper must first receive three coats of thin size, one coat of good white starch, and one coat of a weak solution of gamboge in water. The ingredients are to be applied cold with a sponge, and each coat allowed to dry before the next is applied.

Cork Paper.—A paper under this title was patented in America by Messrs. H. Felt and Co.; it is prepared by coating one side of a thick, soft, and flexible paper with a mixture composed of glue, 20; gelatine, 1; and molasses, 3 parts, and covering with finely-powdered cork, which is afterwards lightly rolled in. The paper thus prepared is said to be used for packing bottles.

New Japanese Paper.—According to the *Bulletin du Musée Commercial*, a native of Japan has recently invented a new process by which paper may be made from seaweed. The paper thus made is said to be very strong, almost untearable, and is sufficiently transparent to admit of its being used as a substitute for window glass; it takes all colours well, and in many respects resembles old window glass.—*Board of Trade Journal*.

Blotting Paper.—This paper, requiring to be very absorbent, is not sized, but is prepared with starch alone, which, while holding the fibres together, does not affect the absorbent property of the paper. Dunbar gives a recipe for making blotting paper which has been found successful, and from which we make a few extracts. In selecting materials for blotting, of high-class, cotton rags of the weakest and tenderest description procurable should be chosen. Boil them with 4 lbs. of caustic soda to the cwt.— that is, if you have no facilities for boiling them in lime alone. When furnished to the breaking-engine, wash the rags thoroughly before letting down the roll; when this is done, reduce them to half-stuff, and as soon as possible convey them to the potcher. When up to the desired colour, drain immediately. The breaker-plate should be sharp for blottings, and the beater-roll and plate also in good order, and the stuff beaten smartly for not more than an hour and a half in the engine. For pink blottings furnish two-thirds white cottons and one-third of Turkey reds if they can be got, or dye with cochineal to desired shade; empty down to the machine before starting, and see that the vacuum pumps are in good condition. Remove weights from couch-roll, and if there are lifting screws raise the top couch-roll a little. Take shake-belt off, as the shake will not be required. Press light with first press, and have the top roll of the second press covered with an ordinary jacket similar to couch-roll jacket. Dry hard, and pass through one calender with weights off, and roll as light as possible, just enough to smooth slightly.

Parchment Paper.—This paper, which is extensively used for covering jars and pots for pickles and jams, is prepared, according to the process of Poumarède and Figuier, as follows:—White unsized paper is dipped for half a minute in strong sulphuric acid, specific gravity 1·842, and afterwards in water containing a little ammonia. By Gaine's process (1857) unsized paper is plunged for a few seconds into sulphuric acid diluted with half to a quarter of its bulk of water (the acid being added to the water), and the solution allowed to cool until of the same temperature as the air. The paper is afterwards washed with weak ammonia. This process, which has been extensively worked by Messrs. De la Rue and Co., produces a far better material than the foregoing.

Mill and Card-board.—In the manufacture of boards refuse materials of all kinds that occur in the paper-mill may be used, and these are sorted according to the quality of boards for which they are best suited. After being well beaten the resulting mass is mixed with suitable proportions of rag pulp, kaolin, chalk, white clays, &c. There are four principal processes by which boards are manufactured, namely,

1. By superposing several sheets of paper and causing them to unite by a sizing material.

2. By superposing several wet leaves at the time of couching.

3. By moulds provided with thick deckles.

4. By special machines similar to those used for making continuous webs of paper, but without a drying cylinder, the sheets being dried in the open air or in a heated room.

The third method is only adopted for boards of moderate thickness, as an excess of pulp would render the draining difficult.

Making Paper or Cardboard with two Faces by Ordinary Machine.—By this process, recently patented by Mr. A. Diana, all kinds of thin or thick paper or cardboard are manufactured with two different faces by means of the ordinary paper-machine, having a single flat table with a single wire-gauze web, without requiring a second metallic web. For this purpose the two pulps are prepared separately, and one is caused to pass on to the web in an almost liquid condition; this is allowed to drain off sufficiently, and the second pulp (also in a liquid condition) is then passed uniformly upon the whole surface of the previous layer. The water drains off from this layer through the first layer, and the paper or cardboard is thus directly formed with two different faces, the subsequent operations being as ordinarily employed in paper-making. The space between two of the suction cases employed for drawing off the water in the pulp is a suitable point for the distribution of the diluted second pulp, which is almost liquid.

Test Papers.—These papers, which are extensively used both in the laboratory and the factory, for determining the presence of acids or alkalies in various liquids, may be prepared as follows:—*Litmus paper*, for detecting the presence of acids, is prepared by first making an infusion of litmus. Reduce to a paste with a pestle and mortar 1 oz. of litmus, adding a little boiling water; then add more boiling water—from 3 to 4 ozs. in all—and put the mixture into a flask and boil for a few minutes; finally, add more boiling water to make up half a pint, and when cold filter the liquor. To prepare the test paper, a sufficient quantity of the liquid being poured into a flat dish, pieces of unsized paper are steeped in the blue liquid, so that all surfaces may be thoroughly wetted; the paper is then to be hung up by one corner to drain, and afterwards dried. As many sheets of paper as may be required should be treated in this way, and the sheets afterwards cut up into convenient strips for use. *Red litmus paper*, for detecting slight traces of alkali in liquids, may be prepared by dipping a glass rod, previously dipped into a very dilute solution of sulphuric acid, into one-half of the above infusion, repeating the operation cautiously until the liquid turns from blue to a slightly red tint. Unsized paper when dipped in this will acquire a reddish colour which is very sensitive to the action of weak alkaline liquors, and the vapour of ammonia restores the blue colour instantly. *Turmeric paper* is prepared by dipping unsized paper in a decoction of turmeric—about 2 ozs. to the pint. Paper steeped in this solution and dried acquires a yellow colour, which turns brown in alkaline solutions.

CHAPTER XVII.

MACHINERY USED IN PAPER-MAKING.

Bentley and Jackson's Drum Washer.—Drying Cylinder.—Self-Acting Dry Felt Regulator.—Paper Cutting Machine.—Single Web Winding Machine. —Cooling and Damping Rolls.—Reversing or Plate Glazing Calender.— Plate Planing Machine.—Roll Bar Planing Machine.—Washing Cylinder for Rag Engine.—Bleach Pump.—Three-roll Smoothing Presses.—Backwater Pump.—Web Glazing Calender.—Reeling Machine.—Web Ripping Machine.—Roeckner's Clarifier.—Marshall's Perfecting Engine.

Apart from the mechanical contrivances which are referred to in various parts of this work, in which their application is explained, it will be necessary to direct attention to certain machines and appliances which are adopted at some of the more advanced paper-mills in this country and in America; but since the various makers of paper-makers' machinery are constantly introducing improvements to meet the requirements of the manufacturer, we must refer the reader to these firms for fuller information than can be given in the limited scope of this treatise. Many of the improvements in paper-making machinery consist in modifications— sometimes of a very important nature—in the construction of certain parts of a machine, whereby the efficiency of the machine as a whole is in some cases considerably augmented. Without offering any critical remarks upon the merits of the respective improvements which have been introduced, it will be sufficient to direct attention to the manufacturer's own description of the principal features of the special mechanical contrivance which he produces for the use of the paper-maker. It may also be said that innumerable patents have been obtained for various improvements in machinery, or parts of machines, engines, etc., which can readily be referred to at the Library of the Patent Office, or any of the public libraries throughout the Kingdom.

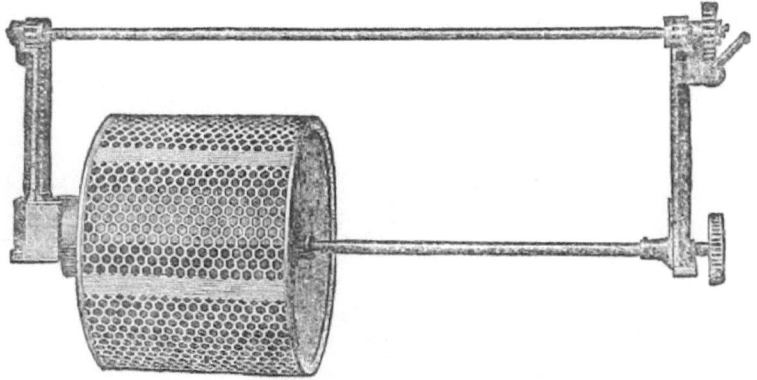

Fig. 43.

Bentley and Jackson's Drum-Washer.—This drum-washer, for use in the rag-engine, is shown in Fig. 43. It has cast-iron ends, strong copper buckets, shaft, stands, lifting-gear, and driving-wheel, but instead of the drum being covered with the ordinary strong brass backing-wire, it is covered with their improved "honey-comb" *backing-plates*, over which the fine wire is wrapped as usual. The honey-comb backing consists of tough rolled brass or copper plates, curved to suit the diameter of the drum, and secured to its ends by cross-bars. It is practicably indestructible, strengthens the drum, and by maintaining its cylindrical form, adds considerably to the durability of the fine covering-wire.

Fig. 44.

Drying Cylinders.—These cylinders, by the same firm, for which patents were obtained in 1872 and 1887, are made with concave and convex ends, the latter type being shown in Fig. 44. The cylinder body is made of hard cast-iron, turned and polished on outside surface. The ends and trunnions are of tough cast iron, turned to fit into their places, and there secured by bolts and nuts by a patented method, whereby no bolts (excepting for the manhole) are put through the metal, an unbroken surface is preserved, and the annoyance of leakage through the bolt-holes is avoided. A manhole and cover is fitted to all cylinders 3 feet in diameter and upwards, and a water-lifter and pipe to remove the condensed steam. The trunnions are bored to receive nozzles or junctions for admitting steam, and the whole, when completed, is carefully balanced and tested by steam pressure to 35 lbs. per square inch. The firm state that they have made cylinders from 2 to 10 feet in diameter by this system.

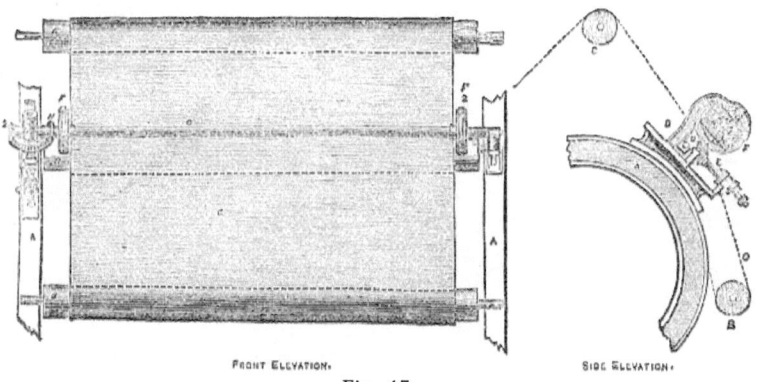
Fig. 45.

Self-acting Dry Felt Regulator.—This contrivance, which is manufactured by Messrs. Bentley and Jackson, is represented in front and side elevation in Fig. 45. A is the framing of the paper-machine, B the felt-rollers, C the dry felt; D is a slide carrying one end of the felt guide-roller B; C is a shaft across the machine, with a pulley F, two-keyed on one end, and a bevel pinion two-keyed on the other end. The pulley F and pinion H are keyed together, and run loose upon the shaft G; I is a bevel-wheel, gearing into the pinions H and 2. The wheel I is connected by a spindle and a pair of bevel-wheels to a screw E, which works through a threaded bush. When the machine is at work, if the felt C should run on one side, it will pass between the pulley F and the guide-roller B, causing the pulley to revolve, and turning the screw E in the threaded bush, thereby moving the slide fixing D and the guide-roller B, which causes the felt to run back. Should the felt run to the other side, it will run in contact with the pulley F 2, and thus reverse the motion of the guide-roller B.

Fig. 46.

Paper-cutting Machine.—This machine (Fig. 46), which is manufactured by the same firm, is constructed to cut from one to eight webs simultaneously, in sheets of any required length, from 8 to 60 inches. It is built on the "Verny" principle, and its operation is as follows:—The webs of paper from the reel-rolls are carried by an endless felt, and the paper is drawn off the rolls by travelling cast-iron gripper beams, which firmly grasp the felt and the webs of paper to be cut, the travel of the beams being equal to the length of the sheet of paper to be cut. When the required length of the sheet is drawn from the rolls, a cast-iron clamp, placed close to the dead cross-cut knife, descends and firmly holds the paper until the movable cross-cut knife has cut off the sheets, which fall on a second endless felt, and are placed by the catchers in the usual manner. As soon as the sheets are cut, the clamp is released, and the travelling-grippers are again ready to seize the paper and repeat the operation.

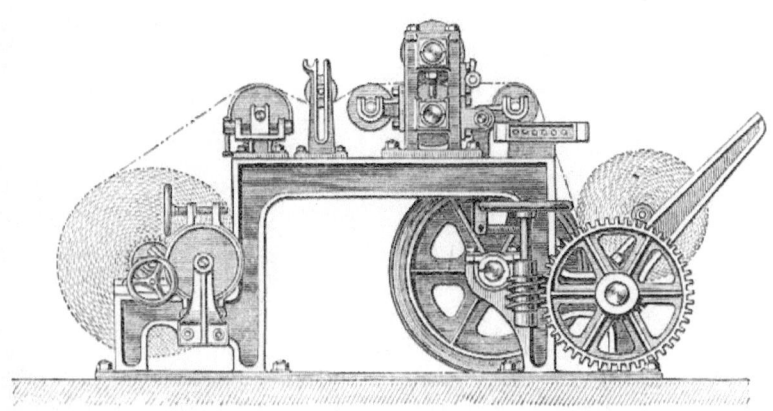

Fig. 47.

Single Web Winding Machine.—This machine (Fig. 47) is constructed for preparing webs of paper for continuous printing-presses. The roll of paper to be prepared is carried by brass bearings having vertical and horizontal screw adjustments attached to standards mounted on a slide, and movable by a screw transversely on the machine to accommodate the deckle edges. The paper web is taken through a pair of iron draw-rolls, carried by brass bearings, fitted in cast-iron stands; there are two pairs of ripping-knives with bosses, springs, and collars, mounted on turned wrought-iron shafts running in brass bearings carried by cast-iron stands; a wrought-iron leading-roll and carrying brackets fitted with brass bushes; a copper measuring roll counter, geared to indicate up to 10,000 yards, with disengaging apparatus to cease measuring when the paper breaks; a friction-drum 2 feet in diameter, made of wood, mounted on cast-iron rings, and a wrought-iron shaft, all carefully turned and balanced; two cast-iron swivelling arms, with brass sliding bearings to carry the mandrel on which the prepared web is to be wound, with screws, struts, wheels and shaft to regulate the angular pressure of the roll of paper against the wood drum, according to its weight and the quantity of paper.

Fig. 48.

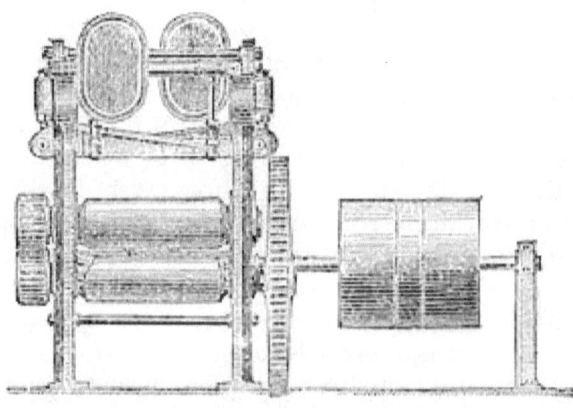

Fig. 49.

Cooling and Damping Rolls.—The illustration (Fig. 48) represents an apparatus, constructed by Messrs. Bentley and Jackson, for cooling and damping paper after leaving the drying cylinders and before passing through the calenders. It consists of two brass rolls bored and fitted with cast-iron ends, brass nozzles, and regulating taps, through which the rolls are supplied with a constant flow of water. The rolls are carried by cast-iron standards, fitted with brass steps and cast-iron caps. Jets of steam are blown on each of the rolls from a perforated copper pipe running parallel with, and at a little distance from, the body of the roll. The steam is condensed on the cold surfaces of the brass rolls, and absorbed by the web of paper, which passes around and in contact with their surfaces, and is consequently damped on *both* sides. The perforated steam-pipes are enclosed by copper hoods, to prevent the steam from spreading, and the supply of steam is regulated by ordinary brass valves or cocks. The rolls are geared together by a pair of spur-wheels, and driven by a pulley of suitable diameter.

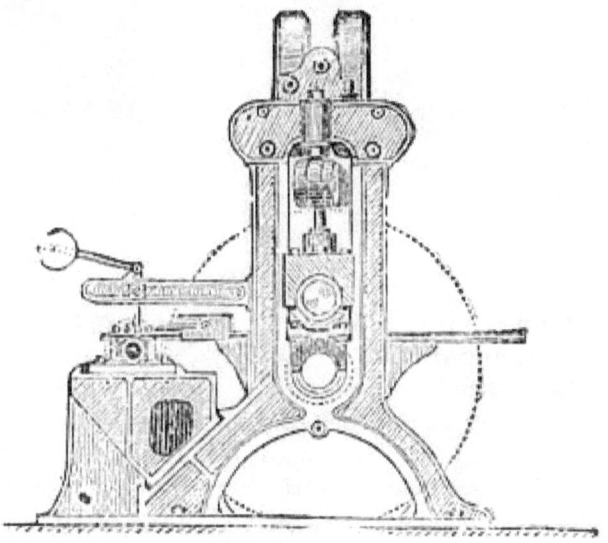

Fig. 50.

Reversing or Plate-glazing Calender.—This machine, which is shown in Figs. 49 and 50, is also made by the firm referred to, and consists of two hammered iron rolls, each about twelve inches in diameter, of any suitable length, carefully turned and carried by strong cast-iron standards, fitted with bell-metal steps. The top roll is provided with setting-down blocks and brasses, compound levers and weights to regulate the pressure required. The two rolls are geared together by strong shrouded wheels, and driven by a strong cast-iron spur-wheel and pinion, a driving-shaft, fast and loose pulleys, carried by cast-iron stands and pedestals fitted with brass steps. The machine is fitted with two metal feed-tables, and a self-acting apparatus for returning the sheets to the rolls, and a handle-lever, slide-bar, and strap-forks for starting and reversing.

Fig. 51.

Plate-planing Machine.—This machine, which is manufactured by Messrs. Bryan Donkin and Co., of Bermondsey, is shown in Fig. 51. By its aid the plates of rag-engines can be sharpened without being taken to pieces. The slide of the machine is made exactly like the roll-bar planing machine (see below), and is so arranged that it can easily be taken off and used for sharpening roll-bars.

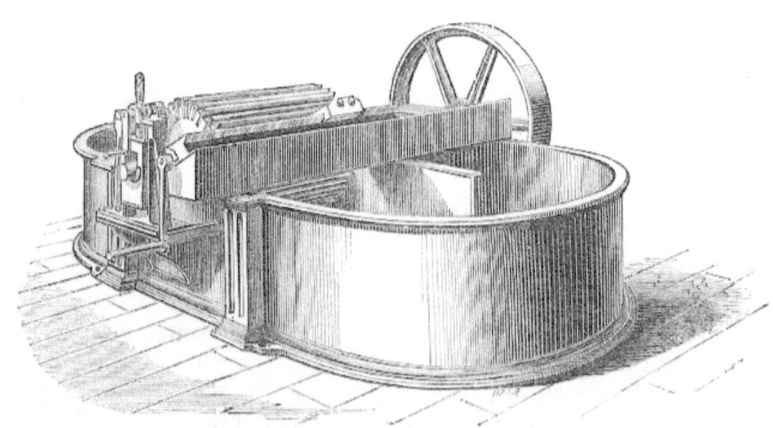

Fig. 52.

Roll-Ear Planing Machine.—In the accompanying engraving (Fig. 52) is shown an apparatus fitted to a rag-engine for sharpening rag-engine roll-bars, and it will be seen that by means of it the operation can be performed without removing the roll from its usual position. The edges of the bars are first planed by a tool supplied by the

Fig. 53.

manufacturers to render the whole cylindrical before sharpening them; the bevelled sides are then planed by suitable tools, two of which accompany the apparatus. This method of sharpening renders the bars uniform in shape, the roll is kept in better working order, and it can be dressed in considerably less time, and at less expense, than can be done by chipping by hand.

Washing-Cylinder for Rag-Engine.—The illustration at Fig. 53 represents the machine as manufactured by Messrs. Bryan Donkin and Co. It is so made that the water is delivered on the driving side of the rag-engine, thus avoiding any trough across the engine, and admitting of the midfeather being thin, as is usual in cast-iron engines. It is all self-contained, and the driving apparatus is wholly on the outside of the engine. The raising and lowering are effected by a worm and worm-wheel, so that the cylinder will stop at any point required.

Bleach Pump.—In the

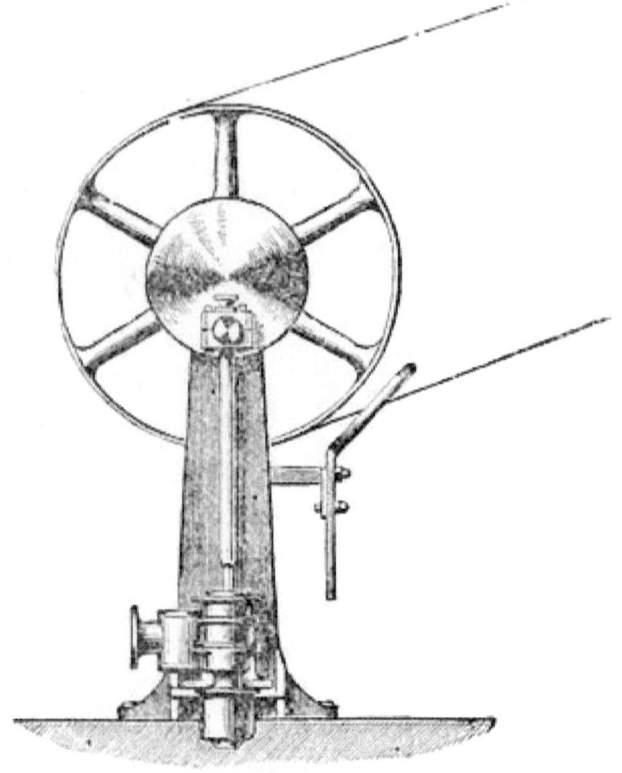

Fig. 54.

accompanying engraving (Fig. 54) is shown a pump, manufactured by Bryan Donkin and Co., which is arranged expressly for the purpose of pumping up bleach-liquor. Each pump is all self-contained, and merely requires a drum and strap to drive it. The live and dead riggers upon the pump allow it to be started and stopped at pleasure. "In all paper-mills," say the manufacturers, "the bleach-liquor should be used over and over again, not only to save bleach, which amounts to a considerable sum in the course of a year, but also to keep the paper clean."

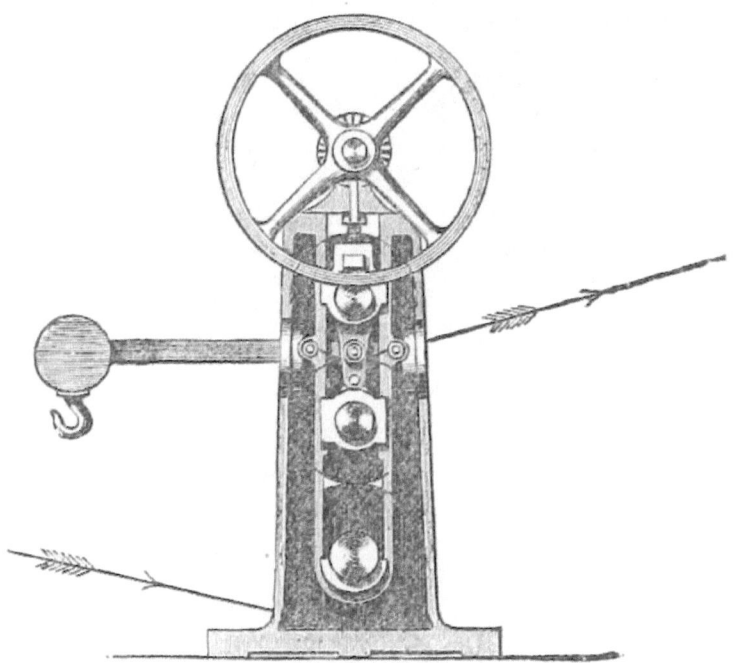

Fig. 55.

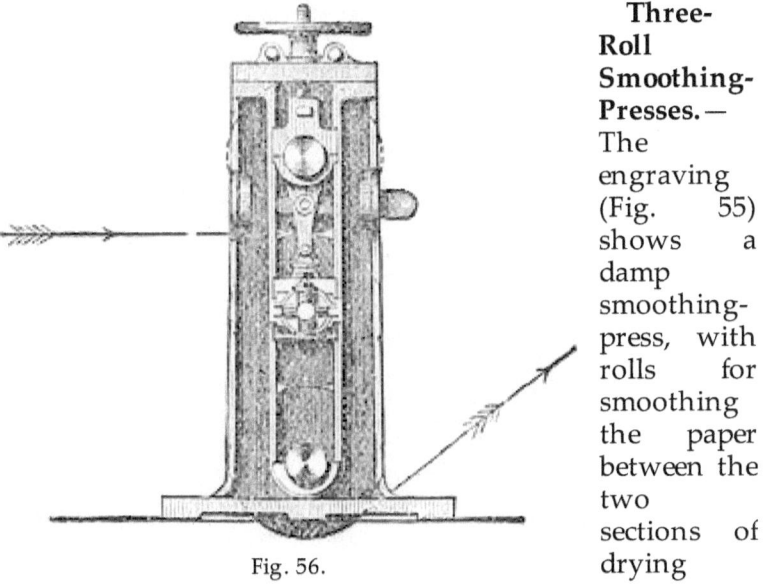

Fig. 56.

Three-Roll Smoothing-Presses. — The engraving (Fig. 55) shows a damp smoothing-press, with rolls for smoothing the paper between the two sections of drying cylinders of a paper-machine. The makers are Messrs. Bryan Donkin and Co. A three-roll smoothing press, for smoothing the paper at the end of a paper-machine, also by the same makers, is shown in Fig. 56.

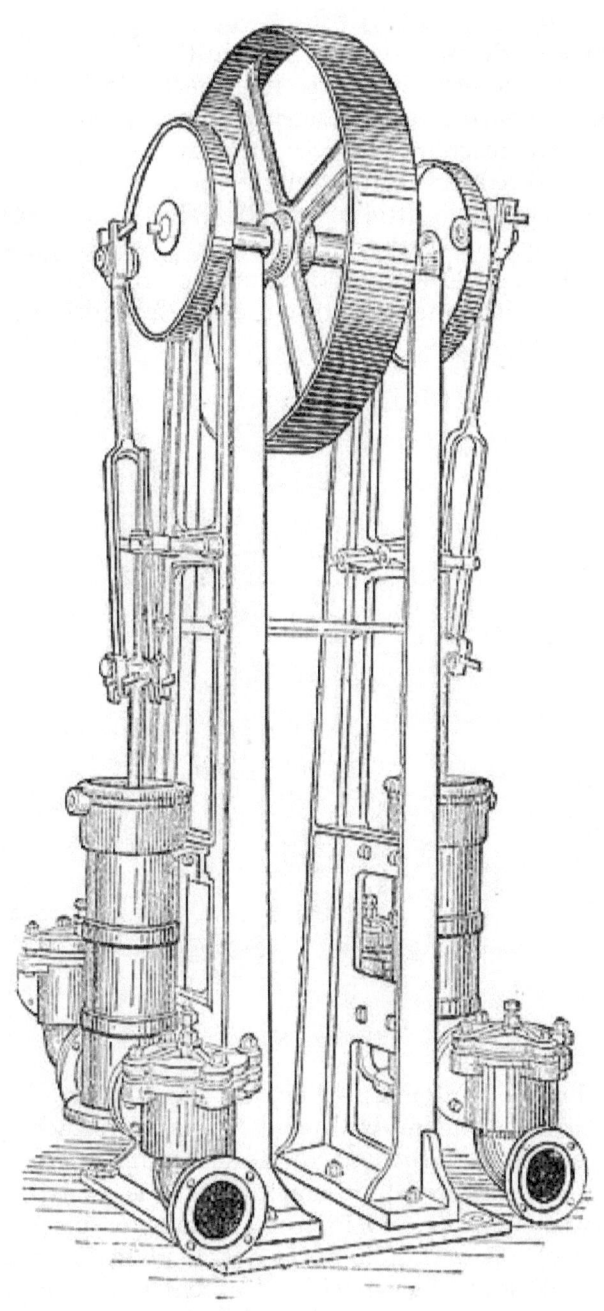

Fig. 57.

Back-water Pump.—The engraving (Fig. 57) shows a pair of back or size-water pumps, manufactured by Bertrams, Limited. The barrels are of cast-iron, lined with copper. The suction and discharge valves are each contained in a chamber with covers, so that every valve could be easily got at by simply releasing the cover. The valve-seats are of brass, with brass guards and rubber clacks. The plungers are of brass, with cup-leathers. All is fitted up on a cast-iron sole-plate, with tall standards, disc-cranks, and driving-pulley between frames.

Fig. 58.

Web-glazing Calender.—Fig. 58 represents Bertrams' web-glazing calender, with steam-engine attached. The illustration shows the machine in front elevation. The steam-engine is specially designed for this class of work, having two cylinders 10 inches in diameter by 16 inches stroke, fitted on a double-hooded sole-plate, with double-throw crank-shaft, fly-wheel, two eccentrics, wrought-iron piston-rods, connecting-rods and valve-rods, steam and exhaust branch pipes with one inlet valve, lubricators, and the cylinders cased with teak legging and brass hoops.

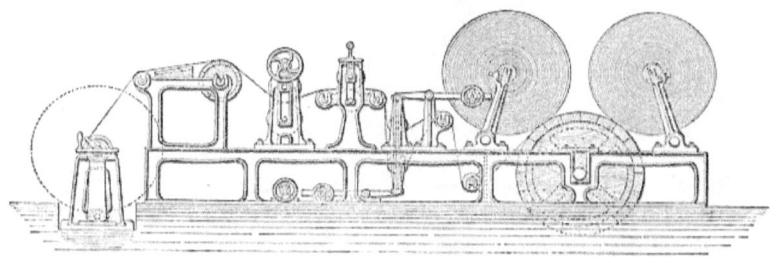

Fig. 59.

Reeling Machine.—One form of reeling machine manufactured by Bertrams, Limited, is shown in Fig. 59, and is used for slitting and re-reeling webs of paper, especially where large webs are requisite for web-calendering, web-printing, and suchlike. The reel of paper from the paper-machine is placed on a sliding-carriage arrangement, the brackets of which are planed and fitted to a planed sole, with wedge or dove-tail corners, and controlled by screws, hand-wheel, etc., so that the reel can quickly and easily be moved forward or backward to suit any unequal reeling that may have taken place on the paper or the machine. A hot cast-iron is provided for mending breaks in the web, and a measuring-roll and counter is also applied. The machine has an important application of drawing-in or regulating rolls of cast iron, with arrangement of expanding pulley for regulating the tension on the paper. Slitting-knives, regulating, dancing, or leading-rolls, of cast iron, etc., are applied for separating the edges and guiding the webs after they are slit. The reeling is performed by a 3-feet diameter drum, cross-shafts, and arms, to which regulating heads are fitted, so that several webs can be run up at one operation.

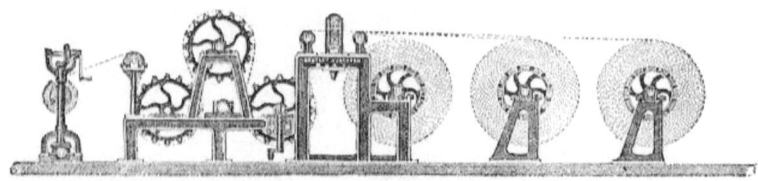

Fig. 60.

Web-Ripping Machine.—This machine, which is manufactured by Messrs. Bentley and Jackson, is shown in Fig. 60, and is constructed to divide webs of paper into two or more widths. It consists of two brass bearings on cast-iron standards, with screw adjustments, a break-pulley and friction-regulator, all mounted on cast-iron slides, movable transversely by means of a screw, geared-wheels, shaft and hand-wheel; a wood guide-roll, about 7 inches diameter, with wrought-iron centres, carried by brass bearings with screw adjustment; three skeleton drums, each 2 feet in diameter, on wrought-iron shafts, carried by brass bearings, and driven by spur-wheels and pinions; two wrought-iron leading-rolls, with brass bearings and cast-iron stands; a pair of strong wrought-iron ripper shafts with circular steel knives, bosses, springs, and collars; cast-iron stands and brass bearings, spur-wheels and driving-pulley; two (or more) changeable wood drums 1 foot 6 inches in diameter, each with wrought-iron shaft and catch-box, carried by brackets fitted with brass steps for easily changing, driven by wrought-iron shafts with pedestals and friction-pulleys, 2 feet in diameter, with regulating screws and lock-nuts, all carried by strong cast-iron framing and standards, and driven by a wrought-iron driving-shaft, with fast and loose driving-pulleys, strap-fork and levers for starting and stopping.

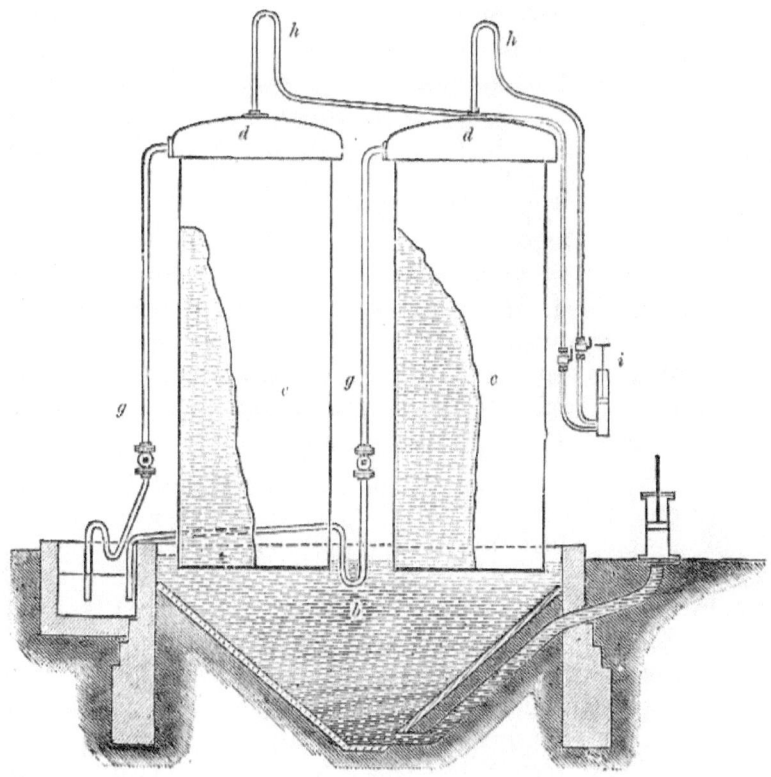

Fig. 61.

Roeckner's Clarifier.—In this apparatus, of which an illustration is given in Fig. 61, Mr. Roeckner has taken advantage of the fact that if a column of liquid is ascending very slowly and quietly within a vessel, it will not be able to carry up with it the solid particles which it contains, which will gradually fall back and sink to the bottom under the action of gravity, without ever reaching the top of the vessel, provided this be of sufficient height. The illustration shows the arrangement of the apparatus on a small scale; the liquor to be clarified is run into a well or reservoir b; into this dip a wrought-iron cylinder c, which is open at the lower end, but hermetically closed at the top by means of the casing d. From this casing air can be withdrawn through a pipe, h, by means of an air-pump i. As soon as this is done the liquid will begin to ascend the cylinder c, and if the height of this is below that to which the water will rise at the atmospheric pressure (say 25 feet), the liquid will ascend until it fills the cylinder and the casing. Into the pocket at the side of the casing there dips a pipe g, which passes out through the opposite side of the casing, descends below the level of the water in the tank, and ends in a discharge-cock. When this cock is opened, the cylinder c and the pipe g form between them a syphon, of which, however, the descending leg is of very small diameter compared with the ascending leg. In consequence, the liquid will rise in the cylinder c very slowly. The sediment it contains will sink back and collect in the bottom of the tank b, and clear water will flow out at the outlet. A sludge-cock at the bottom of the tank allows the solid matter to be drawn off at intervals and conveyed to any convenient place for drying, etc.[30] For drawing clear water from a river, the clarifier would simply be placed in the river, dipping 2 or 3 inches into it below the lowest water-level. The clear water will then be drawn through the clarifier, while the heavier matters will fall down and be carried away by the river current. It is stated that this has proved a great advantage to a paper-mill which used a river, and had, prior to its use, been much troubled through the dirt being pumped with the water. The clarifier to receive the waste from paper-machinery, or from washings in the engines, can be placed in any convenient corner, and by its action the water can be re-used, and the otherwise lost fibres collected, without its action ever being stopped.

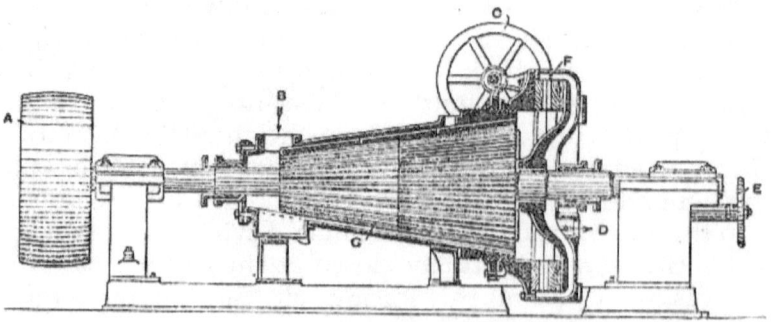

Fig. 62.

Marshall's Perfecting Engine.—This engine, a longitudinal section of which is shown in Fig. 62, has been introduced into this country by Messrs. Bentley and Jackson, and is described in *Industries*[31] as follows:—"The machine, which is the invention of Mr. F. Marshall, of Turner's Falls, Mass., U.S.A., is used in one of the processes of paper manufacture, and has for its purpose the more effectual drawing of the pulp fibre, the clearance of knots from the pulp previous to its delivery on to the paper-making machine, and the saving of time in the treatment of the material. As will be seen in the illustration (Fig. 62), the machine consists essentially of a cast-iron conical casing, bored, and fitted with about two hundred elbowed steel knives, G, placed in sections. At the large end of this conical casing is placed a movable disc, also fitted with about two hundred and ten steel knives, F, and capable of adjustment by means of a screw, worm, worm-wheel, and hand-wheel, E. The revolving cone and disc are of cast iron, fitted with straight steel knives firmly keyed upon a hammered iron shaft, and carefully balanced to prevent vibration. The knives of the revolving cone and disc are brought into contact with the stationary knives by means of the hand-wheel, E, and the disc-knives can be independently adjusted by means of the hand-wheel C, which actuates a screw on the conical casing by means of the worm and worm-wheel shown. The machine is driven by means of a pulley A, and the whole machine is mounted on a cast-iron base-plate. The pulp material enters the engine in the direction indicated by the arrow, B, at the small end of the cone, and is by the rotary and centrifugal action of the revolving cone, propelled to its large end, and during its passage is reduced to a fine pulp by the action of the knives. It then passes through the knives, F, of the stationary and rotating discs, by which the fibres are further crushed or split up, all knots or strings rubbed out, and the pulp effectually cleared previous to its exit through the passage D." We are informed that the machine is capable of treating from 900 lbs. to 1,200 lbs. of pulp per hour. The power required to drive it is estimated at from 40 i.h.p. to 50 i.h.p. when making 300 revolutions per minute. This, however, is dependent on the amount of friction caused between the surfaces of the fixed and revolving knives. The flow space occupied is 12ft. 6in. in length, and 4ft. in width. The perfecting machine, in its complete form, is shown in Fig. 63.

Fig. 63.

CHAPTER XVIII.

RECOVERY OF SODA FROM SPENT LIQUORS.

Recovery of Soda.—Evaporating Apparatus.—Roeckner's Evaporator.—Porion's Evaporator—American System of Soda Recovery.—Yaryan Evaporator.

Recovery of Soda.—Probably one of the most important improvements in modern paper-making, at least from an economical point of view, is the process of recovering one of the most costly, and at the same time most extensively used, materials employed in the manufacture—soda. While not a great many years since (and in some mills is still the case even now), it was customary to allow the spent soda liquors resulting from the boiling of various fibres to run into the nearest rivers, thus not only wasting a valuable product, but also polluting the streams into which they were allowed to flow, means are now adopted by which a considerable proportion of the soda is recovered and rendered available for further use. The means by which this is effected are various, but all have for their object the expulsion of the water and the destruction of the organic matters dissolved out of the fibrous substances in the process of boiling with caustic soda solutions. One of the main objects of the various methods of recovering the soda from spent liquors is to utilise, as far as practicable, all the heat that is generated from the fuel used, whereby the process of evaporation may be effected in the most economical way possible. The principle upon which the most successful methods are based is that the flame and heat pass over and under a series of evaporating pans, and through side flues, by which time the heat has become thoroughly utilised and exhausted. When all the water has been expelled, the resulting dry mass is ignited and allowed to burn out, when the black ash that remains, which is carbonate of soda, is afterwards dissolved out, and the alkaline liquor causticised with lime in the usual manner. According to Dunbar, 8 cwt. of recovered ash and 4½ cwt. of good lime will produce 900 gallons of caustic ley at 11° Tw. The liquor is then pumped into settling tanks, from which it is delivered to the boilers when required.

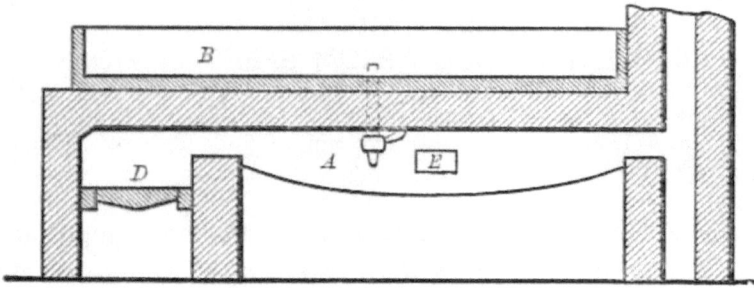

Fig. 64.

Evaporating Apparatus.—An ordinary form of evaporator for the recovery of the soda is shown in Fig. 64. It consists of a chamber A, of the nature of a reverberatory furnace, lined with fire-brick, the bottom of which is slightly hollowed. Above this is a tank B containing the liquor, which is run down into the chamber as required by means of a pipe C, provided with a tap. At one end of the chamber is a furnace D, the flame of which passes through the chamber and over the surface of the liquor lying upon the floor, heating the chamber, evaporating, and at last incinerating, its contents, and at the same time warming the liquor in the tank above, and evaporating some of its water. The products of the combustion in the furnace, and of evaporation, pass by the flue into a chimney, and escape thence into the air. There is a door E in the side of the furnace near the level of the floor of the chamber, and this is opened from time to time to enable the workmen to stir and move about the contents of the chamber, and finally, when the process is sufficiently advanced, to draw out the residue. The first effect produced is the reduction of the liquor to the consistence of tar. Later on, a white crust, which is the incinerated material, forms on the surface, and is drawn on one side by the workmen, so as to allow of fresh crust being formed. When all the charge has become solid it is drawn. The charge is usually withdrawn before the conversion into carbonate is completed; it is then raked out into barrows and placed in a heap, generally in a shed or chamber, open on one side, but sometimes in a closed brick-chamber or den, where the combustion continues for several weeks. The result is the fusion of the material into a grey rocky substance, which consists chiefly of carbonate and silicate of soda.

Various modifications of the esparto evaporator and calciner have, however, been introduced since the recovery of soda has become more general, and are in use at various works, all having for their main object the economising of fuel and the utilising of the waste heat of the fire, which in the old-fashioned calciner goes up the chimney and is lost. The leading principle, of all of them is to use the waste heat in concentrating the liquor preparatory to its being run into the part where the calcination is to be effected. This is done by so extending and widening out the flue as to cause the heated air and flame, after they have performed their function in the calcination, to pass over or under their layers of liquor, lying upon shelves or floors in such a way that the liquor shall become more and more concentrated as it approaches the calciner by successive steps or gradations. [32] —*Dr. Ballard.*

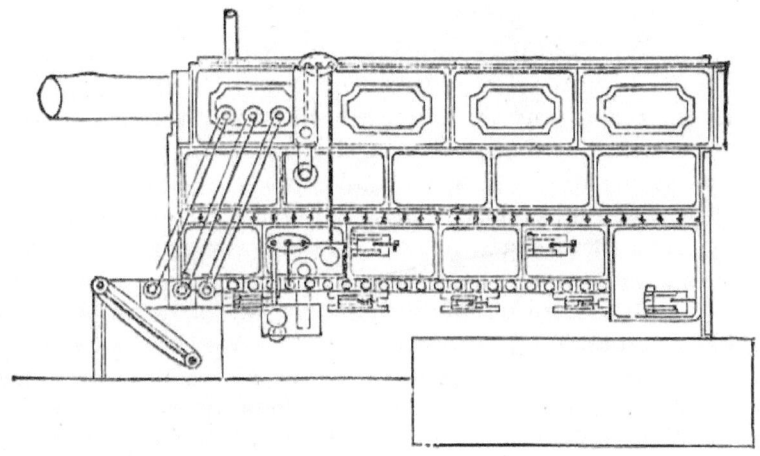

Fig. 65.

Roeckner's Evaporator.—This apparatus, an illustration of which is shown in Fig. 65, is thus described by Dr.

Ballard, medical officer of the Local Government Board, who was specially appointed by the board to investigate the effluvium nuisances which arise in connection with certain manufacturing industries. "In this apparatus there is above the calcining floor a series of shelves or shallow pans, alternating in such a manner that the liquor flowing from the tank above into the uppermost of them, flows, after a partial evaporation, over the edge of the shelf into the shelf or shallow pan next below, and in this way from shelf to shelf, still becoming more and more concentrated until it reaches the final floor, over which the flame from the actual fire plays, and where the first part of the calcination is effected. The heated air, in passing to the chimney, passes over each of these shelves in succession, heating them and concentrating the liquor upon them. There is between the lower shelves an arrangement for causing the liquor to pass from the upper to the lower by means of a pipe, instead of its running over the edge. At the top of all is a covered tank, where the temperature of the liquor is raised before it is run into the evaporator. In order to promote the heating of the liquor in this tank, the lower part of the tank is made to communicate by side pipes with tubes passing across the evaporator near the fire, as, for instance, at the bridge and at the further end of the calcining floor. In this way a circulation of liquor is set up which serves to heat the liquor in the tank more effectually. A pipe from the top of the tank leads to the chimney-shaft, conducting any vapours into it. As the incinerated crust forms it is raked on one side, and when sufficient of it has accumulated it is drawn to an opening (provided with a damper) at the side or end of the floor, and discharged down this opening into a brick chamber below, which is inclosed by iron doors, and from which a flue conducts the vapours that arise during the final fusion through the fire in such a way as to consume them." By recent improvements Mr. Roeckner has constructed an apparatus for condensing and rendering inoffensive the vapours eliminated from the liquor during its evaporation on the successive shelves of his evaporator.

Porion's Evaporator.—This evaporator and incinerating furnace much resembles in principle an ordinary reverberatory furnace, except that it is provided with paddle agitators, which project the liquid upwards, causing it to descend in a spray, thus increasing the surface of the liquid coming in contact with the hot air and current of smoke traversing the furnace. By this method the expense of fuel is greatly reduced. The residue is in a state of ignition when it is withdrawn from the furnace, and is piled in heaps so that it may burn slowly. When the combustion is complete, the resulting calcined mass is treated with water, and the carbonate of soda formed is afterwards causticised in the usual way. About two-thirds of the soda is thus recovered.

The Yaryan Evaporator.—Mr. Homer T. Yaryan, of Toledo, Ohio, U.S.A., has introduced some important improvements in evaporating apparatus, which have been fully recognised in America, and appear to have been attended with success. The principle involved is that of multiple effects, in which the evaporation takes place while the liquid is flowing through heated coils of pipe or conduits, and in which the vapour is separated from the liquid in a chamber, at the discharge end of the coils, and is conducted to the heating cylinder surrounding the evaporating coils of the next effect, from the first to the last effect. The objects of the invention are: (1) to provide extended vaporising coils or conduits and increased heating surface for each liquid feed supply in the heating cylinders, and provide improved means for feeding the liquid, whereby each set or coil of vaporising tubes will receive a positive and uniform supply of liquid without danger of the feed ducts being clogged by extraneous matter; (2) to positively control the amount of liquid fed by the pump to the evaporating coils, and make it more uniform than heretofore, regardless of the speed of the pump; (3) to provide improved separating chambers at the discharge ends of the vaporising coils so as to better free liquid and solid particles from the vapours; (4) to provide for the successful treatment of the most frothy liquids by causing the vapours carrying solid and liquid particles to pass through catch-all chambers, where they are arrested and precipitated and then returned to the evaporating coils; (5) to secure a more positive flow and circulation of liquid from the evaporating cylinder of one effect to another, under the influence of a better vacuum than heretofore in multiple-effect vacuum evaporating apparatus; (6) to provide for transferring a better concentrated liquid into the separating chamber containing cooler concentrated liquid in direct connection with the condenser and vacuum pump, so as to equalise the temperature of the two liquids, and then draw off both by one tail pump.

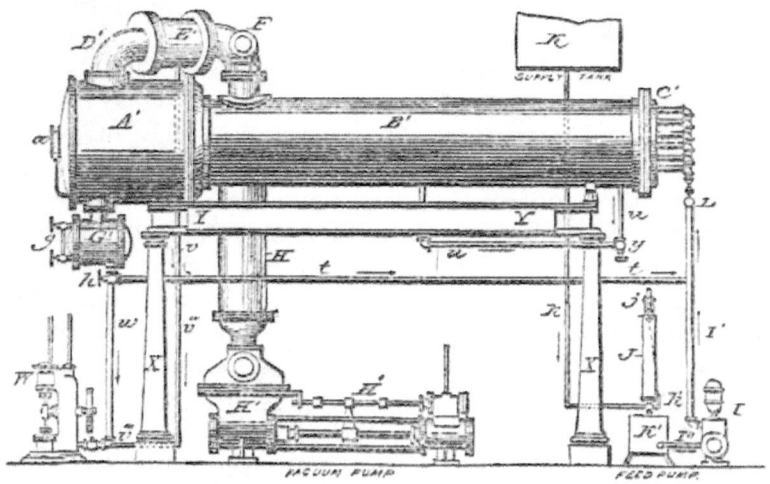

Fig. 66.

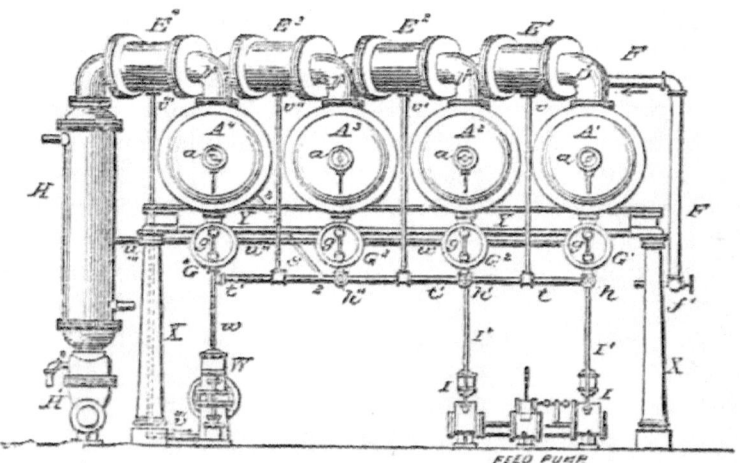

Fig. 67.

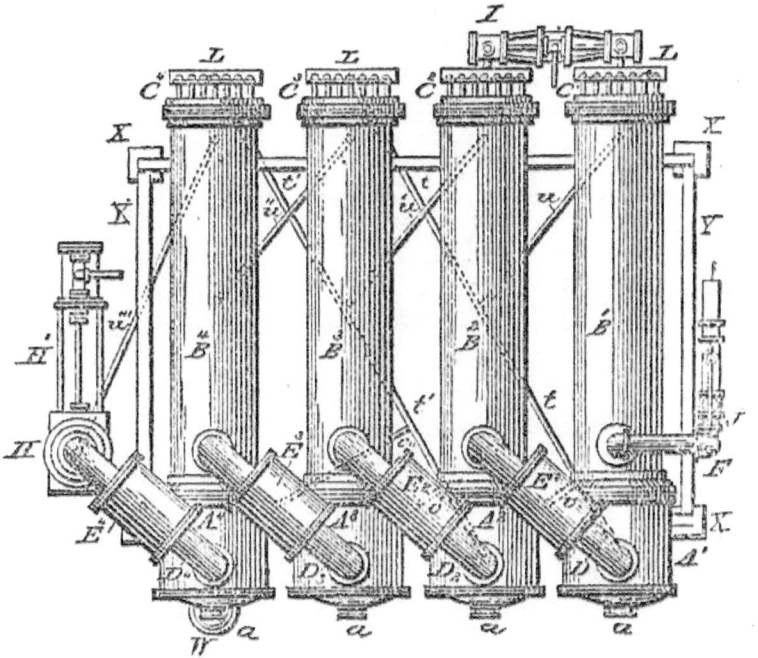

Fig. 68.

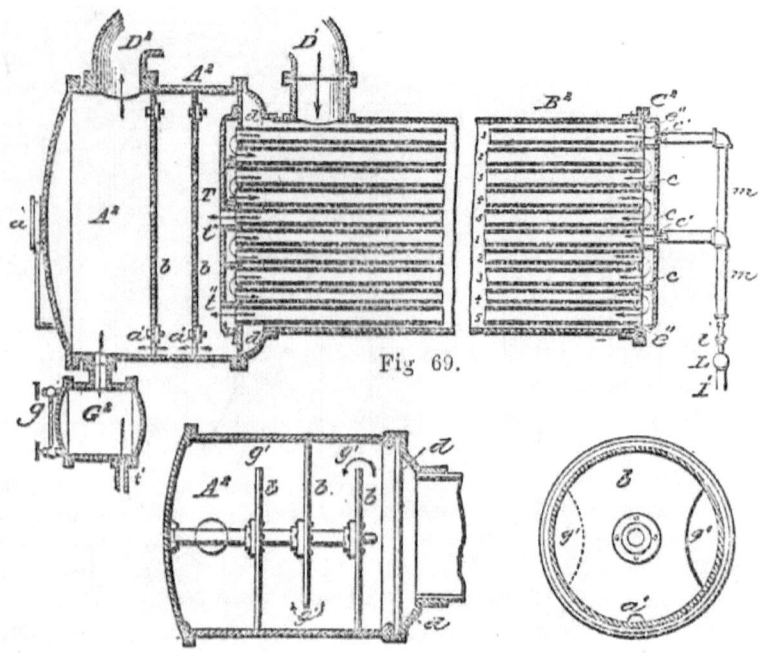

Fig. 69. Fig. 70. Fig. 71.

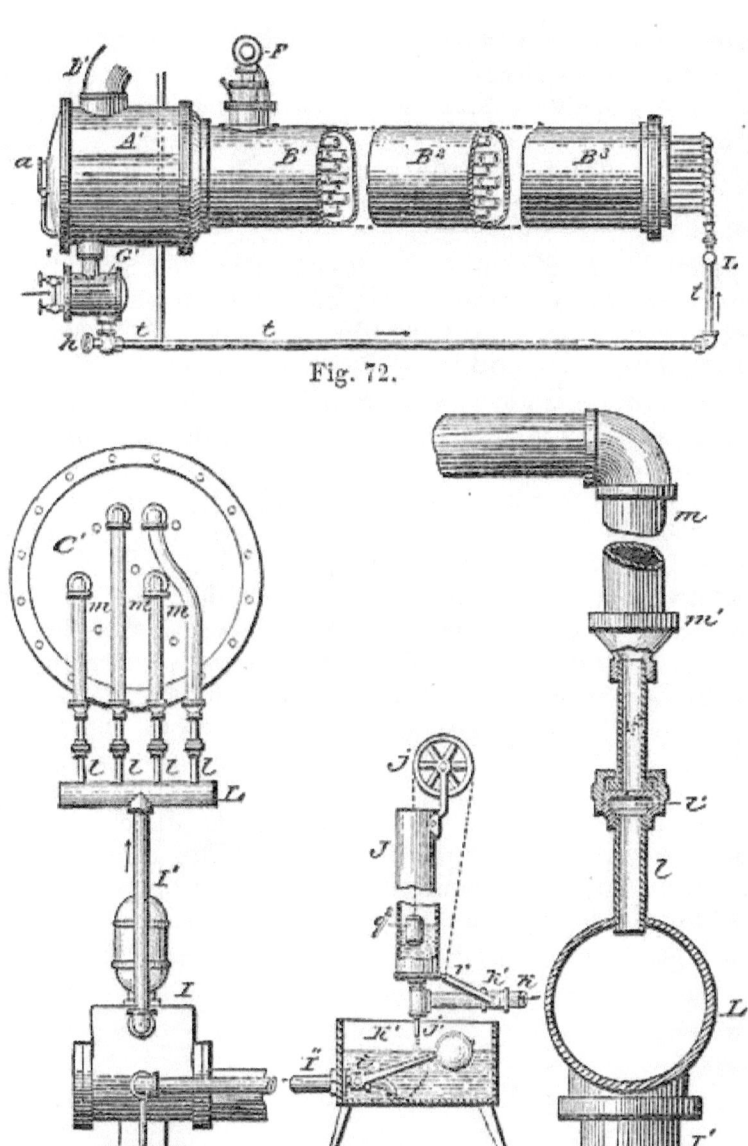

Fig. 72. Fig. 73. Fig. 74.

The present invention comprises a series of important improvements on an apparatus described by Mr. Yaryan in a former English patent, No. 14,162 (1886), and covers a number of important modifications in construction, whereby improved results are secured. It is only necessary, therefore, to give the details of the new patent, No. 213 (1888), since it embodies the latest improvements which practical working of the apparatus has suggested. In reference to the accompanying illustrations the following details are given: Fig. 66 represents a side elevation of the apparatus; Fig. 67, the front elevation; Fig. 68, a top plan view; Fig. 69, a vertical section of a cylinder showing the evaporating coils and separating chamber; Fig. 70 is a horizontal section; and Fig. 71, a vertical section of the separating chamber shown in Fig. 69, both on reduced scale; Fig. 72 is a broken section of the cylinders for showing the connections of the liquid pipe from the first to the third effect evaporator; Fig. 73 is a rear end view of a cylinder with manifold, the feed pump and a sectional view of the feed box and supply devices; Fig. 74 represents a sectional view, on enlarged scale, of the manifold and a feed duct; Fig. 75 is an inside view of a return bend-head; Fig. 76 an inside view of a section of the head; Fig. 77, a vertical cross section thereof on enlarged scale, and showing the partitions forming cells for connecting the ends of the evaporating tubes; Fig. 78 is a vertical longitudinal section of a catch-all chamber; Fig. 79, a cross section thereof; Fig. 80 is a vertical longitudinal section of new form of separating chamber; and Fig. 81 represents a side view and Fig. 82 an end view of the cylinders for showing the pipe connection between the separating chambers of the third and fourth effect evaporators.

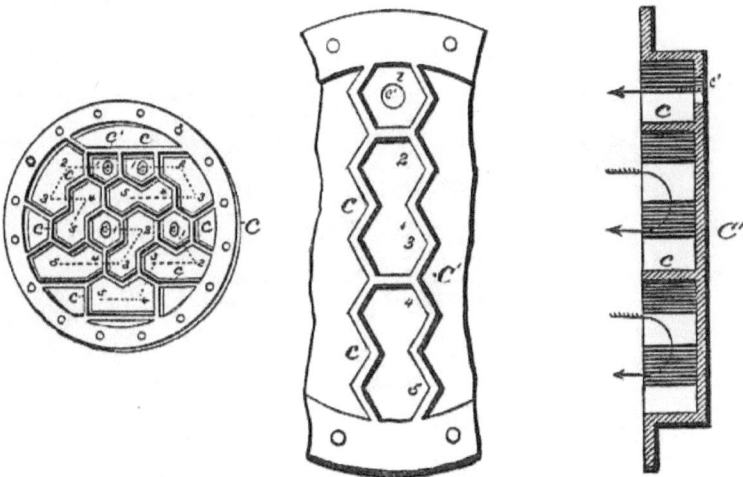

Fig. 75. Fig. 76. Fig. 77.

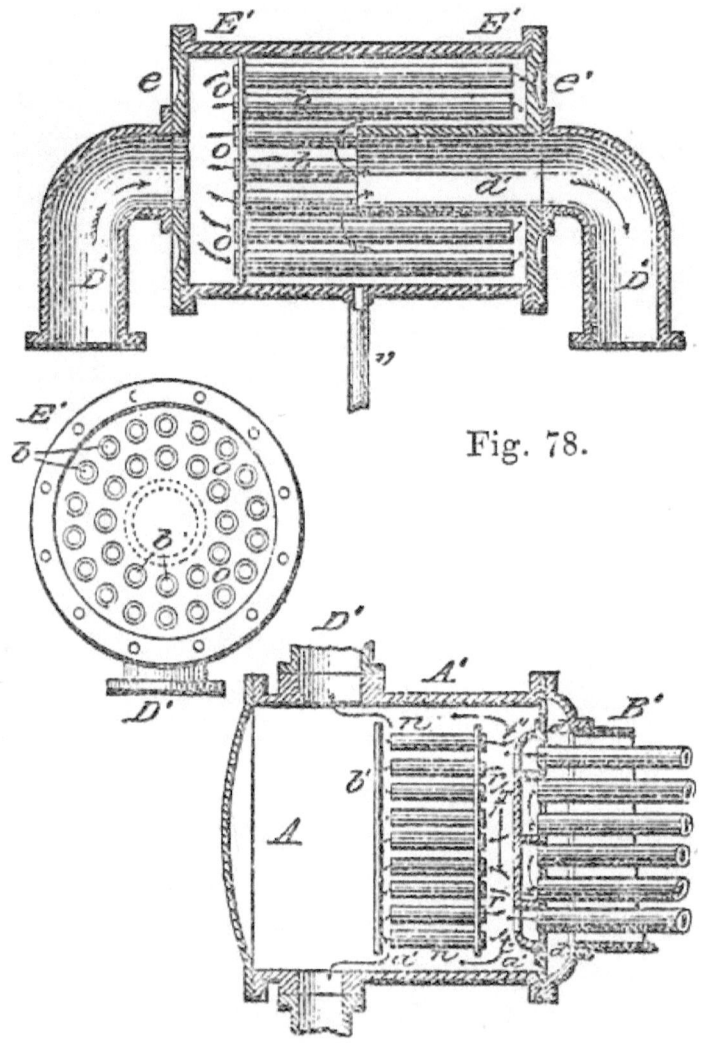

Fig. 78. Fig. 79. Fig. 80.

The evaporating cylinders are mounted upon a framework Y, supported upon columns X X, or other suitable supports. The apparatus is shown arranged as quadruple effect, with four connected cylinders, but multiple effect apparatus may be constructed with an increased number of cylinders up to ten or twelve. The heating cylinders B^1 B^2 B^3 B^4, containing the evaporating tubes or coils, are preferably arranged in the same horizontal plane, and are provided at the discharge ends of the evaporating coils with separating chambers, A^1 A^2 A^3 A^4, of enlarged diameter, and at the supply ends of the coils with the coils with return bend ends, C^1 C^2 C^3 C^4. From each separating chamber, A^1, A^2, valve pipe D^1 D^2 D^3 leads into the shell of the next heating cylinder, as B^2, B^3, B^4, and vapour pipe D^4 leads from the last separator A^4 to the condenser H, and the vacuum pump H^1. A cylindrical catch-all chamber E^1, E^2, E^3, E^4, is connected in each vapour pipe between each separator and each successive heating cylinder, as shown in Figs. 66, 67, and 68, and in detail in Fig. 75. Gauge glass and liquid receiving chambers, G^1, G^2, G^3, G^4, connect with the bottom of each separating chamber for receiving the liquid as it is separated from the vapour, and a gauge glass g is applied to each of such chambers. Liquid discharge and transfer pipes t, t^1, having valves h, h^1, as best shown in Figs. 66, 68, and 72, lead respectively from chambers G^1, G^2, of the first and second effect to the manifold feed pipes leading into the cylinders B^3, B^4, of the third and fourth effect for the purpose hereafter described. The main steam supply pipe F, having a safety valve f and stop valve f^1, Figs. 66, 67, and 68, connects with the heating cylinder B^1 of the first effect. The evaporating tubes 1, 2, 3, 4, 5, are expanded or otherwise secured in the tube sheets d and e'' at opposite ends of the cylinders, and are properly connected at the ends in sets of five to form coils. The outer rear return-bend head C^1 C^2, etc., are provided on their insides with numerous short intersecting partition plates c, forming single and double cells, properly arranged for connecting the evaporating tubes in sets of five, as shown in Figs. 75, 76, 77.

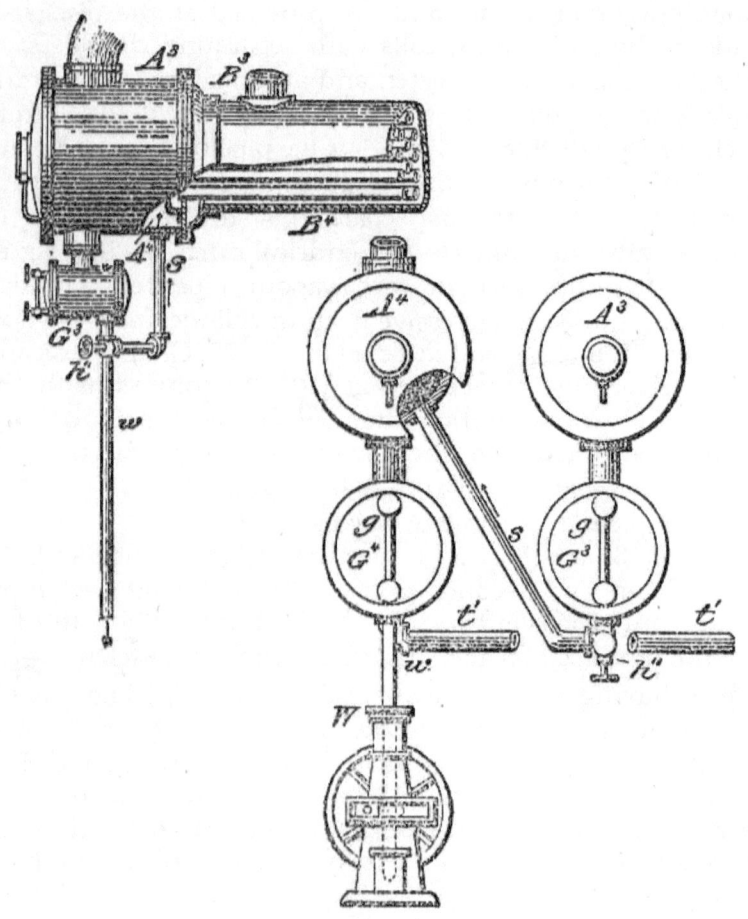

Fig. 81. Fig. 82.

The heads are pierced with holes c' for connecting the liquid supply pipes M of the manifolds L. The inner return-bend head T in the separating chambers are formed like heads $C^1 C^2$, etc., with intersecting partition plates x, and are provided with discharge openings t'' for every fifth tube, as shown in Fig. 69. Tube sheet d is made of considerably larger diameter than cylinders $B^1 B^2$, etc., and acts as a vibrating diaphragm, to accommodate the expansion and contraction of the tubes. The separating chambers may be constructed with dash plates $b\ b$, two or more in number, having openings $g'\ g'$ alternately upon opposite sides for the passage of vapour, and opening a' at the bottom for the passage of liquid, as shown in Fig. 80. Here a tube sheet z is provided near the openings of the evaporating tubes, and in such sheet are set numerous small horizontal tubes n, which discharge against a vertical arresting plate b' set near their open ends. Water and solid matter are impelled against the plate and thereby arrested and caused to flow down to the bottom of the chamber. The liquid feed apparatus consists of a supply tank K, stand-pipe J, feed box K^1, double pump I, manifold L, and connecting pipes and valves. The liquid to be evaporated flows from tank K, through pipe k, to stand-pipe J and box K^1, the flow being constant and uniform, and of the desired quantity, by means of a valve k' having a lever handle r' which is connected by a cord or chain passing over a pulley j with float q in stand-pipe J. The valve opening in pipe k being properly adjusted by means of the float, etc., the liquid is admitted to the stand-pipe J while the column of liquid is automatically maintained at any desired height and pressure regardless of the quantity in the supply tank, by means of the float q, which, as it rises, tends to close valve k', and as it falls, to open the valve. From the bottom of the stand-pipe J, nozzle j' discharges a constant and uniform stream of liquid into feed box K^1. The suction pipe I" of pump I extends into box K^1, where it terminates in a turned-down nozzle provided with valve i having a lever handle and float z. As a given amount of liquid is constantly running into the box, should the pump run too fast the float lowers, partially closing the valve and lessening the amount of liquid drawn at each stroke of the pump, and preventing air from being drawn in, since the end of the suction pipe is always sealed by the liquid. The liquid is forced by pump I into the manifolds L, from which it flows through the contracted ducts l into the enlarged feed pipes m, as shown in Figs. 73 and 74. Ducts l are of about one-half inch diameter, and the upper and lower sections thereof are connected by a union coupling, one portion of which l' has a reducer with opening one-quarter inch diameter, more or less, according to the amount of liquid it is desired to feed.

The catch-all chambers E^1 E^2, etc., Figs. 66, 78, and 79, are provided each at its inlet end *e*, with tube sheet *o* extending across its diameter a short distance in front of the opening of vapour pipe D^1, and in such sheet are fixed numerous longitudinal tubes *p* extending to near the opposite head *e'*, so that vapours carrying watery or solid particles are impelled against the head and arrested. Liquid and solid matter, arrested in the catch-all chambers, flow through pipes *v v' v"* down into the fluid transfer pipe *t t'* (Figs. 67, 68, and 72), and thence into the evaporating coils and through pipe *v'''* directly to the tail pump W, Fig. 67. By use of the catch-all chambers the most frothy liquids can he readily and economically managed. A liquid transfer pipe *s*, having a valve *h"*, leads directly from receiving chamber G^3 of the third effect to the separating chamber A^4 of the fourth effect, the latent heat being carried off in the vapours drawn by the vacuum pump H^1 into the chamber H, and the finished liquid of both effects is drawn off through pipe *w* by one and the same tail pipe pump W. The water of condensation accumulating in the heating cylinders B^1 B^2, etc., is transferred from one to the other through connecting pipes *u u' u"* having valves *y*, shown in Figs. 66, 67, and 68; and finally from cylinder B^4 through pipe *u'''* directly into condenser H. The specification of the patent, which those interested will do well to consult, next describes the operation of the apparatus.

American System of Soda Recovery.—Mr. Congdon gives an exhaustive description[33] of the method of recovering soda in the United States, from whose interesting paper we extract the following:—The spent liquors are delivered to the Yaryan evaporator from the pans at a density of 6° to 7° B. at 130° F. Here they are concentrated to 34° to 42° at 140° F. At this density they are fed into furnaces of a reverberatory type, where they are burnt to a cherry-red heat; and the ash then raked out. This ash, which averages 50 per cent. of soda, is weighed in iron barrows on suitable scales, and wheeled into the leaching-room for lixiviation. The system of leaching, as it is termed in the States, is conducted as follows:—Iron tanks are used, with suitable piping, that allows pumping from one tank to another, and also to pump from any one of them up to the causticising tanks in the alkali-room. There is also a water-line by which water may be pumped into any of the tanks, and there is a spout used in washing away the black ash sludge. The leaching-tanks have false bottoms of 2in. by 2in. stuff, placed crosswise, over which is a layer of gravel, on which lies a layer of straw, by which the liquor is filtered. The gravel is removed every few days, and the straw with every charge. When one of the tanks is filled with black ash, it is "wet down" with the stored liquor (the strongest of the stored weak liquors), and also with the strongest weak liquors from the tanks, and with weak liquors obtained from these tanks by pumping water upon them and keeping them full. This is all pumped up to the causticising-tank until the strength is reduced to 2° or 1½° B. The remaining liquor is then drained into a tank known as the "clear-liquor" tank, owing to there being no black ash in it. The liquor from the next weakest pan is then pumped upon the pan containing the black ash, and the next weakest liquor pumped upon this. The weaker pans are then in succession pumped upon the stronger, and the water pumped upon these, and thus a very perfect washing is obtained. The sludge left behind is nothing but charcoal, with a slight trace of carbonate of soda. Mr. Congdon illustrates the above system thus. The tanks stand as follows:—

No. 1. Clear liquor, 1° to 2° B. (strongest).

No. 2. Black ash sludge (weaker than No. 3).

No. 3. Black ash, after sending up to causticising-tank (strongest sludge).

No. 4. Fresh black ash.

No. 5. Weaker than No. 2 (sludge only).

No. 6. Weaker than No. 5 (sludge and weakest liquor).

The method of procedure is as follows:—

Liquor from No. 3 drained into No. 1 (now full).

No. 6 pumped on to No. 2 (No. 6 sludge thrown away).

Liquor from No. 2 drained upon No. 3.

Water put on No. 5.

No. 5 pumped upon No. 2 (No. 5 sludge thrown away).

The black ash is treated thus:—

No. 4, full of black ash, is wet down with Nos. 1, 2, and 3, and pumped up to the causticising-tank.

Water is pumped out to Nos. 2 and 3, and then drained upon No. 4, the liquor still being pumped up from No. 4 while the water is being pumped upon Nos. 2 and 3, which are kept full. This is continued until the liquor tests only 2° to 1° B.

No. 4 is now drained upon No. 1.

No. 3 pumped upon No. 4, and this drained into No. 1 (now full).

No. 3 pumped upon No. 5.

Water pumped upon No. 2 (No. 2 the next to be thrown away).

No. 5 is by this time full of fresh black ash, and the same process is carried out with No. 4.

CHAPTER XIX.

DETERMINING THE REAL VALUE OR PERCENTAGE OF COMMERCIAL SODAS, CHLORIDE OF LIME, ETC.

Examination of Commercial Sodas.—Mohr's Alkalimeter.—Preparation of the Test Acid.—Sampling Alkalies.—The Assay.—Estimation of Chlorine in Bleaching Powder.—Fresenius' Method.—Gay-Lussac's Method.—The Test Liquor.—Testing the Sample.—Estimation of Alumina in Alum Cake, etc.

In a manufacture such as paper-making, which involves the consumption of enormous quantities of materials of variable quality, as soda ash, caustic soda, and bleaching powder, for example, it will be readily seen that some means should be at the command of the consumer who does not avail himself of the services of a practical chemist at his works, by which he can ascertain the *actual* value of the various substances he uses. An art which, up to a certain point in its progress, is mainly a chemical operation, it would undoubtedly be more safely and economically conducted when supervised by persons well acquainted with chemical principles and reactions, and less dependent upon individual judgment, than is, perhaps, too frequently the case. Under such supervision more perfect uniformity of results—a consideration of the greatest importance in a manufacture of this kind—would be ensured.

Fig. 83. Fig. 84. Fig. 85.

Examination of Commercial Sodas.—The methods of determining the percentage of real alkali in the commercial products which have received the name of *Alkalimetry* are fortunately of a simple character, and such as a person of ordinary intelligence and skill can readily manipulate and render thoroughly reliable by exerting the necessary care. He must, however, be provided with a few indispensable appliances, which will be described, and with these he should make several trials upon various samples until he finds that his results are uniform and his manipulation easy and reliable. He will require a chemical balance,[34] capable of weighing to the tenth of a grain; a few glass "beakers" (Fig. 83) of various sizes, capable of holding from four to eight or ten ounces of fluid; several glass stirrers; a bottle of litmus solution, made by dissolving litmus in hot water; books of litmus and turmeric papers; and several glass flasks (Fig. 84) of various sizes, capable of holding from four to eight ounces. Besides these accessories, certain measuring instruments, termed *alkalimeters* or *burettes*, are employed, of which either of the two following may be employed. These instruments are of glass, and hold up to 0 or zero exactly 1,000 grains. The scale is graduated in a hundred divisions, which are again subdivided into tenths. Bink's burette is shown in Fig. 85, and Mohr's burette in Fig. 86. The latter, being provided with a stand, enables the operator to add the test liquor—with, which the burette is charged—drop by drop, when the alkaline solution to be tested is near the point of saturation, without engaging the hands.

Mohr's Alkalimeter.—This useful instrument (Fig. 86) and the method of using it is thus described by Mohr:—"I have succeeded in substituting for expensive glass stop-cocks an arrangement which may be constructed by any person with ease, which remains absolutely air and water-tight for an indefinite period, which may be opened and regulated at will by the pressure of the fingers, and which costs almost nothing. It consists of a small piece of vulcanized indiarubber tube, which is closed by a clamp of brass wire (Fig. 87). The ends of this clamp, which I call a pressure-cock, are bent laterally at right angles in opposite directions and furnished with knobs, so that when both ends are pressed the clamp is opened, and a single drop or a continuous current of liquid may be allowed to escape at pleasure. The measuring-tube is a straight glass cylinder, as uniform as possible, graduated to 0·2 or 0·1 cubic centimètres, and somewhat contracted at its lower end, so as to fit into the indiarubber tube. A small piece of glass tube

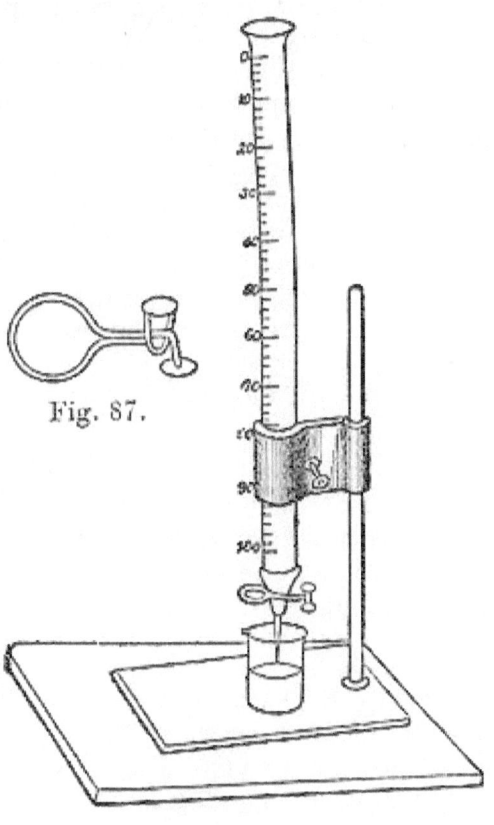

Fig. 86. Fig. 87.

inserted below the pressure-cock forms the spout. The pressure-cock has the advantage of not leaking, for it closes itself when the pressure of the fingers is removed. The measure, furnished with the pressure-cock, is fastened upon an appropriate stand, which can be placed at any required height. When used, it is filled above the zero point with test liquor, the cock opened for an instant, so as to let the air escape from the spout, and the level of the solution is then adjusted. This is done by bringing the eye level with the zero point, and applying a gentle pressure to the cock until the liquid has sunk so low that the inferior curve of the liquid touches the graduation like the circle of a tangent; the cock is then closed, and at the same moment the liquid remains at zero, and continues to do so for weeks if evaporation is prevented. The test-measure being normally filled, the experiment may be commenced; this is done sitting, while the filling of the measure is done standing.

"The weighed sample of alkali is first placed in a beaker-glass, and the test-liquor is allowed to flow into it by gently pressing the cock. Both hands are set at liberty, for when the pressure-cock is released it closes of itself. The volumetric[35] operation may be interrupted at pleasure, in order to heat the liquid, shake it, or do whatever else may be required. The quantity of liquid used may be read off at any moment, and in repeating an experiment, the limit of the quantity used before may be approached so near that the further addition of liquid may be made drop by drop." The test-acid to be used *volumetrically*—that is, with the alkalimeter, has a specific gravity of 1·032 at 60° F., and 1,000 grains by measure contain exactly 40 grains of real or anhydrous (that is, without water) sulphuric acid.

The chemical principles involved in the process of alkali-testing may be thus briefly stated:—According to the laws of chemical combination defined by the atomic theory of Dalton, all substances combine in *definite* proportions or "equivalents"; thus, 1 part by weight of *hydrogen* combines with 8 parts by weight of *oxygen* to form water. The equivalent number of hydrogen, therefore, is 1, and of oxygen 8, and that of water 9. Again, 3 equivalents of oxygen combine with 1 equivalent of sulphur (16) to form sulphuric acid; thus, sulphur 16, oxygen 24, equals anhydrous sulphuric acid 40; therefore 40 is the *equivalent* or combining number of this acid, and it cannot be made to unite with alkalies or other bases in any other proportion. For example, 40 *grains* by weight of *pure* sulphuric acid will neutralise exactly 53 grains of *dried carbonate of soda*, 31 grains of *pure anhydrous soda*, or 40 grains of *hydrate of soda* (caustic soda). This being so, it is only necessary to have exactly 40 grains of *real* sulphuric acid in 1,000 grains of water to form a *test-acid*, which, when employed to neutralise an alkaline solution, will show, by the proportion of dilute acid used to saturate the alkali, the absolute percentage present in the sample.

Preparation of the Test-Acid or Standard Solution.—As there is some trouble involved in the preparation of the test-liquor, it is advisable to prepare a sufficient quantity at a time to last for many operations. It may be readily made by mixing 1 part of concentrated sulphuric acid with 11 or 12 parts of *distilled water*, the mixture being made in what is termed a "Winchester" bottle, which holds rather more than half a gallon, and is provided with a glass stopper. The acid solution must be *adjusted* or brought to the proper strength after it has cooled down to 60° F.; and it should be *faintly tinged* with litmus, which will give it a pinkish hue. The acid, to be of the proper strength, should *exactly* neutralise 53 grains of pure carbonate of soda, previously calcined at a red heat, or 31 grains of pure anhydrous soda. To prepare the anhydrous carbonate of soda, a few crystals of carbonate of soda are placed in a Berlin porcelain crucible, and this must be heated over a spirit-lamp or Bunsen burner. When all the water of crystallisation has become expelled, the calcination is continued until the mass is at a bright red heat, when the vessel may be allowed to cool. 53 grains of the calcined carbonate are now to be carefully weighed, and next dissolved in a glass beaker, in about 2 ounces of distilled water. The alkalimeter is now to be charged with the test-acid to the level of zero, and (if Mohr's burette be used) the beaker containing the alkaline solution is to be placed upon the stand immediately beneath the exit-tube. Now press the knobs of the pressure-cock, and allow a portion of the liquor to flow into the beaker. When the effervescence which immediately sets up subsides, make further additions of the test-liquor from time to time, until the effervescence becomes sluggish, at which period the acid must be added with greater caution. When the solution approaches saturation it acquires a purplish tint (due to the litmus with which the acid is tinged), which it retains until the point of saturation is reached, when it suddenly changes to a pink colour. After each addition of the acid the solution should be stirred with a thin and clean glass rod; and before the final change from purple to pink, the end of the glass rod should be applied to a strip of blue litmus paper, when, if the moistened spot touched assumes a red colour, the saturation is complete; if, on the contrary, the paper is unchanged, or has a violet or reddish hue, add the test-liquor, one or two drops at a time, with continued stirring, until a drop of the solution applied with a glass rod reddens litmus paper, when the saturation is finished. If any test-liquor remain in the burette, this indicates that there is excess of acid in the test-liquor; consequently more distilled water must be added to the bulk, the burette emptied and refilled with the reduced liquor, and another 53 grains of anhydrous carbonate of soda treated as before, until 1,000 grains of the acid liquor *exactly* neutralise the solution. Should the whole contents of the burette in the first trial be used before saturation is complete, a little more sulphuric acid must be put into the Winchester or test-acid bottle, and a 53-grain solution of carbonate of soda treated as before. A very little practice will enable the operator to adjust his test-liquor with perfect accuracy; and, to prevent mistakes, the bottle should be labelled "Test-acid," and always be kept closed by its stopper.

Sampling Alkalies.—Soda-ash of commerce is usually packed in wooden casks, and in order to obtain a fair average sample from a large number of these casks, which may represent one consignment, it is important to take small samples, as near the centre of each cask as possible, from as many of the casks as time will permit. Each sample, as drawn from the cask, should be at once placed in a rather wide-mouthed bottle furnished with a well-fitting cork. Each sample should be numbered and marked with the brand which distinguishes the cask from which it was taken. The duty of sampling should be placed in the hands of a person of known integrity and intelligence.

When about to test a sample of soda-ash, the contents of the bottle should first be emptied upon a sheet of dry paper, and the larger lumps then crushed to reduce the whole to a coarse powder, and this must be done as quickly as possible to prevent absorption of moisture from the atmosphere. 100 grains of the alkali must now be accurately weighed and put into a glass flask (Fig. 84), and the remainder of the alkali returned to the bottle and the vessel securely corked. About half an ounce of distilled water is then to be put into the flask and gentle heat applied, with an occasional shaking, until the alkali is all dissolved. The flask is then to be set aside for a few minutes, until any insoluble matter present has subsided, when the clear liquor is to be carefully poured into a beaker glass; the sediment must be washed several times with small quantities of distilled water, and the washings added to the solution in the beaker. This washing is of great importance and must be performed several times, or until the last washing liquor produces no effect upon yellow turmeric paper, which even slight traces of alkali will turn a brown colour. So long as this brown tint is given to the turmeric paper the presence of alkali is assured, and the washing must be continued. It is important, after each washing, to pour off the last drop of the liquor above the sediment, by which the operation is more effectual, and is effected with less water than when this precaution is not observed. In order to ensure perfect accuracy in the result, every particle of the washings must be added to the contents of the beaker-glass in which the assay is to be made.

The Assay.—The alkalimeter is first to be filled with the test-acid exactly to the line 0 or zero of the scale as described, and the beaker containing the solution to be tested then placed immediately beneath the dropping tube of the instrument; a thin glass rod should be placed in the beaker as a stirrer. The acid liquor is then allowed to flow gradually into the alkaline solution (which should be repeatedly stirred with the glass rod), by pressing the knobs of the pressure-cock, until the solution assumes a purple tint, which it will retain until the exact point of saturation has been arrived at, when, as before stated, it will suddenly change to a pink colour. Before the latter stage is reached the beaker should be placed over a spirit lamp or Bunsen burner, and the liquid heated to expel the carbonic acid which is evolved, and partly absorbed by the solution during the process of saturation. When the neutralisation is complete, the alkalimeter is allowed to repose for a few moments, so that the acid liquor may drain from the interior of the glass tube into the bulk of the fluid, and the quantity of test-acid used is then determined by reading off the number of divisions of the alkalimeter that have been exhausted, every one of which represents 1/100th part, or 1 per cent. of *alkali*, whenever the *equivalent weight* is taken for assay. Every 1/10th part of an alkalimeter division represents 1/10th of 1 per cent., and the result is thus obtained without the necessity of any calculation. The following table shows the *equivalent* or combining proportions of soda with 40 grains of real (that is, anhydrous) sulphuric acid:—

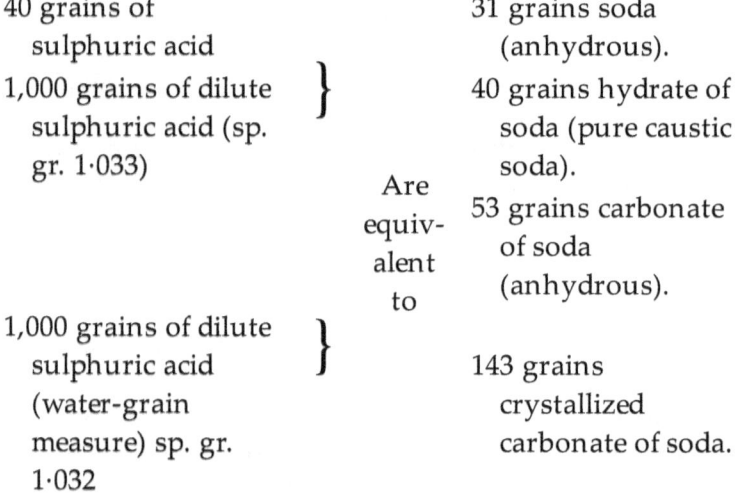

40 grains of sulphuric acid		31 grains soda (anhydrous).
1,000 grains of dilute sulphuric acid (sp. gr. 1·033)	Are equivalent to	40 grains hydrate of soda (pure caustic soda).
		53 grains carbonate of soda (anhydrous).
1,000 grains of dilute sulphuric acid (water-grain measure) sp. gr. 1·032		143 grains crystallized carbonate of soda.

Mr. Arnot recommends the following method for alkali testing: "The sample, which should be a fair average of the drum or cask from which it is drawn, should, in the case of caustic soda, be quickly crushed into small fragments, and returned to the stoppered bottle in which it was collected for testing. It need not be finely ground, and, indeed, should not be, as it very readily attracts moisture from the air. The contents of the drum are usually pretty uniform, and the crushing recommended will give the operator a sample quite fit to work upon. Samples of soda-ash and soda crystals will, of course, be fairly representative of the casks from which they are drawn. One hundred grains of the prepared sample must be weighed out upon a watch-glass or slip of glazed paper, and transferred to a porcelain basin, with at least half a pint of boiling water. The watch-glass is preferable for caustic soda, and the weighing in the case of that agent must be done expeditiously. While the sample is dissolving the burette will be charged with the standard acid. To the soda solution a few drops of solution of litmus, sufficient to colour it distinctly, will be added. The acid will then be run into the blue soda liquor; at first, within reasonable limits, this may be done rapidly, but towards the close of the operation the acid must be added cautiously, and the solution kept well stirred. In the case of caustic, when the blue has distinctly changed to red, the operation may be considered completed, and the measures may be read off the burette; and this is, without calculation, the result required. When the soda in the sample is a carbonate, the blue colour of the litmus will be changed to pink before all the soda is neutralised, owing to a portion of the liberated carbonic acid remaining in the solution; this must be eliminated by placing the basin over a Bunsen burner and boiling the solution. The blue colour will thus be restored, and more acid must be added, repeating the boiling from time to time, until the red colour becomes permanent. It is sometimes necessary to filter the soda solution before testing; this applies specially to recovered soda, and, although in a less degree, to soda-ash." When the soda solution is filtered, it will be necessary to thoroughly wash out the liquor absorbed by the filtering paper, the washings being added to the bulk of the liquor as before. The best plan is to allow the soda solution to stand for some time until all the sediment has deposited, and then to pour off as much of the liquor as possible, and then to wash the sediment into a very small filter, in which it will receive further washing, until no trace of alkali can be detected in the last wash water.

Estimation of Chlorine in Bleaching Powder.—It is desirable that the manager or foreman of a paper-mill should have at his command some ready means by which he may test the percentage of chlorine in samples of bleaching powder, or chloride of lime, delivered at the mill, not alone to enable him to determine the proportions to be used in making up his bleaching liquors, but also to ensure his employers against possible loss in case of inferior qualities being delivered at the mill. Bleaching powders being purchased according to percentage, it is absolutely necessary that the purchaser should have this determined to his own satisfaction before either using or paying for the material. Good chloride of lime should contain 35 per cent. of available chlorine, but the powder should not be accepted which contains less than 32 per cent. There are several methods of estimating the percentage of chlorine in bleaching powder, which is composed of hypochlorite of lime, chloride of calcium, and hydrate of lime, the latter substances being of no service in the bleaching process.

According to Fresenius, in freshly prepared and perfectly normal chloride of lime, the quantities of hypochlorite of lime and chloride of calcium present stand to each other in the proportion of their equivalents. When such chloride of lime is brought into contact with dilute sulphuric acid, the whole of the chlorine it contains is liberated in the elementary form. On keeping chloride of lime, however, the proportion between hypochlorite of lime and chloride of calcium gradually changes: the former decreases, the latter increases. Hence from this cause alone, to say nothing of original difference, the commercial article is not of uniform quality, and on treatment with acid gives sometimes more, and sometimes less, chlorine. As the value of bleaching powder depends entirely upon the amount of chlorine set free on treatment with acids, chemists have devised very simple methods of determining the available amount of chlorine in any given sample, these methods having received the name of *chlorimetry*. The method of Fresenius is generally considered both practicable and reliable.

Fresenius' Method of preparing the solution of bleaching powder to be tested is as follows:—Carefully weigh out 10 grains of the sample, and finely triturate it in a mortar with a little cold water, gradually adding more water; next allow the liquor to settle, then pour the liquid into a litre flask, and triturate the residue again with a little water, and rinse the contents of the mortar carefully into the flask, which should then be filled with water up to the graduated mark. Now shake the milky fluid and proceed to examine it while in the turbid state; and each time, before measuring off a fresh portion, the vessel must be again shaken to prevent the material from depositing. The results obtained with the solution in its turbid condition are considered more accurate and reliable than when the clear liquid alone is treated, even though the deposit be frequently washed. This may be proved, Fresenius says, by making two separate experiments, one with the decanted clear liquor, and another with the residuary turbid mixture. In an experiment made in his own laboratory the decanted clear fluid gives 22·6 of chlorine, the residuary mixture 25·0, and the uniformly mixed turbid solution 24·5. One cubic centimètre of the solution of chloride of lime so prepared corresponds to 0·01 gramme of chloride of lime.

Gay-Lussac's Method.—This method, which is known as the *arsenious acid process*, has been much adopted for the determination of chlorine in bleaching powders, and is conducted as follows:—

The Test-liquor.—This is prepared by dissolving 100 grains of *pure* arsenious acid in about 4 ounces of pure hydrochloric acid, and the solution is to be diluted with water until, on being poured into a graduated 10,000 grains measure-glass, it occupies the volume of 700 grains measure marked on the scale. Each 1,000 grains measure of this liquid now contains 14·29 grains of arsenious acid, corresponding to 10 grains of chlorine, or 1/10 grain of chlorine for every division or degree of the scale of the chlorimeter, for which purpose a Mohr's burette of the above capacity may be used, or a graduated tube of the form shown in Fig. 85 may be employed.

Testing the Sample.—100 grains of the chloride of lime to be tested are next dissolved in water, and poured into a tube graduated up to 2,000 grains measure. The whole must be well shaken in order to obtain a uniformly turbid solution, and half of it (1,000 grains measure) transferred to a graduated chlorimeter, which is, therefore, thus filled up to 0°, or the zero of the scale, and contains exactly 50 grains of the chloride of lime under examination, whilst each degree or division of the scale contains only ½ grain. 1,000 grains measure of the arsenious acid test-liquor are now poured into a glass beaker, and a few drops of a solution of sulphate of indigo added, in order to impart a faint, but distinct, blue colour to it; the glass is then to be shaken so as to give a circular movement to the liquid, and whilst it is whirling round the chloride of lime solution from the chlorimeter is gradually and cautiously added until the blue tinge given to the arsenious acid test-liquor is destroyed, care being taken to stir the mixture well with a glass rod during the whole process, and to stop as soon as the decoloration is complete. We will assume that in order to destroy the blue colour of 1,000 grains measure of the arsenious acid test-liquor 90 divisions or degrees of the chloride of lime solution have been employed. These 90 divisions, therefore, contained the 10 grains of chlorine required to destroy the colour of the test solution; and since each division represents ½ grain of chloride of lime, 45 grains of chloride of lime (10 grains of chlorine) were present in the 90 divisions so employed, from which the percentage strength may be ascertained:—

For 45 : 10 :: 100 : 22·22.

The chloride of lime examined, therefore, contained 22¼ per cent. (nearly) of chlorine. This method is extremely simple and trustworthy when properly employed, but to ensure accuracy certain precautions must be adopted. Instead of pouring the test liquor into the solution of the sample (as in alkalimetry), the solution of the sample must be poured into the test-liquor. If the contrary plan were adopted the hydrochloric acid of the test-liquor would liberate chlorine gas so fast that much would be lost, and the result rendered incorrect. By pouring, on the contrary, the chloride of lime solution into the arsenious acid solution the chlorine is disengaged in small portions at a time, and meets with an abundance of arsenious acid to react on. The mixture of chloride of lime should also be employed turbid.

Estimation of Alumina in Alum Cake, etc.—Mr. Rowland Williams, F.C.S., in a paper read before the Chemical Society in June, 1888, describes a method of estimating the alumina in alums, alum cakes, and sulphate of alumina, by which he obtained more accurate results than are obtained by the ordinary ammonia method of estimation. After pointing out several objections to the method of precipitating the alumina by ammonia, he proceeds:—"There is another method for the estimation of alumina which is not so well known as the above. This is by means of sodium thiosulphate. Having had a very extensive and successful experience of this process, I can recommend it with confidence. Considerable practice is, however, necessary in order to secure good results, as certain conditions must be carefully attended to, otherwise the precipitation will be incomplete. The estimation is made in a moderately dilute solution. In the case of alum cake and sulphate of alumina I dissolve 400 grains in water, filter, dilute to 10,000 grains. I use 1,000 grains of this solution (equal to 40 grains of the sample) for estimating the alumina. If any free acid is present it is neutralised by a few drops of carbonate of soda solution, and the whole diluted to about 8 ounces measure. A large quantity of crystallized thiosulphate of soda is then added, and the liquid boiled for at least half-an-hour, constantly replacing the water lost by evaporation. By the end of that time all the alumina will be precipitated in a finely-divided form, along with more or less free sulphur. The precipitate is then filtered off and washed well with boiling water. The filtration and washing take place very rapidly, and may generally be accomplished in about twenty minutes, this being a great saving of time in comparison with the long and tedious washing by decantation, which is necessary in the case of gelatinous alumina. Before filtration, it is advisable to add a drop or two of carbonate of soda solution, lest the liquid should have become slightly acid during boiling."

CHAPTER XX.

USEFUL NOTES AND TABLES.

Preparation of Lakes.—Brazil-wood Lake.—Cochineal Lake.—Lac Lake.—Madder Lake.—Orange Lake.—Yellow Lake.—Artificial Ultramarine.—Twaddell's Hydrometer.—Dalton's Table showing the proportion of Dry Soda in Leys of Different Densities.—Table of Strength of Caustic Soda Solutions at 59° F.—Table showing the Specific Gravity corresponding with the degrees of Baumé's Hydrometer.—Table of Boiling Points of Alkaline Leys.—Table showing the Quantity of Caustic Soda in Leys of Different Densities.—Table showing the Quantity of Bleaching Liquid at 6° Twaddell required to be added to Weaker Liquor to raise it to the given Strength.—Comparative French and English Thermometer Scales.—Weights and Measures of the Metrical System.—Table of French Weights and Measures.—List of Works relating to Paper Manufacture.

Preparation of Lakes.—These are prepared by either of the following processes:—1. By adding a solution of alum, either alone or partly saturated with carbonate of potassa, to a filtered infusion or decoction of the colouring substance, and after agitation precipitating the mixture with a solution of carbonate of potash ("salt of tartar"). 2. By precipitating a decoction or infusion of the colouring substance made with a weak alkaline ley, by adding a solution of alum. 3. By agitating recently precipitated alumina with a solution of the colouring matter, prepared as before, until the liquid is nearly discoloured, or the alumina acquires a sufficiently dark tint. The first method is usually employed for aciduous solutions of colouring matter, or for those whose tint is injured by alkalies; the second for those that are brightened, or at least uninjured, by alkalies; the third, those colouring matters that have a great affinity for gelatinous alumina, and readily combine with it by mere agitation. By attention to these general rules, lakes may be prepared from almost all animal and vegetable colouring substances that yield their colour to water, many of which will be found to possess great beauty and permanence.

The precise process adapted to each particular substance may be easily ascertained by taking a few drops of its infusion or decoction, and observing the effects of alkalies and acids on the colour.

The quantity of alum or of alumina employed should be nearly sufficient to decolour the dye-liquor, and the quantity of carbonate of potassa should be so proportioned to the alum as to exactly precipitate the alumina, without leaving free or carbonated alkali in the liquid. The first portion of the precipitate has the deepest colour, and the shade gradually becomes paler as the operation proceeds.

A beautiful "tone" of violet, red, and even purple may be communicated to the colouring matter of cochineal by the addition of perchloride of tin; the addition of arseniate of potassa (neutral arsenical salt) in like manner gives shades which may be sought for in vain with alum or alumina. After the lake is precipitated it must be carefully collected, washed with cold distilled water, or the purest rain-water, until it ceases to give out colour.

Brazil-wood Lake.—1. Take of ground Brazil wood 1 lb., water 4 gallons; digest for 24 hours, then boil for 30 or 40 minutes, and add of alum 1½ lb., dissolved in a little water; mix, decant, strain, and add of solution of tin ½ lb.; again mix well and filter; to the clear liquid add, cautiously, a solution of salt of tartar or carbonate of soda, as long as a deep-coloured precipitate forms, carefully avoiding excess. 2. Add washed and recently precipitated alumina to a strong and filtered decoction of Brazil wood. Inferior to the last.

Cochineal Lake.—1. Cochineal (in coarse powder) 1 oz.; water and rectified spirit, of each, 2½ ozs.; digest for a week; filter and precipitate the tincture with a few drops of solution of tin, added every 2 hours, until the whole of the colouring matter is thrown down; lastly, wash the precipitate in distilled water and dry it; very fine. 2. Digest powdered cochineal in ammonia water for a week, dilute the solution with a little water, and add the liquid to a solution of alum, as long as a precipitate falls, which is the lake. Equal to the last. 3. Coarsely powdered cochineal 1 lb., water 2 gallons; boil 1 hour, decant, strain, add a solution of salt of tartar, 1 lb., and precipitate with a solution of alum. By adding the alum first, and precipitating the lake with the alkali, the colour will be slightly varied. All the above are sold as carminated or Florence lake, to which they are often superior.

Lac Lake.—Boil fresh stick-lac in a solution of carbonate of soda, filter the solution, precipitate with a solution of alum, and proceed as before. A fine red.

Madder Lake.—1. Take of Dutch grappe or crop madder 2 oz., tie it in a cloth, beat it well in a pint of water in a stone mortar, and repeat the process with fresh water (about 5 pints) until it ceases to yield colour; next boil the mixed liquor in an earthen vessel, pour it into a large basin, and add of alum 1 oz., previously dissolved in boiling water, 1 pint; stir well, and while stirring, pour in gradually of a strong solution of carbonate of potassa (salt of tartar) 1½ oz.: let the whole stand until cold, then pour off the supernatant liquor, drain, agitate the residue with boiling water, 1 quart (in separate portions), decant, drain, and dry. Product, ½ oz. The Society of Arts voted their gold medal to the author of the above formula. 2. Add a little solution of acetate of lead to a decoction of madder, to throw down the brown colouring matter, filter, add a solution of tin or alum, precipitate with a solution of carbonate of soda or of potassa, and otherwise proceed as before. 3. Ground madder, 2 lbs.; water, 1 gallon; macerate with agitation for 10 minutes, strain off the water, and press the remainder quite dry; repeat the process a second and a third time; then add to the mixed liquors, alum, ½ lb., dissolved in water, 3 quarts; and heat in a water-bath for 3 or 4 hours, adding water as it evaporates: next filter, first through flannel, and when sufficiently cold, through paper; then add a solution of carbonate of potassa as long as a precipitate falls, which must be washed until the water comes off colourless, and lastly, dry. If the alkali be added in 3 successive doses, 3 different lakes will be obtained, successively diminishing in beauty.

Orange Lake.—Take of the best Spanish annotta 4 ozs.; pearlash, ¾ lb.; water, 1 gallon; boil it for half an hour, strain, precipitate with alum, 1 lb., dissolved in water, 1 gallon, observing not to add the latter solution when it ceases to produce an effervescence or a precipitate. The addition of some solution of tin turns this lake a lemon yellow; acids redden it.

Yellow Lake.—1. Boil French berries, quercitron bark, or turmeric, 1 lb., and salt of tartar, 1 oz., in water, 1 gallon, until reduced to one half; then strain the decoction and precipitate with a solution of alum. 2. Boil 1 lb. of the dye-stuff with alum, ½ lb.; water, 1 gallon, as before, and precipitate the decoction with a solution of carbonate of potash.

Artificial Ultramarine.—This is obtained by several processes, of which the following are examples:—1. Take kaolin, 37 parts; sulphate of soda, 15; carbonate of soda, 22; sulphur, 18; and charcoal, 8 parts; mix these intimately, and heat in large covered crucibles for twenty-four to thirty hours. The resulting product is then to be again heated in cast-iron boxes at a moderate temperature, until the required tint is obtained; it is finally pulverised, washed in a large quantity of water, and the floating particles allowed to subside in a separate vessel; the deposited colour is now collected and dried. 2. Expose to a low red heat, in a covered crucible as long as fumes are given off, a mixture composed of: kaolin, 2 parts; anhydrous carbonate of soda and sulphur, of each 3 parts. Some persons use one-third less carbonate of soda.

Twaddell's Hydrometer, which is much employed for ascertaining the strength of soda and chloride of lime solutions, etc., is so graduated and weighted that the 0 or zero mark is equal to 1,000, or the specific gravity of distilled water at the temperature of 60° F., and each degree on the scale is equal to ·005; so that by multiplying this number by the number of degrees marked on the scale, and adding 1·, the real specific gravity is obtained. Thus 10° Twaddell indicates a specific gravity of 1050, or 1·05, and so on.

Imitation Manilla Pulp from Wood.—Mr. George E. Marshall, of Turner's Falls, Mass., patented a process some years back by which wood, under the action of hot water, and under a heavy pressure, acquires the characteristic colour of manilla. The wood, having been cut as usual, is placed in a closed vessel or tank capable of resisting high pressure, if necessary, of 450 lbs. to the square inch, the material being closely packed. At the bottom of this tank is an opening with a valve, through which the water, previously heated to a point above boiling, and below 280°, is forced by a hydraulic press to such an extent as to saturate and to completely permeate the wood, and to soften and drive out of the pores the gum, resins, and acids; and if the temperature is kept sufficiently hot, it gives the pulp the desired colour belonging to a finely-made manilla paper. This may be aided somewhat by the introduction of a small quantity of some alkaline substance to act on the acids. The water may be heated in a coil outside, and forced into the tank by a hydraulic press. The water thus heated and forced in leaves the wood or the pulp in the most desirable condition for work and for colour. Pulp made from wood treated below the boiling point will be white; but this process is said to secure the desired manilla colour by raising the temperature to 240° or 250° for a light pulp, and as high as 280° for a dark pulp. No pressure is required from the steam above three atmospheres, but the press may give from 450 to 500 lbs. to the square inch, and practice has shown that the greater the pressure the more speedy is the operation on the wood.[36]

Testing Ultramarines.—The sample of ultramarine should be examined as to its power of resisting the action of alum solutions, which may readily be done by the method suggested by Mr. Dunbar:—"Dissolve the same amount of each sample in water, and mix in this water about ½ lb. of pulp. When thoroughly mixed, and each lot of pulp is well and evenly coloured, add one glassful of the ordinary mill alum liquor, either from pure alum, or aluminous cake to each, losing no time over the operation. Stir each well and continuously with a glass rod, and note the glasses carefully as to the length of time each sample keeps its colour." To ascertain the *staining power*, so called, of the ultramarine, and at the same time the tone, or tint, which it will impart when mixed with pulp, 25 grains of each sample should be mixed with 100 of kaolin or sulphate of lime (pearl hardening) and the several mixtures then worked up into a paste with a little water by means of a spatula, when the differences in the staining power of the respective samples will at once become apparent if either be of inferior quality. To make the test more complete, a like amount of commercially pure ultramarine should be mixed with 100 grains of kaolin for the purpose of comparison. In this way a ready judgment may be formed as to the quality of the sample under examination.

Strength of Paper.—The comparative strength of samples of paper may he determined by cutting strips an inch in width from each sample, and suspending these from a rigid iron bar. Weights are then cautiously attached to each until the sample breaks, when the difference in the weights sustained by the respective samples before the breaking point is reached will determine the comparative strength of the samples tested. Mr. Parkinson, of St. George's Road, Preston, furnishes a simple contrivance for determining the breaking points of paper, and so comparing their value.

TABLES.

I.—DALTON'S TABLE SHOWING THE PROPORTION OF DRY SODA IN LEYS OF DIFFERENT DENSITIES.

Specific gravity of solution.	Dry Soda per cent. by weight.	Boiling points.	Specific gravity of solution.	Dry Soda per cent. by weight.	Boiling points.
1·85	63·6	600°	1·36	26·0	235°
1·72	53·8	400°	1·32	23·0	228°
1·63	46·6	300°	1·29	19·0	224°
1·56	41·2	280°	1·23	16·0	220°
1·50	36·8	265°	1·18	13·0	217°
1·47	34·0	255°	1·12	9·0	214°
1·44	31·0	248°	1·06	4·7	213°
1·40	29·0	242°			

II.—Table of Strength of Caustic Soda Solutions at 59° F. = 150° C. (Tünnerman).

Specific Gravity (Water 1,000).	Degrees Twaddell.	Per cent. of Soda.	Equivalent per cent. of 60 per cent. Caustic Soda.
1·0040	0·80	0·302	0·503
1·0081	1·62	0·601	1·001
1·0163	3·26	1·209	2·015
1·0246	4·92	1·813	3·021
1·0330	6·60	2·418	4·030
1·0414	8·28	3·022	5·037
1·0500	10·00	3·626	6·043
1·0587	11·74	4·231	7·051
1·0675	13·50	4·835	8·059
1·0764	15·28	5·440	9·067
1·0855	17·10	6·044	10·073
1·0948	18·96	6·648	11·080
1·1042	20·84	7·253	12·090
1·1137	22·74	7·857	13·095
1·1233	24·66	8·462	14·103
1·1330	26·60	9·066	15·110
1·1428	28·56	9·670	16·117
1·1528	30·56	10·275	17·125
1·1630	32·60	10·879	18·131
1·1734	34·68	11·484	19·140

1·1841	36·82	12·088	20·147
1·1948	38·96	12·692	21·153
1·2058	41·16	13·297	22·161
1·2178	43·56	13·901	23·170
1·2280	45·60	14·506	24·177
1·2392	47·84	15·110	25·170

III.—Table showing the Specific Gravity corresponding with the Degrees of Baumé's Hydrometer.

Liquids denser than Water.

Degrees.	Specific Gravity.	Degrees.	Specific Gravity.	Degrees.	Specific Gravity.
0	1·0000	26	1·2063	52	1·5200
1	1·0066	27	1·2160	53	1·5353
2	1·0133	28	1·2258	54	1·5510
3	1·0201	29	1·2358	55	1·5671
4	1·0270	30	1·2459	56	1·5833
5	1·0340	31	1·2562	57	1·6000
6	1·0411	32	1·2667	58	1·6170
7	1·0483	33	1·2773	59	1·6344
8	1·0556	34	1·2881	60	1·6522
9	1·0630	35	1·2992	61	1·6705
10	1·0704	36	1·3103	62	1·6889
11	1·0780	37	1·3217	63	1·7079
12	1·0857	38	1·3333	64	1·7273
13	1·0935	39	1·3451	65	1·7471
14	1·1014	40	1·3571	66	1·7674
15	1·1095	41	1·3694	67	1·7882
16	1·1176	42	1·3818	68	1·8095
17	1·1259	43	1·3945	69	1·8313
18	1·1343	44	1·4074	70	1·8537

19	1·1428	45	1·4206	71	1·8765
20	1·1515	46	1·4339	72	1·9000
21	1·1603	47	1·4476	73	1·9241
22	1·1692	48	1·4615	74	1·9487
23	1·1783	49	1·4758	75	1·9740
24	1·1875	50	1·4902	76	2·0000
25	1·1968	51	1·4951		

IV.—Table of Boiling Points of Alkaline Leys.

Alkaline Ley.	Specific Gravity.	Percentage of Alkali.	Boils at degrees Fahrenheit.
Soda	1·18	13	217°
Potash	1·23	19·5	220
Soda	1·23	16	220
Potash	1·28	23·4	224
Soda	1·29	19	224
Soda	1·32	23	228
Potash	1·33	26·3	229
Soda	1·36	26	235
Soda	1·40	29	242
Potash	1·42	34·4	246
Soda	1·47	34	255
Potash	1·44	36·8	255
Soda	1·5	36·8	265
Potash	1·52	42·9	276
Potash	1·6	46·7	290
Soda	1·63	46·6	300
Potash	1·68	51·2	329

V.—Table showing the Quantity of Caustic Soda in Leys of different Densities (Water 1,000).

Specific gravity.	Soda per cent.	Specific gravity.	Soda per cent.
1·00	0·00	1·22	20·66
1·02	2·07	1·24	22·58
1·04	4·02	1·26	24·47
1·06	5·89	1·28	26·33
1·08	7·69	1·30	28·16
1·10	9·43	1·32	29·96
1·12	11·10	1·34	31·67
1·14	12·81	1·35	32·40
1·16	14·73	1·36	33·08
1·18	16·73	1·38	34·41
1·20	18·71		

VI. — TABLE SHOWING THE QUANTITY OF BLEACHING LIQUID AT 6° TWADDELL (SPECIFIC GRAVITY 1·030) REQUIRED TO BE ADDED TO WEAKER LIQUOR TO RAISE IT TO THE GIVEN STRENGTHS.

Strength of Sample in $1/12°$.	Required Strength.	Proportions Required.	
		Given Sample.	Liquor at 6°.
		parts.	part.
Water	$8/12°$	8	1
1	"	9¼	1
2	"	11	1
3	"	13½	1
4	"	17	1
5	"	23	1
6	"	35	1
7	"	71	1
Water	$6/12°$	11	1
1	"	13½	1
2	"	17	1
3	"	23	1
4	"	35	1
5	"	71	1
Water	$4/12°$	17	1
1	"	23	1
2	"	35	1
3	"	71	1
Water	$3/12°$	23	1

| 1 | " | 35 | 1 |
| 2 | " | 71 | 1 |

VII.—Comparative French and English Thermometer Scales.

French or Centigrade.		English or Fahrenheit.	
0	Cent. or C.	equals 32	Fahr. or F.
5	"	" 41	"
10	"	" 50	"
15	"	" 59	"
20	"	" 68	"
25	"	" 77	"
30	"	" 86	"
35	"	" 95	"
40	"	" 104	"
45	"	" 113	"
50	"	" 122	"
55	"	" 131	"
60	"	" 140	"
65	"	" 149	"
70	"	" 158	"
75	"	" 167	"
80	"	" 176	"
85	"	" 185	"
90	"	" 194	"
95	"	" 203	"

100	"	(Water boils)	"	212 "	(Water boils)
200	"		"	392 "	
300	"		"	572 "	
356	"	(Mercury boils)	"	662 "	(Mercury boils)

VIII.—Weights and Measures of the Metrical System.
(From the British Pharmacopœia.)

WEIGHTS.

1 Milligramme	=	the thousandth part of one gramme, or	0·001
1 Centigramme	=	the hundredth "	0·01 "
1 Décigramme	=	the tenth "	0·1 "
1 Gramme	=	weight of a cubic centimètre of water at 4° C.	1·0
1 Décagramme	=	ten grammes	10·0
1 Hectogramme	=	one hundred grammes	100·0
1 Kilogramme	=	one thousand grammes	1,000·0

MEASURES OF CAPACITY.

1 Millilitre	=	1	cubic centimètre,	or the measure of	
1 Centilitre	=	10	"	"	1
1 Décilitre	=	100	"	"	10
1 Litre	=	1,000	"	"	1,00

MEASURES OF LENGTH.

1 Millimètre	=	the thousandth part of		one mètre, or	0·001
1 Centimètre	=	the hundredth	"	"	0·01
1 Décimètre	=	the tenth	"	"	0·1
1 Mètre	=	the ten-millionth part of a quarter of th(			

1 Mètre   =   meridian of the earth.

IX.—TABLE OF FRENCH WEIGHTS AND MEASURES.

Kilogramme, 1,000 grammes, equals 2 lbs. 3¾ ozs. nearly.
Gramme (the unit) equals 15·432 grains.

FRENCH MEASURE OF VOLUME.

1 Litre (the unit) equals 34 fluid ozs. nearly.

LONG MEASURE.

Mètre (the unit)	equals	39·371	inches.
Décimètre (10th of a mètre)	"	3·9371	"
Centimètre (100th of a mètre)	"	0·3937	"
Millimètre (1,000th of a mètre)	"	0·0393	"

LIST OF WORKS RELATING TO PAPER MANUFACTURE.

"Practical Remarks on Modern Paper." J. Murray. Edinburgh, 1829.
"Manuel du Fabricant des Papiers." L. S. Le Normand. Paris, 1834.
"L'Industrie de la Papetrie." G. Planche. Paris, 1853.
"Die Fabrikation des Papiers." L. Müller. Berlin, 1855.
"Manufacture of Paper and Boards." A. Proteaux. Philadelphia, 1866.
"Manufacture of Paper." C. Hofmann. Philadelphia, 1873.
"Pflanzenfasir." Hugo Müller. Leipzig, 1873.
"Bamboo Considered as a Paper-making Material." London, 1875.
"Etudes sur les Fibres Végétales." Vétillart. Paris, 1876.
"Technology of the Paper Trade" (Cantor Lectures). Arnot. Journal Society of Arts, 1877.
"The Practical Paper-maker." J. Dunbar. London, 1881.
"Forestry and Forest Products." Edinburgh, 1884.
"A Treatise on Paper." R. Parkinson. Preston, 1886.
"Manufacture of Paper." C. T. Davis. Philadelphia, 1887.
"Manufacture of Paper." Tomlinson.
"Text Book of Paper-making." C. F. Cross and E. J. Bevan.

Articles on paper-making will also be found in the following encyclopædias, journals, etc:—

"Encyclopædia Britannica," vol. xvii.; "Encyclopædia Metropolitana," 1845; "Tomlinson's Cyclopædia;" "New American Cyclopædia;" "British Manufacturing Industries;" "English Cyclopædia;" "Encyclopædia Americana;" "Penny Cyclopædia;" *Paper Makers' Monthly Journal*; *Paper Makers' Circular*; *Paper Trade Journal*; *American Paper Trade Journal*.

INDEX.

Acetic acid, 64, 98

Acid, arsenious, process, 231
  or bisulphite processes, objections to, 74
  boracic, 46
  carbonic, 97
  fluo-silicic, 175
  hydrochloric, 55, 232
  hypochlorous, 98
  nitric, 66
  nitrous, 66
  nitro-hydrochloric, 64
  oxalic, 98
  processes, McDougall's boiler for, 72
  sulphuric, 47, 99
    anhydrous, 225
  sulphurous, 55, 175
  test, 224
  test, preparation of, 225
  treatment of wood, 64

Acids, action of, on cellulose, 2

Acicular fibres, 3

Action of acids on cellulose, 2

Adamsonia, 85

Adamson's process, 77

African esparto, 47

Agalite, 115

Agar-agar, 178

Agave Americana, 8

Alexandria rags, 21

Algerian esparto, 47

Alkali, caustic, 48
  testing, 224

Alkalimeter, Mohr's, 223

Alkalimeters, 222

Alkalimetry, 221

Alkaline leys, boiling points of, 243

Alkalis, sampling, 227

Alum, 116

Alum, bleach liquor, 100
  cake, estimation of alumina in, 233
  concentrated, 119
  crystallised, 119
  liquor, 240
  pearl, 119
  porous, 167

Alumina, estimation of, in alum, &c., 233
  sulphate of, 100

Aluminium, chloride of, 100
  hypochlorite of, 100

Aluminous cake, 119

American combinations for colouring, 167
  method of sizing, 123
  ochre, 167
  refining engines, Mr. Wyatt on, 103
  system of soda recovery, 218
  wood pulp, 60

Ammonia, 233

Ammoniacal water, 6

Andreoli's electrolytic bleaching process, 96

Anhydrous soda, 225
  sulphuric acid, 225

Aniline blues, 166
  reds, 166
  sulphate of, 8
  triethyl rose, 98

Animal size, preparation of, 120, 122
  sized papers, 123
  or tub-sizing, 122

Annotta, Spanish, 238

Antichlor, 109

Antique paper, 157

Apparatus, disintegrating, 72
  evaporating, 205

Aqua regia, 66

Arnot, Mr., on beating-engines, 102
  on finishing, 160

Arnot's method of alkali testing, 229

Artificial flowers, colouring paper for, 168
  ultramarine, 238

Arsenious acid process, 231

Asbestos, 73, 115

Ash, black, 219

Aussedat's process, 63

Azure blue, 170

Back-water pump, Bertrams', 195

Bagging, old, 10

Balsam, Canada, 179

Baltic rags, 21

Bamboo cane, 10, 18

Bambusa vulgaris, 18

Banana fibre, 10

Bank-notes, water-marking, 147

Baobab, 85

Bark fibres, 6
  oak, 166
  paper mulberry, 10

Barre and Blondel's process, 66

Bast bagging, 10

Baumé's hydrometer, 242

Beakers, 222, 224

Beater, 37
  Jordan, 103, 104
  Kingsland, 104

Beating, 101

   Dunbar's observations on, 102
   engine, 103
     Bertrams', 105
     Forbes', 105
     Umpherston's, 105
   engines, Arnot on, 102
   operations of, 107
   or refining, 101

Belgian rags, 20

Bentley and Jackson's boiler, 80
   cooling and damping rolls, 189
   drum-washer, 185
   dry felt self-acting regulator, 186
   glazing calender, 155
   rag-cutter, 24
     engine, 38
   single-cylinder machine, 153
   web-ripping machine, 198

Benzine, 5, 77

Berlin blue, 168

Bertrams' back-water pump, 195
   beating-engine, 105
   conical pulp-saver, 144
   damping-rolls, 155
   edge-runner, 82
   esparto-cleaner, 40
   large paper machine, 134
   rag boiler, 29
     cutting-machine, 23
     engine, 37
   revolving strainer and knotter, 137
   revolving knife-cutter, 162
   reeling machine, 197
   single-sheet cutter, 162
   web-glazing calender, 196
   willowing and dusting machine, 26

Beetroot refuse, 10

Beyrout rags, 21

Bichromate of potassa, 165

Binders' clippings, 10

Birch, 60

Bisulphite of lime, 71
   magnesium, 70
  process, Blitz's, 72

    Francke's, 68
    Graham's, 73
    Mitscherlich's, 71
    objections to, 74

Black ash, 219
  calicoes, 20
  cotton, 20
  Frankfort, 171
  lamp, 166

Blacks, 20

Bleach, 93
  liquor, alum, 100
    Wilson's, 100
    zinc, 99
  mixer, 92
  pump, Donkin's, 193

Bleaching, 89
  agent, 90
  with chloride of lime, 92
    chlorine gas, Glaser's process, 93
  C. Watt, jun.'s, electrolytic process, 94
  electrolytic, Andreoli's process, 96
  Hermite's process, 96
  esparto, 50
  liquid, table showing quantity to be used, 244
  liquor, 50, 91
    preparation of, 92

Bleaching liquors, 3
  Lunge's process of, 98
  new method of, 100
  operations, 89
  powder, 92
    estimation of chlorine in, 230
    Fresenius' method, 231
    Gay-Lussac's method, 231

Bleaching, sour, 91
  Thompson's process, 97
  Young's method, 100

Blending, 112

Blitz's process, 72

Blotting-papers, 21, 181

Blue, 166
  azure, 170
  Berlin, 168
  Bremen, 170

cottons, 20
  dark, 170
  indigo, 166
  linens, 20
  mineral, 171
  pale, 170
  paper, 19
  Paris, 169
  Prussian, 165
  rags, 19
  smalts, 165

Blues, 20
  aniline, 166

Boiler, Bentley and Jackson's, 80
  Roeckner's, 45

Boiling, American, 60
  esparto, 41
  rags, 29
  straw, 81
  waste paper, 86

Boracic acid, 46

Borax, 169

Boxes, suction, 148

Brazil wood, 166
  lake, 236

Breaking half-stuff, 39
  points of paper, method of determining, 240

Breaking and washing, 34

Breast-roll, 149

Bremen blue, 170

"Broke" paper, 85

Bromine, 6
  water, 6

Broom, 10

Broussonetia papyrifera, 18

Brown, 167

brown, dark, 170
  reddish, 172

275

Bucking-keir, 88

Buckwheat straw, 10

Buff envelope, 167

Bunsen burner, 225

Burettes, 222

Calcined soda, 93

Calciner, 206

Calcium, acetate of, 98
   chloride of, 109, 230
   hypochlorite of, 3
   salts, 99

Calender, glazing, 154

Calendering, 154
   super, Mr. Wyatt on, 158

Calicoes, black, 20

Canada balsam, 179

Cane, bamboo, 10
   rattan, 10

Caoutchouc, 73

Carbonate of lime, 119
   magnesia, 46
   potassa, 235, 236
   soda, 31

Carbonell's esparto process, 46

Carbonic acid, 97

Carbonisation, 75

Cardboard, 182
   with two faces by ordinary machinery, 182
   work, 179

Carminated lake, 237

Carrageen moss, 178

Carrying tubes, 143

Castile soap, 121

Caustic alkali, 48
  potash, 3, 7
  soda, 31
    ley, 31
    table showing quantities of, in leys of different densities, 243

Causticising soda, 32, 205
  tanks, 218

Cellulose, 1
  action of acids on, 2
  determination of, 5
  of flax, 4
  physical characteristics of, 3
  white, 76

Chemical combination, 224
  processes, 55
  wood pulp, 54

Chilled-iron glazing-rolls, 156

China clay, 114
  grass, 10

Chloride of aluminium, 100
  calcium, 101, 230
  lime, 47, 230
    bleaching with, 92
    testing samples of, 232
  magnesium, 96
  potassium, 95
  sodium, 95, 109
  zinc, 99

Chlorimeter, 232

Chlorimetry, 231

Chlorine, 2, 90, 232
  gas, bleaching with, 93
  in bleaching powder, estimation of, 230
  test for, 110

Chrome, lemon, 170
  orange, 166
  yellow, 166

Cinnabar, 171

Citrate of tin, 169

Clarifier, Roeckner's, 199

Clay, China, 114

Clogging, 116

"Close" paper, 112

Cobalt, oxide of, 165

Cochineal, 121, 166
   lake, 236

Colcothar, 170

Coloured cotton, 20
   papers, 165

Colouring, 121
   American combinations for, 167
   materials, mixing, with pulp, 168
   matters used in paper making, 166
   paper for artificial flowers, 168

Commercial sodas, examination of, 221

Comparative cost of animal and engine sizing, estimate of, 128
   French and English thermometer scales, 244

Composition for waterproof paper, 177

Concentrated alum, 119

Conical pulp-saver, 144

Cooling and damping rolls, Bentley and Jackson's, 189

Copal, white, 179

Copper, green, 170
   hydrated oxide of, 175
   sulphate, 146

Copperas, 165

Copying-paper, 120

Corchorus capsularis, 4

Cork, 180
   paper, 180

Cost of animal and engine sizing, comparative estimate of, 128

Cotton fibre, 3

   filaments of, 7
   pieces, 20
   rags, 10
   seed waste, 10
     oil soap, 121
   superfine whites, 20
   waste, 10
   wool, 10

Cottons, blue, 20

outshot, 20
unbleached, 20

Coucher, 130

Couch-rolls, 149

Coupier and Mellier's process, 80, 84

Crop madder, 237

Crystallised alum, 119

Cupro-ammonium, 2, 174
  Wright's process of preparing, 175

Cutting, 22, 161
  machine, 23
    Verny's, 187

Cutter, single-sheet, 162

Cutters, 22

Cylinder, drying, 185
  machine, single, 152
  washing, 193

Cylinders, drying, 151

Dalton's table showing proportion of dry soda in leys of different densities, 241

Damping-rolls, Bertrams', 155

Dandy-roll, 144

Deckle, 130
  frame, 143
  strap, 143

De la Rue's improvements in water-marks, 147

Determination of cellulose, 5

Determining the real value or percentage of commercial sodas, chloride of lime, &c., 221

Devil, Donkin's, 27

Dextrin, 2

Diana's process for making paper or cardboard with two faces by ordinary machinery, 182

Digester, 65

Disinfecting machine, 12

Disintegrating apparatus, 79

Doctor, the, 150

Donkin's bleach-mixer, 92
    pump, 193
  glazing machine, 157
    press, 157
  plate-planing machine, 191
  rag boiler, 30
    dusting machine, 26
  washing cylinder for rag-engine, 193

Double crown, 164
  demy, 164
  royal, 164

Double-sized paper, 126

Drab, 167

Drainers, 39

Draining, 39

Dr. Mitscherlich's process, 71

Drum-washer, 34
  Bentley and Jackson's, 185

Dry-felt regulator, self-acting, 186

Drying cylinder, 185
  cylinders, 151

Dunbar's method of treating esparto, 48
  observations on beating, 102

Duster, 26

Dusting, 26

Dutch grappe madder, 237

Dyers' wood waste, 10

Edge-runner, Bertrams', 82

Ekman's process, 70

Elastic fibres, 3
  packing, 72

Electrolytic bleaching process, Andreoli's, 96
    Hermite's, 96
    C. Watt's, 94

Electrotypes for water-marking, 146

Engine, beating, 103
    Bertrams', 105
    Forbes', 105
    Umpherston's, 105
  Marshall's perfecting, 201
  size, French method of preparing, 120
  sizing, 115

Engines, beating, Mr. Arnot on, 102
  refining, American, Mr. Wyatt on, 103

English green, 172
  pink, 172

Envelope, buff, 167
  orange-red gold, 167
  yellow gold, 167

Eosine, 166

Equivalents, chemical, 224

Esparto, African, 47
  Algerian, 47
  bleaching, 50
  boiler, Sinclair's, 42, 43
  boiling, 41
  cleaner, Bertrams', 40
  Dunbar's treatment of, 48
  fibre, 4
  Gabes, 47
  grass, 10, 16
  Mallary's process for, 46
  Oran, 47
  picking, 40
  preliminary treatment of, 40
  Carbonell's process for, 46
  Sfax, 47
  Spanish, 47
  Susa, 47
  Tripoli, 47
  washing boiled, 49
  willowing, 41
  Young's process for boiling, 50

Estimation of alumina in alum cake, &c., 233
   of chlorine in bleaching powder, 230
   of commercial sodas, 221

Eucalyptus, oil of, 178

Evaporating apparatus, 205

Evaporator, esparto, 206
  Porion's, 208
  Roeckner's, 206
  Yaryan's, 208

Evaporators, American, 61, 208

Examination of commercial sodas, 221

Feebly-ribbed, or smooth fibres, 5

Felt, 72, 101

Felting, 131

Fern leaves, 10

Ferrocyanide of potassium, 165

Fibre, banana, 10
  cotton, 3
  esparto, 4
  flax, 7
  hemp, 8
  jute, 4, 8
  linen, 4
  Manilla, 4
  sulphite, and resin, 76
  yellow pine, 4

Fibres, acicular, 3
  bark, 6
  elastic, 3
  round-ribbed, 5
  smooth, or feebly-ribbed, 5
  spiral, 8
  straw, 4
  various, treatment of, 80
  vegetable, micrographic examination of, 5
  vegetable, recognition of, by the microscope, 6

Fibrous waste, 11

Finished paper, packing the, 163

Finishing, 157

  Arnot on, 160
  house, 163
  and sizing, 132

First press-roll, 150

Flask, 227

Flax, cellulose of, 4
  fibre, or linen, 7
  New Zealand, 8, 10
  tow, 11
  waste, 10

Flocks, 73

Florence lake, 237

Foolscap, 164

Forbes' beating-engine, 105

Foreign rags, 20

Fourdrinier machine, 133

Francke's bisulphite process, 68

Frankfort black, 169

French and English thermometer scales, comparative, 244
  measure of volume, 245
  rags, 20
  weights and measures, table of, 245

Fresenius' method of estimating bleaching powder, 231

Friction-glazing, 157

Fridet and Matussière's process, 66

Furnace, incinerating, 208

Fustians, 20

Fustic, 169

Gabes esparto, 47

Gaine's process for making parchment paper, 182

Gamboge, 169

Gas, chlorine, bleaching with, 93

receiver, 65

Gay-Lussac's method of estimating bleaching powder, 231

German rags, 21

Glaser's process for bleaching with chlorine gas, 93

Glauber's salt, 109

Glazing calender, 154
   press, Donkin's, 157
   rolls, chilled-iron, 156
   web, 154

Glucose, 2

Glue pieces, 122
   stock, 124

Glycerin, 120

Graham's process, 73

Grass, China, 10
   esparto, 10, 16
   sea, 11

Green, copper, 170
   English, 172
   pale, 170
   Schweinfurth, 171

Grey linens, 20

Ground madder, 237
   wood pulp, 85

Guillotine rag-cutter, 24

Gum arabic, 169
   sandarac, 179
   tragacanth, 168

Gunny, 20
   bags, 10

Gutta-percha, 147

Half jute and linen, 20
   stuff, 39, 101
     breaking, 39

Hemp fibre, 8

Manilla, 4, 10
sizal, 8
tarred, 20
waste, 10
white, 20

Hermite's electrolytic bleaching process, 96

High-pressure boiler, 63

Hollander, or rag-engine, 34, 129

Home rags, 20

Hop-bines, 10

Hydrate of soda, 225

Hydrated oxide of copper, 175

Hydro-cellulose, 1

Hydrochloric acid, 55, 232

Hydro-extractor, 94

Hydrometer, Baumé's, 242
Twaddell's, 238

Hypochlorite of aluminium, 100
calcium, 3
lime, 92, 98, 230
soda, 8
sodium, 96

Hypochlorous acid, 98

Hyposulphite of soda, 110

Iodide of potassium, 111

Imitation Manilla pulp from wood, 239

Imperial, 164

Incinerating furnace, 208

Indiarubber, vulcanised, 223

Indigo, 98, 166
sulphate of, 232

Ink, lithographic, 180

Introduction of wood pulp, 17

Irish moss, 178

Iron, oxide of, 34

Iron, pernitrate of, 165
  sulphate of, 170

Isinglass, 179

Japanese paper, new, 180

Jordan's beating engine, 103, 104

Jouglet's process for waterproof paper, 177

Jute fibre, 4, 8
  Manilla, &c., 84
  spinners' waste, 20
  waste, 10, 20

Kaolin, 114, 182

Keegan's process, 59

Killing the colour, 121

Kingsland beating-engine, 104

Knife, revolving, 161

Knotter and strainer, revolving, 137

Kollergang, or edge-runner, 82

Lac lake, 237

Laid paper, 130

Lake, Brazil-wood, 236
  carminated, 237
  cochineal, 236
  Florence, 237
  lac, 237
  madder, 237
  orange, 238
  scarlet, 171

Lakes, preparation of, 235

Lamp-black, 166, 169

287

Leaching, 218
  tanks, 218

Lead, nitrate of, 167
  white, 171

Leather waste, 11

Leghorn rags, 21

Lemon chrome, 170

Leys, alkaline, boiling point of, 243
  of different densities, table showing quantities of caustic soda in, 243

Lime, bisulphite of, 71
  carbonate of, 119
  chloride of, 23, 47, 110
    bleaching with, 92
    testing, 232
  hypochlorite of, 92, 98, 230
  milk of, 33, 72, 110
  sulphate of, 100

Limed skins, 122

Linen, 4
  fibre, 4
  or flax fibre, 7
  pieces, 20
  rags, 10
  waste, 10

Linens, blue, 20
  extra fine, 20
  grey, 20
  strong, 20
  white, 20

Liquor, bleaching, preparation of, 92

Liquors, bleaching, 3
  spent, recovery of soda from, 218

Lithographic ink, 180
  paper, 180

Litmus paper, 183

Lixiviation, 75

Loading, 114

Logwood, 166

Long measure, French, 246

Lunge's bleaching process, 9

Machine, Bentley and Jackson's perfecting, 201
   web-ripping, 198
  Bertrams' large paper, 13
    rag-cutting, 23
    reeling, 197
    web-glazing, 196
    willowing and dusting, 26
  disinfecting, 12
  Donkin's plate-planing, 191
    rag-dusting, 23
  Fourdrinier, 133
  rag-cutting, 23
  roll-bar planing, 191
  single-cylinder, 152
    web-winding, 188
  sizing, 126
  Verny's paper-cutting, 187
  wire and its accessories, 142
  Yankee, 152

Machinery, making paper by, 133
  used in paper-making, 184

Machines, wet,

57

Madder, Dutch, 237
  ground, 237
  lake, 237

Magnesia, carbonate of, 46
  sulphate of, 46

Magnesian limestone, 69

Magnesite, 46, 70

Magnesium, bisulphite of, 70
  chloride of, 96

Maize husks and stems, 10

Making the paper, 130
  paper or cardboard with two faces by ordinary machinery, 182
  paper by hand, 129
    machinery, 133

Mallary's process for esparto, 46

Manganese, peroxide of, 94

Manilla fibre, 4
  hemp, 4, 10
  jute, &c., 84
  paper, 85

Manilla, imitation, from wood pulp, 239

Manning winder, 159

Maori-prepared phormium, 8

Materials, raw, 10
  used in paper-making, 9

Marking, water, 146

Marshall's perfecting engine, 201

McDougall's boiler for acid processes, 72

Mechanical processes, 78
  wood pulp, 113
    Voelter's process of preparing, 78

Megass, or cane trash, 10

Mellier's process, 84

Method of sizing, American, 123

Metrical system, weights and measures of, 245

Micrographic examination of vegetable fibres, 5

Microscope, recognition of vegetable fibres by, 6

Midfeather, 35

Milk of lime, 33, 72, 110

Millboard, 175, 182

Mincing the fibre, 102

Mineral blue, 171
  orange, 166

Miscellaneous papers, 174

Mixed fines, 20
  prints, 20

Mixing colouring materials with pulp, 168

Mohr's alkalimeter, 223

Molasses, 180

Morfit's process for toughening paper, 178

Morocco papers, stains for, 171

Mucilage, 94

Mustard oil, 46
  stems, 10

Nascent chlorine, 96

Netting, old, 11

New Japanese paper, 180
  method of bleaching, 100

New rags, 20

New Zealand flax, 8, 10

Nitric acid, 66

Nitro-hydrochloric acid, 64

Nitrous acid, 66

Notes and tables, 235

Nutgalls, 166

Nuttall's rag-cutter, 24

Oak-bark, 166

Oakum, 11

Objections to the acid or bisulphite process, 74

Ochre, American, 167
   yellow, 165, 166

Oil, boiled, 179
   cotton-seed, soap, 121
   of eucalyptus, 178
   linseed, 179
   mustard, 46
   resin, 178
   of turpentine, 179
   of vitriol, 100

Oiled paper, 180

Old bagging, 10
   bast bagging, 10
   canvas, 10
   netting, 11
   rope, 10
   style, 157

Operation of beating, 107

Oran esparto, 47

Orange chrome, 166
   lake, 238
   mineral, 166
   red gold envelope, 167
   yellow, 171

Organic acid, 99

Outshot cottons, 20

Outshots (whites), 20

Overhaulers, 22

Oxalic acid, 98

Oxide of cobalt, 165
  iron, 34
  zinc, 99

Packing the finished paper, 163

Pale blue, 170

Panels, millboard, 175

Pasteboard, 179

Paper, animal-sized, 123
  antique, 157
  blotting, 21, 181
  blue, 19
  breaking points of, method of determining, 240
  "broke," 85
  or cardboard with two faces made by ordinary machinery, 182
  colouring, for artificial flowers, 168
  copying, 120
  cork, 180
  cutting machine, Verny's, 187
  double sized, 126
  hand-made, 129
  new Japanese, 180
  machine, Bertrams' large, 134
    Fourdrinier's, 133
    Yankee, 152
  making by hand, 129
    by machinery, 133
    machinery used in, 184
    materials used in, 9
  manilla, 85
  imitation manilla, from wood, 239
  Morfit's process for toughening, 178
  mulberry, 18
  bark, 10
  oiled, 180
  old style, 157
  parchment, 181
  shavings, 58
  sizes of, 164
  strength of, 240
    Parkinson's contrivance for determining, 240
  toned, 165
  toughening, 178
  tracing, 179
  transparent, 179
  turmeric, 183
  varnished, 179
  vegeto-mineral, 115
  waste, 85

   boiling, 86
     Ryan's process for treating, 87
  water-marked, 130
  waterproof, 174
     Jouglet's process, 177
  for windows, 181
  coloured, 165
  miscellaneous, 174
  Morocco, stains for, 171
  printing, 164
  satin, stains for, 172
  test, 183
  wrapping, 178
  writing, 164

Parchment liquor, 171
  paper, 181
  shavings, 171

Paris blue, 169

Parker and Blackman's disinfecting machine, 12

Parting, 131

Partington's process, 71

Pearl alum, 119

Pearlash, 238

Pearl hardening, 114

Peat, 10

Pectin, 6

Pectose, 6

Perchloride of tin, 236

Perfecting engine, Marshall's, 201

Pernitrate of iron, 165

Peroxide of manganese, 94

Petroleum, 178

Phormium tenax, 8

Physical characteristics of cellulose, 3

Picking esparto, 40

Pictet and Brélaz's process, 64

Pieces, cotton, 20
  linen, 20

Pink, 166
  English, 172

Plate-glazing, 157
  calender, reversing, 191
  planing machine, 190

Poplar, 10, 60

Porion's evaporator, 208

Porous alum, 167

Potash, 74
  carbonate of, 235
  caustic, 3, 7
  yellow prussiate of, 165

Potassa, carbonate of, 235

Potassium, chloride of, 95
  iodide of, 111
  ferrocyanide of, 165

Potcher, 37

Poucher, 39

Poumarède and Figuier's process for parchment paper, 181

Preliminary operations, 19
  treatment of esparto, 40

Preparation of animal size, 122
  bleaching liquor, 92
  lakes, 235
  test acid, 225

Press, glazing, Donkin's, 157

Press-rolls, 150

Presse-pâte, 51

Printing-paper, 103
  papers, 164

Prints, light, 20
  mixed, 20

Process, Adamson's, 77
  American wood pulp, 60
  Andreoli's electrolytic bleaching, 96
  arsenious acid, 231
  Aussedat's, 63
  Barre and Blondel's, 66
  Blitz's, 72
  Carbonell's esparto, 46
  Coupier and Mellier's, 80
  C. Watt's electrolytic bleaching, 94
  Diana's, for making paper with two faces by ordinary machinery, 182
  Dr. Mitscherlich's, 71
  Eckman's, 70
  Francke's bisulphite, 68
  Fridet and Matussière's, 66
  Gaine's, for making parchment paper, 182
  Graham's, 73
  Hermite's electrolytic bleaching, 96
  Jouglet's, for preparing waterproof paper, 177
  Keegan's, 59
  Lunge's bleaching, 98
  Mallary's esparto, 46
  Mellier's, 84
  Morfit's, 178
  Partington's, 71
  Pictet and Brélaz's, 64
  Poumarède and Figuier's, 181
  retting, 129
  Ritter and Kellner's, 71
  Ryan's, 87
  Scoffern and Tidcombe's, 174
  Sinclair's, 58
  Thompson's, 97
  Thune's, 79
  Voelter's, 78
  Watt and Burgess's, 55
  Wright's, 175
  Young's, 50
  Young and Pettigrew's, 66

Processes, acid or bisulphite, objections to, 74
  McDougall's boiler for, 72
  chemical, 55
  mechanical, 78
  sulphide, 77
  sulphite, 68

Prussian blue, 165

Prussiate of potash, 165

Pulp, ground wood, 85
  long-fibred, 111
  mechanical wood, 113
  mixing colouring matter with, 168

rag, 72

Pulp saver, 143
    conical, 144
  strainers, 137
    Bertrams' revolving, 137
    Roeckner's, 140

Pulp, sulphite, 68, 160
  wood, American, 60
    first introduced by Mr. C. Watt, 17
    imitation Manilla from, 239

Pump, vacuum, 149

Quercitron, 166

Rag bagging, 11
  boiler, Bertrams', 29
    Donkin's, 30
  cutter, Nuttall's, 24
  cutting-machine, Bertrams', 23
    Donkin's, 26
  engine, 34
    Bentley and Jackson's, 38
    Bertrams', 37
  pulp, 72

Rags, 11
  Alexandria, 21
  Baltic, 21
  Belgian, 20
  Beyrout, 21
  blue, 19
  boiling, 29
  cotton, 10
  country, 21
  disinfecting, 12
  foreign, 20
  French, 20
  German, 21
  home, 20
  Leghorn, 21
  linen, 10
  new, 20
  Russian, 21
  sorting, 19
  treatment of, 19
  Trieste, 21
  Turkey, 21
  woollen, 21

Rattan cane, 10

Raw materials, 10

Recognition of vegetable fibres by the microscope, 6

Recovery of soda, American system, 218
   from spent liquor, 204

Red, cherry, 170
   dark, 170
   litmus paper, 183
   ochre, 172
   pale, 171
   Turkey, 170
   Venetian, 166

Reds, aniline, 166

Reeds, 10

Reeling machine, Bertrams', 197

Refining or beating, 101
   engine, 159
      Jordan's, 103
   engines, American, Mr. Wyatt on, 103

Regulating box, 136

Resin, 6, 115
   oil, 178
   size, 118
   soap, 116

Resinous soaps, 179

Retree, 85, 164

Retting, 4
   process of, 129

Reversing or plate-glazing calender, 190

Revolving knife, 161
   cutter, 162
   strainer and knotter, 137

Rhamnus catharticus, 169

Ritter and Kellner's process, 71

Roeckner's boiler, 45
   clarifier, 199
   evaporator, 206
   pulp strainers, 140

Roll-bar planing machine,

191

Rolls, couch, 149
  press, 150
  smoothing, 151, 152

Rope, 20
  bagging, 20
  hard, 20
  tarred, 20
  white, 20

Round-ribbed fibres, 5

Royal, 164

Russian rags, 21

Ryan's process for treating waste paper, 87

Sailcloth, 11

Salt of tartar, 235

Sampling alkalies, 227

Sandarac, gum, 179

Sand-table, 136
  tables, 149
  trap, 50, 136

Sap green, 169

Satin papers, stains for, 172

Save-all, 143

Sawdust, 10

Scarlet lake, 171

Schweinfurth green, 171

Scoffern and Tidcombe's process for waterproof paper, 174

Sea grass, 11

Seaweeds, 178

Second press-roll, 150

Seconds rags, 20

Seconds, whites, 20

Self-acting dry felt regulator, 186
  cleansing strainer, 139

Separating tank, 61

Setting, 174

Settling of the pulp, 131

Sfax esparto, 47

Shavings, paper, 58
  parchment, 171
  wood, 10, 55

Shoddy, 11

Silk cocoon waste, 11

Silver white, 173

Sinclair's esparto boiler, 42, 43
  process, 58

Single-cylinder machine, 152

Single-sheet cutter, 162
  web-winding machine, 188

Sizal, or sisal hemp, 8

Size, animal, preparation of, 122
  engine, French method of preparing, 120
  resin, 118

Sizes of paper, 164

Sizing, 115
  American method of, 123
  and finishing, 132
  machine, 126
  tub or animal, 122
  Mr. Wyatt's remarks on, 127
  zinc soaps in, 121

Skip, 153

Small post, 164

Smalts blue, 121, 165

Smoothing presses, three-roll, 194
  rolls, 151, 152

Soap, Castile, 121
  cotton-seed oil, 121
  resin, 116

Soaps, zinc, in sizing, 121

Soda, anhydrous, 225
  ash, 31, 227
  calcined, 93
  carbonate, 31
  caustic, 31
    table showing the quantities of leys of different densities, 243
  dry, Dalton's table, showing the proportion of, in leys of different densities, 241
  hydrate of, 225
  hypochlorite of, 8
  hyposulphite of, 110
  ley, caustic, 31
  recovery of, 104
  recovery of, American system of, 218
  solutions, caustic, table showing strength of, 241
  sulphite of, 110
  thiosulphite of, 110, 233

Sodas, commercial, examination of, 221

Sodium, chloride of, 95, 109
  hypochlorite of, 96
  thiosulphite of, 233

Sorting rags, 19, 22

Sour bleaching, 91

Souring, 99

Spanish annotta, 238
  esparto, 47

Spent liquors, recovery of soda from, 204
  liquors, 218

Spiral fibres, 8

Spruce, 60

Stable manure, 11

Staining power of ultramarines, 240

Stains for Morocco papers, 171
  satin papers, 172

Standard test-acid solution, 225

Starch paste, 117

Strainer and knotter, Bertrams' revolving, 137
   self-cleansing, 139

Strainers, 57, 137
   Roeckner's pulp, 140

Straw, 16
   boiling, 81
   buckwheat, 10
   fibres, 4
   wheat, 10

Strength of paper, determination of, 240

Strings, 20

Strong linens, 20

Stuff-chests, 57, 112, 136
   pump, 136

Sturtevant blower, 60

Suction boxes, 148

Sulphate of alumina, 100
   aniline, 8
   copper, 146
   indigo, 232
   iron, 170
   lime, 100
   magnesia, 46
   zinc, 99, 119

Sulphide processes, 77

Sulphite fibre, 76
   and resin, 76
   processes, 68
   pulp, 68
   of soda, 110
   wood pulp, 160

Sulphur, 72, 225

Sulphuric acid, 47, 91, 99
   anhydrous, 225

Sulphurous acid, 175
   gas, 55

Super-calendering, 157

American, Mr. Wyatt on, 157

Superfine white cotton, 20

Superfines, white, 20

Supply-box, 136

Surface-sizing, 122

Susa esparto, 47

Table of boiling points of alkaline leys, 243
    French and English thermometer scales, 244
    French weights and measures, 245
    showing proportion of dry soda in leys of different densities, 241
    showing the quantity of bleaching liquid to be used, 244
    showing the quantity of bleach liquor required to be added to weaker liquors, 244
    showing the quantity of caustic sodas in leys of different densities, 243
    showing the specific gravity corresponding with the degrees of Baumé's hydrometer, 242
    of strength of caustic soda solutions, 241
    of weights and measures of the metrical system, 245

Tables and notes, 235
    sand, 149

Tan waste, 10

Tarpaulin, 11, 77

Tarred hemp, 20
    rope, 20
    string, 20

Tartar, salts of, 235

Tea colour, 167

Test acid, preparation of, 224, 225
    for chlorine, 110
    liquor, 232
    papers, 183

Testing chloride of lime, 232
    ultramarines, 239

Thermometer scales, comparative French and English, 244

Thiosulphite of soda, 110
    sodium, 233

Thirds, whites, 20

Thompson's bleaching process, 97

Three-roll smoothing process, 194

Thune's process, 79

Tiles, paper, 175

Tin, citrate of, 169
 perchloride of, 236

Tobacco stalks, 10

Toned paper, 165

Torrance's drainer, 39

Toughening paper, 178

Tracing paper, 179

Tragacanth, gum, 168

Transparent paper, 179

Treatment of esparto, 40
 rags, 19, 29
 various fibres, 80
 wood, 53, 68

Triethyl rose aniline, 98

Tripoli esparto, 47

Tub-sizing, 122

Turmeric paper, 183

Turkish minium, 170

Turkey rags, 21
 red, 170

Turpentine, oil of, 179
 Venice, 179

Twaddell's hydrometer, 238

Ultramarine, 121, 165
 artificial, preparation of, 238

Ultramarines, staining power of, 240
 testing, 239

Umpherston's beating-engine, 105

Unbleached cottons, 20

Vacuum pumps, 149

Vanadate of ammonia, 72

Various fibres, treatment of, 80

Varnished paper, 179

Varrentrapp's zinc bleach liquor, 100

Vat for hand paper-making, 129

Vegetable fibres, micrographic examination of, 5

Vegetable fibres, recognition of, by the microscope, 6

Vegeto-mineral paper, 115

Venetian red, 166

Venice turpentine, 179

Verdigris, 169

Verny's paper-cutting machine, 187

Violet, 171
  dark, 172
  light, 171

Vitriol, oil of, 57, 90, 106

Voelter's process for preparing mechanical wood pulp, 78

Volumetric assaying, 224

Vulcanised india-rubber, 223

Vulcanite, 148

Washing, American, 61
  boiled esparto, 49
  and breaking, 34
    engine, 37
  cylinder for rag-engine, 193

Waste, cotton, 10
  cotton-seed, 10

flax, 10
hemp, 10
jute, 10
linen, 10
liquors, recovery of soda from, 204
paper, 10, 85
   boiling, 86
   Ryan's process for, 87
tan, 10

Water-marked paper, 130

Water-marking, 146

Water-marks, De la Rue's improvements in, 147

Waterproof composition for paper, 177
paper, 174
   for flooring, 177
   Jouglet's process, 177
   for roofing, 177

Watt and Burgess's wood-paper process, 55

Watt's electrolytic bleaching process, 94

Wax, 6, 120
  soap, 169

Web-glazing, 154
  calender, Bertrams', 196

Web-ripping machine, 198

Weights and measures, French table of, 245

Weights and measures of the metrical system, 245

Wet machines, 57

White cellulose, 76
  copal, 179
  hemp, 20
  lead, 171
  linens, 20

Willow and duster, Bertrams', 25
  Masson, Scott, and Co.'s, 40

Willowing, 24
  esparto, 41

Wilson's bleach liquor, 100

Winding machine, single-web, 188

Wood, acid treatment of, 64
  fibre, 53
  paper, Watt's patent for, 17
  pulp, American method of preparing, 60
  pulp, chemical, 54
  mechanical, 113
  shavings, 10, 55, 77
  pulp, sulphite, 160
  treatment of, 53, 68
  pulp, Voelter's mechanical process for preparing, 78
  waste, dyers', 10

Woollen rags, 21

Wrapping papers, 178

Wright's process for preparing cupro-ammonium, 175

Writing papers, 164

Wyatt, Mr., on American refining engines, 103
  on American super-calendering, 157

Wyatt, Mr., on sizing, 127

Xyloidin, 67

Yankee machine, 152

Yaryan evaporator, 208

Yellow chrome, 166
  gold envelope, 167
  lake, 238
  ochre, 165, 166
  pale, 172, 173
  pine fibre, 4

Young's method of bleaching, 100

Young and Pettigrew's process, 66

Young's process for cleaning esparto, 50

Zinc bleach liquor, 99
  chloride of, 99
  oxide of, 99, 100
  salts, 100
  soaps in sizing, 121
  sulphate of, 99, 119

Zostera marina, 11

PRINTED BY J. S. VIRTUE AND CO., LIMITED, CITY ROAD, LONDON

FOOTNOTES:

[1] Cantor Lectures, *Journal of Society of Arts*, vol. xxvi. p. 74.
[2] Needle-shaped, slender and sharp-pointed.
[3] Manilla hemp.
[4] For this purpose, a microscope having a magnifying power of 120 to 150 diameters will be found efficient.
[5] "Commercial Organic Analysis." By A. H. Allen, F.C.S., vol. i. p. 316.
[6] For Table of French Measures see end of this work.
[7] *Pectous*, pertaining to or consisting of *pectose* or *pectin*. Pectose is a substance contained in the pulp of unripe fleshy fruit, also in fleshy roots and other vegetable organs. It is insoluble in water, but under the influence of acids is transformed into *pectin*.
[8] A *litre* equals 34 fluid ounces *nearly*.
[9] "Commercial Organic Analysis." By A. H. Allen, F.C.S., vol. i.
[10] *Septa*, plural of *septum*, a partition, as the partitions of an orange, for example.
[11] "Manufacture of Paper." By C. T. Davis, Philadelphia, 1887.
[12] Patent dated 16th December, 1884, No. 539.
[13] "Forestry and Forest Products," p. 501, and Cross and Bevan's "Text Book of Paper-making," p. 65.
[14] "Practical Paper Maker," by James Dunbar. Mackenzie and Storrie, Leith, 1887.
[15] "Practical Treatise on the Manufacture of Paper." By Carl Hofmann, Philadelphia, 1873.
[16] *The Chemist*. Edited by Charles and John Watt, p. 552; 1855.
[17] *School of Mines Quarterly, a Journal of Applied Science*. Jan., 1889.
[18] The *cord* is a pile containing 128 cubic feet, or a pile 8 feet long, 4 feet high, and 4 feet broad.
[19] Wagner's "Jahresb." 1860, p. 188.
[20] *Paper-Makers Monthly Journal*, March 15th, 1889.
[21] Sometimes also called *thiosulphite of soda*.
[22] "The Art of Soap-making." By Alexander Watt. London, Crosby Lockwood and Son, 4th edition, 1890.
[23] Sometimes called "concentrated alum," "pearl alum," etc.

[24] Muspratt's "Chemistry Applied to the Arts."

[25] "Art of Leather Manufacture." By Alexander Watt. Crosby Lockwood and Son, 1885.

[26] "Proceedings of the Society of Civil Engineers," vol. lxxix. p. 245.

[27] *Paper-Makers' Monthly Journal,* April 15th, 1889.

[28] The berries of *Rhamnus catharticus* made into a decoction by boiling.

[29] *Paper Trade Journal,* New York, April 20th, 1889.

[30] *Sanitary World,* March 29th, 1884.

[31] *Industries,* January 25th, 1889.

[32] "Seventh Annual Report of Local Government Board," 1877-8.

[33] School of Mines *Quarterly Journal of Applied Science,* January, 1889, New York.

[34] These balances may be obtained from Mr. Oertling, Coppice Row, London, or of any philosophical instrument maker.

[35] There are two principal methods of analysing or assaying alkalies by means of the test-acid, namely, *volumetric,* or by volume, and *gravimetric,* or by weight, in which a specific gravity bottle, capable of holding exactly 1,000 grains of distilled water, is used.

[36] New York *Paper Trade Journal,* 1878.

7, Stationers' Hall Court, London, E.C.
May, 1894.

A CATALOGUE OF BOOKS

INCLUDING NEW AND STANDARD WORKS IN
ENGINEERING: CIVIL, MECHANICAL, AND MARINE;
ELECTRICITY AND ELECTRICAL ENGINEERING;
MINING, METALLURGY; ARCHITECTURE,
BUILDING, INDUSTRIAL AND DECORATIVE ARTS;
SCIENCE, TRADE AND MANUFACTURES;
AGRICULTURE, FARMING, GARDENING;
AUCTIONEERING, VALUING AND ESTATE AGENCY;
LAW AND MISCELLANEOUS.
PUBLISHED BY

CROSBY LOCKWOOD & SON.

MECHANICAL ENGINEERING, etc.

D. K. Clark's Pocket-Book for Mechanical Engineers.

THE MECHANICAL ENGINEER'S POCKET-BOOK OF TABLES, FORMULÆ, RULES AND DATA. A Handy Book of Reference for Daily Use in Engineering Practice. By D. KINNEAR CLARK, M. Inst. C. E., Author of "Railway Machinery," "Tramways," &c. Second Edition, Revised and Enlarged. Small 8vo, 700 pages, 9s. bound in flexible leather covers, with rounded corners and gilt edges.

SUMMARY OF CONTENTS.

MATHEMATICAL TABLES.—MEASUREMENT OF SURFACES AND SOLIDS.—ENGLISH WEIGHTS AND MEASURES.—FRENCH METRIC WEIGHTS AND MEASURES.—FOREIGN WEIGHTS AND MEASURES.—MONEYS.—SPECIFIC GRAVITY, WEIGHT AND VOLUME—MANUFACTURED METALS.—STEEL PIPES.—BOLTS AND NUTS.—SUNDRY ARTICLES IN WROUGHT AND CAST IRON, COPPER, BRASS, LEAD, TIN, ZINC.—STRENGTH OF MATERIALS.—STRENGTH OF TIMBER.—STRENGTH OF CAST IRON.—STRENGTH OF WROUGHT IRON.—STRENGTH OF STEEL.—TENSILE STRENGTH OF COPPER, LEAD, ETC.—RESISTANCE OF STONES AND OTHER BUILDING MATERIALS.—RIVETED JOINTS IN BOILER PLATES.—BOILER SHELLS—WIRE ROPES AND HEMP ROPES.—CHAINS AND CHAIN CABLES.—FRAMING.—HARDNESS OF METALS, ALLOYS AND STONES.—LABOUR OF ANIMALS.—MECHANICAL PRINCIPLES.—GRAVITY AND FALL OF BODIES.—ACCELERATING AND RETARDING FORCES.—MILL GEARING, SHAFTING, ETC.—TRANSMISSION OF MOTIVE POWER.—HEAT.—COMBUSTION: FUELS.—WARMING, VENTILATION, COOKING STOVES.—STEAM.—STEAM ENGINES AND BOILERS.—RAILWAYS.—TRAMWAYS.—STEAM SHIPS.—PUMPING STEAM ENGINES AND PUMPS.—COAL GAS, GAS ENGINES, ETC.—AIR IN MOTION.—COMPRESSED AIR.—HOT AIR ENGINES.—WATER POWER.—SPEED OF CUTTING TOOLS.—COLOURS.—ELECTRICAL ENGINEERING.

⁂ OPINIONS OF THE PRESS.

"Mr. Clark manifests what is an innate perception of what is likely to be useful in a pocket-book, and he is really unrivalled in the art of condensation. Very frequently we find the information on a given subject is supplied by giving a summary description of an experiment, and a statement of the results obtained. There is a very excellent steam table, occupying five and-a-half pages; and there are rules given for several calculations, which rules cannot be found in other pocket-books, as, for example, that on page 497, for getting at the quantity of water in the shape of priming in any known weight of steam. It is very difficult to hit upon any mechanical engineering subject concerning which this work supplies no information, and the excellent index at the end adds to its utility. In one word, it is an exceedingly handy and efficient tool, possessed of which the engineer will be saved many a wearisome calculation, or yet more wearisome hunt through various text-books and treatises, and, as such, we can heartily recommend it to our readers, who must not run away with the idea that Mr. Clark's Pocket-book is only Molesworth in another form. On the contrary, each contains what is not to be found in the other; and Mr. Clark takes more room and deals at more length with many subjects than Molesworth possibly could."—*The Engineer.*

"It would be found difficult to compress more matter within a similar compass, or produce a book of 650 pages which should be more compact or convenient for pocket reference.... Will be appreciated by mechanical engineers of all classes."—*Practical Engineer.*

"Just the kind of work that practical men require to have near to them."—*English Mechanic.*

MR. HUTTON'S PRACTICAL HANDBOOKS.

Handbook for Works' Managers.

THE WORKS' MANAGER'S HANDBOOK OF MODERN RULES, TABLES, AND DATA. For Engineers, Millwrights, and Boiler Makers; Tool Makers, Machinists, and Metal Workers; Iron and Brass Founders, &c. By W. S. HUTTON, Civil and Mechanical Engineer, Author of "The Practical Engineer's Handbook." Fourth Edition, carefully Revised and partly Re-written. In One handsome Volume, medium 8vo, price 15s. strongly bound.

☞ *The Author having compiled Rules and Data for his own use in a great* variety of modern engineering work, and having found his notes extremely useful, decided to publish them—revised to date—believing that a practical work, suited to the DAILY REQUIREMENTS OF MODERN ENGINEERS, would be favourably received.

In the Fourth Edition the First Section has been re-written and improved by the addition of numerous Illustrations and new matter relating to STEAM ENGINES and GAS ENGINES. The Second Section has been enlarged and Illustrated, and throughout the book a great number of emendations and alterations have been made, with the object of rendering the book more generally useful.

∴ OPINIONS OF THE PRESS.

"The author treats every subject from the point of view of one who has collected workshop notes for application in workshop practice, rather than from the theoretical or literary aspect. The volume contains a great deal of that kind of information which is gained only by practical experience, and is seldom written in books."—*Engineer.*

"The volume is an exceedingly useful one, brimful with engineers' notes, memoranda, and rules, and well worthy of being on every mechanical engineer's bookshelf."—*Mechanical World.*

"The information is precisely that likely to be required in practice.... The work forms a desirable addition to the library not only of the works' manager, but of anyone connected with general engineering."—*Mining Journal.*

"A formidable mass of facts and figures, readily accessible through an elaborate index.... Such a volume will be found absolutely necessary as a book of reference in all sorts of 'works' connected with the metal trades."—*Ryland's Iron Trades Circular.*

"Brimful of useful information, stated in a concise form, Mr. Hutton's books have met a pressing want among engineers. The book must prove extremely useful to every practical man possessing a copy."—*Practical Engineer.*

New Manual for Practical Engineers.

THE PRACTICAL ENGINEER'S HAND-BOOK. Comprising a Treatise on Modern Engines and Boilers: Marine, Locomotive and Stationary. And containing a large collection of Rules and Practical Data relating to recent Practice in Designing and Constructing all kinds of Engines, Boilers, and other Engineering work. The whole constituting a comprehensive Key to the Board of Trade and other Examinations for Certificates of Competency in Modern Mechanical Engineering. By WALTER S. HUTTON, Civil and Mechanical Engineer, Author of "The Works' Manager's Handbook for Engineers," &c. With upwards of 370 Illustrations. Fourth Edition, Revised, with Additions. Medium 8vo, nearly 500 pp., price 18s. Strongly bound.

☞ *This work is designed as a companion to the Author's* "WORKS' MANAGER'S HAND-BOOK." *It possesses many new and original features, and contains, like its predecessor, a quantity of matter not originally intended for publication, but collected by the author for his own use in the construction of a great variety of* MODERN ENGINEERING WORK.

The information is given in a condensed and concise form, and is illustrated by upwards of 370 Woodcuts; and comprises a quantity of tabulated matter of great value to all engaged in designing, constructing, or estimating for ENGINES, BOILERS, *and* OTHER ENGINEERING WORK.

∴ OPINIONS OF THE PRESS.

"We have kept it at hand for several weeks, referring to it as occasion arose, and we have not on a single occasion consulted its pages without finding the information of which we were in quest."—*Athenæum.*

"A thoroughly good practical handbook, which no engineer can go through without learning something that will be of service to him."—*Marine Engineer.*

"An excellent book of reference for engineers, and a valuable text-book for students of engineering."—*Scotsman.*

"This valuable manual embodies the results and experience of the leading authorities on mechanical engineering."—*Building News.*

"The author has collected together a surprising quantity of rules and practical data, and has shown much judgment in the selections he has made.... There is no doubt that this book is one of the most useful of its kind published, and will be a very popular compendium."—*Engineer.*

"A mass of information, set down in simple language, and in such a form that it can be easily referred to at any time. The matter is uniformly good and well chosen and is greatly elucidated by the illustrations. The book will find its way on to most engineers' shelves, where it will rank as one of the most useful books of reference."—*Practical Engineer.*

"Full of useful information and should be found on the office shelf of all practical engineers."—*English Mechanic.*

Practical Treatise on Modern Steam-Boilers.

STEAM-BOILER CONSTRUCTION. A Practical Handbook for Engineers, Boiler-Makers, and Steam Users. Containing a large Collection of Rules and Data relating to Recent Practice in the Design, Construction, and Working of all Kinds of Stationary, Locomotive, and Marine Steam-Boilers. By WALTER S. HUTTON, Civil and Mechanical Engineer, Author of "The Works' Manager's Handbook," "The Practical Engineer's Handbook," &c. With upwards of 300 Illustrations. Second Edition. Medium 8vo, 18s. cloth.

☞ *This work is issued in continuation of the Series of Handbooks written by the Author, viz:*—"THE WORKS' MANAGER'S HANDBOOK" *and* "THE PRACTICAL ENGINEER'S HANDBOOK," *which are so highly appreciated by Engineers for the practical nature of their information; and is consequently written in the same style as those works.*

The Author believes that the concentration, in a convenient form for easy reference, of such a large amount of thoroughly practical information on Steam-Boilers, will be of considerable service to those for whom it is intended, and he trusts the book may be deemed worthy of as favourable a reception as has been accorded to its predecessors.

⁂ OPINIONS OF THE PRESS.

"Every detail, both in boiler design and management, is clearly laid before the reader. The volume shows that boiler construction has been reduced to the condition of one of the most exact sciences; and such a book is of the utmost value to the *fin de siècle* Engineer and Works' Manager."—*Marine Engineer.*

"There has long been room for a modern handbook on steam boilers; there is not that room now, because Mr. Hutton has filled it. It is a thoroughly practical book for those who are occupied in the construction, design, selection, or use of boilers."—*Engineer.*

"The book is of so important and comprehensive a character that it must find its way into the libraries of everyone interested in boiler using or boiler manufacture if they wish to be thoroughly informed. We strongly recommend the book for the intrinsic value of its contents."—*Machinery Market.*

"The value of this book can hardly be over-estimated. The author's rules, formulæ, &c., are all very fresh, and it is impossible to turn to the work and not find what you want. No practical engineer should be without it."—*Colliery Guardian.*

Hutton's "Modernised Templeton."

THE PRACTICAL MECHANICS' WORKSHOP COMPANION. Comprising a great variety of the most useful Rules and Formulæ in Mechanical Science, with numerous Tables of Practical Data and Calculated Results for Facilitating Mechanical Operations. By WILLIAM TEMPLETON, Author of "The Engineer's Practical Assistant," &c. &c. Sixteenth Edition, Revised, Modernised, and considerably Enlarged by WALTER S. HUTTON, C.E., Author of "The Works' Manager's Handbook," "The Practical Engineer's Handbook," &c. Fcap. 8vo, nearly 500 pp., with 8 Plates and upwards of 250 Illustrative Diagrams, 6s., strongly bound for workshop or pocket wear and tear.

⁂ OPINIONS OF THE PRESS.

"In Its modernised form Hutton's 'Templeton' should have a wide sale, for it contains much valuable information which the mechanic will often find of use, and not a few tables and notes which he might look for in vain in other works. This modernised edition will be appreciated by all who have learned to value the original editions of 'Templeton'."—*English Mechanic.*

"It has met with great success in the engineering workshop, as we can testify; and there are a great many men who, in a great measure, owe their rise in life to this little book."—*Building News*.

"This familiar text-book—well known to all mechanics and engineers—is of essential service to the every-day requirements of engineers, millwrights, and the various trades connected with engineering and building. The new modernised edition is worth its weight in gold."—*Building News*. (Second Notice.)

"This well-known and largely-used book contains information, brought up to date, of the sort so useful to the foreman and draughtsman. So much fresh information has been introduced as to constitute it practically a new book. It will be largely used in the office and workshop."—*Mechanical World*.

"The publishers wisely entrusted the task of revision of this popular, valuable, and useful book to Mr. Hutton, than whom a more competent man they could not have found."—*Iron*.

Templeton's Engineer's and Machinist's Assistant.

THE ENGINEER'S, MILLWRIGHT'S, and MACHINIST'S PRACTICAL ASSISTANT. A collection of Useful Tables, Rules and Data. By WILLIAM TEMPLETON. 7th Edition, with Additions. 18mo, 2s. 6d. cloth.

⁂ OPINIONS OF THE PRESS.

"Occupies a foremost place among books of this kind. A more suitable present to an apprentice to any of the mechanical trades could not possibly be made."—*Building News*.

"A deservedly popular work. It should be in the 'drawer' of every mechanic."—*English Mechanic*.

Foley's Office Reference Book for Mechanical Engineers.

THE MECHANICAL ENGINEER'S REFERENCE BOOK, for Machine and Boiler Construction. In Two Parts. Part I. GENERAL ENGINEERING DATA. Part II. BOILER CONSTRUCTION. With 51 Plates and numerous Illustrations. By NELSON FOLEY, M.I.N.A. Folio, £5 5s. half-bound.

SUMMARY OF CONTENTS.

PART I.

MEASURES.—CIRCUMFERENCES AND AREAS, &C., SQUARES, CUBES, FOURTH POWERS.—SQUARE AND CUBE ROOTS.—SURFACE OF TUBES—RECIPROCALS.—LOGARITHMS.—MENSURATION.—SPECIFIC GRAVITIES AND WEIGHTS.—WORK AND POWER.—HEAT.—COMBUSTION.—EXPANSION AND CONTRACTION.—EXPANSION OF GASES.—STEAM.—STATIC FORCES.—GRAVITATION AND ATTRACTION.—MOTION AND COMPUTATION OF RESULTING FORCES.—ACCUMULATED WORK.—CENTRE AND RADIUS OF GYRATION.—MOMENT OF INERTIA.—CENTRE OF OSCILLATION.—ELECTRICITY.—STRENGTH OF MATERIALS.—ELASTICITY.—TEST SHEETS OF METALS.—FRICTION.—TRANSMISSION OF POWER.—FLOW OF LIQUIDS.—FLOW OF GASES.—AIR PUMPS, SURFACE CONDENSERS, &C.—SPEED OF STEAMSHIPS.—PROPELLERS.—CUTTING TOOLS.—FLANGES.—COPPER SHEETS AND TUBES.—SCREWS, NUTS, BOLT HEADS, &C.—VARIOUS RECIPES AND MISCELLANEOUS MATTER.

WITH DIAGRAMS FOR VALVE-GEAR, BELTING AND ROPES, DISCHARGE AND SUCTION PIPES, SCREW PROPELLERS, AND COPPER PIPES.

PART II.

Treating of, Power of Boilers.—Useful Ratios.—Notes on Construction.—Cylindrical Boiler Shells.—Circular Furnaces.—Flat Plates—Stays.—Girders.—Screws.—Hydraulic Tests.— Riveting.—Boiler Setting, Chimneys, and Mountings.—Fuels, &c.—Examples of Boilers and Speeds of Steamships.—Nominal and Normal Horse Power.

With DIAGRAMS for all Boiler Calculations and Drawings of many Varieties of Boilers.

⁂ Opinions of the Press.

"This appears to be a work for which there should be a large demand on the part of mechanical engineers. It is no easy matter to compile a book of this class, and the labour involved is enormous, particularly when—as the author informs us—the majority of the tables and diagrams have been specially prepared for the work. The diagrams are exceptionally well executed, and generally constructed on the method adopted in a previous work by the same author.... The tables are very numerous, and deal with a greater variety of subjects than will generally be found in a work of this kind; they have evidently been compiled with great care and are unusually complete. All the information given appears to be well up to date.... It would be quite impossible within the limits at our disposal to even enumerate all the subjects treated; it should, however, be mentioned that the author does not confine himself to a mere bald statement of formulæ and laws, but in very many instances shows succinctly how these are derived.... The latter part of the book is devoted to diagrams relating to Boiler Construction, and to nineteen beautifully-executed plates of working drawings of boilers and their details. As samples of how such drawings should be got out, they may be cordially recommended to the attention of all young, and even some elderly, engineers.... Altogether the book is one which every mechanical engineer may, with advantage to himself add to his library."—*Industries.*

"Mr. Foley is well fitted to compile such a work.... The diagrams are a great feature of the work.... Regarding the whole work, it may be very fairly stated that Mr. Foley has produced a volume which will undoubtedly fulfil the desire of the author and become indispensable to all mechanical engineers."—*Marine Engineer.*

"We have carefully examined this work, and pronounce it a most excellent reference book for the use of marine engineers."—*Journal of American Society of Naval Engineers.*

"A veritable monument of industry on the part of Mr. Foley, who has succeeded in producing what is simply invaluable to the engineering profession."—*Steamship.*

Coal and Speed Tables.

A POCKET BOOK OF COAL AND SPEED TABLES, for Engineers and Steam-users. By Nelson Foley, Author of "The Mechanical Engineer's Reference Book." Pocket-size, 3s. 6d. cloth.

"These tables are designed to meet the requirements of every-day use; they are of sufficient scope for most practical purposes, and may be commended to engineers and users of steam."—*Iron.*

"This pocket-book well merits the attention of the practical engineer. Mr. Foley has compiled a very useful set of tables, the information contained in which is frequently required by engineers, coal consumers and users of steam."—*Iron and Coal Trades Review.*

Steam Engine.

TEXT-BOOK ON THE STEAM ENGINE. With a Supplement on Gas Engines, and Part II. on Heat Engines. By T. M. Goodeve, M.A., Barrister-at-Law, Professor of Mechanics at the Royal College of Science, London; Author of "The Principles of Mechanics," "The Elements of Mechanism," &c. Twelfth Edition, Enlarged. With numerous Illustrations. Crown 8vo, 6s. cloth.

"Professor Goodeve has given us a treatise on the steam engine which will bear comparison with anything written by Huxley or Maxwell, and we can award it no higher praise."—*Engineer.*

"Mr. Goodeve's text-book is a work of which every young engineer should possess himself."—*Mining Journal.*

Gas Engines.

ON GAS-ENGINES. With Appendix describing a Recent Engine with Tube Igniter. By T. M. Goodeve, M.A. Crown 8vo, 2s. 6d. cloth.

[*Just published.*]

"Like all Mr. Goodeve's writings, the present is no exception in point of general excellence. It is a valuable little volume."—*Mechanical World.*

Steam Engine Design.

A HANDBOOK ON THE STEAM ENGINE, with especial Reference to Small and Medium-sized Engines. For the Use of Engine-Makers, Mechanical Draughtsmen, Engineering Students and Users of Steam Power. By HERMAN HAEDER, C.E. English Edition, Re-edited by the Author from the Second German Edition, and Translated, with considerable Additions and Alterations, by H. H. P. POWLES, A.M.I.C.E., M.I.M.E. With nearly 1,100 Illustrations. Crown 8vo, 9s. cloth.

"A perfect encyclopædia of the steam engine and its details, and one which must take a permanent place in English drawing-offices and workshops."—*A Foreman Pattern-maker.*

"This is an excellent book, and should be in the hands of all who are interested in the construction and design of medium sized stationary engines.... A careful study of its contents and the arrangement of the sections leads to the conclusion that there is probably no other book like it in this country. The volume aims at showing the results of practical experience, and it certainly may claim a complete achievement of this idea."—*Nature.*

"There can be no question as to its value. We cordially commend it to all concerned in the design and construction of the steam engine."—*Mechanical World.*

Steam Boilers.

A TREATISE ON STEAM BOILERS: Their Strength, Construction, and Economical Working. By ROBERT WILSON, C.E. Fifth Edition. 12mo, 6s. cloth.

"The best treatise that has ever been published on steam boilers."—*Engineer.*

"The author shows himself perfect master of his subject, and we heartily recommend all employing steam power to possess themselves of the work."—*Ryland's Iron Trade Circular.*

Boiler Chimneys.

BOILER AND FACTORY CHIMNEYS: Their Draught-Power and Stability. With a Chapter on *Lightning Conductors.* By ROBERT WILSON, A.I.C.E., Author of "A Treatise on Steam Boilers," &c. Second Edition. Crown 8vo, 3s. 6d. cloth.

"A valuable contribution to the literature of scientific building."—*The Builder.*

Boiler Making.

THE BOILER-MAKER'S READY RECKONER & ASSISTANT. With Examples of Practical Geometry and Templating, for the Use of Platers, Smiths and Riveters. By JOHN COURTNEY, Edited by D. K. CLARK, M.I.C.E. Third Edition, 480 pp., with 140 Illusts. Fcap. 8vo, 7s. half-bound.

"No workman or apprentice should be without this book."—*Iron Trade Circular.*

Locomotive Engine Development.

THE LOCOMOTIVE ENGINE AND ITS DEVELOPMENT. A Popular Treatise on the Gradual Improvements made in Railway Engines between 1803 and 1893. By CLEMENT E. STRETTON, C.E., Author of "Safe Railway Working," &c. Second Edition, Revised and much Enlarged. With 95 Illustrations. Crown 8vo, 3s. 6d. cloth.

[*Just published.*]

"Students of railway history and all who are interested in the evolution of the modern locomotive will find much to attract and entertain in this volume."—*The Times.*

"The author of this work is well known to the railway world, and no one probably has a better knowledge of the history and development of the locomotive. The volume before us should be of value to all connected with the railway system of this country."—*Nature.*

Fire Engineering.

FIRES, FIRE-ENGINES, AND FIRE-BRIGADES. With a History of Fire-Engines, their Construction, Use, and Management; Remarks on Fire-Proof Buildings, and the Preservation of Life from Fire; Statistics of the Fire Appliances in English Towns; Foreign Fire Systems Hints on Fire-Brigades, &c. &c. By CHARLES F. T. YOUNG, C.E. With numerous Illustrations. 544 pp., demy 8vo, £1 4s. cloth.

"To those interested in the subject of fires and fire apparatus, we most heartily commend this book. It is the only English work we now have upon the subject."—*Engineering*.

"It displays much evidence of careful research; and Mr. Young has put his facts neatly together. His acquaintance with the practical details of the construction of steam fire engines, old and new, and the conditions with which it is necessary they should comply, is accurate and full."—*Engineer*.

Estimating for Engineering Work, &c.

ENGINEERING ESTIMATES, COSTS AND ACCOUNTS: A Guide to Commercial Engineering. With numerous Examples of Estimates and Costs of Millwright Work, Miscellaneous Productions, Steam Engines and Steam Boilers; and a Section on the Preparation of Costs Accounts. By A GENERAL MANAGER. Demy 8vo, 12s. cloth.

"This is an excellent and very useful book, covering subject-matter in constant requisition in every factory and workshop.... The book is invaluable, not only to the young engineer, but also to the estimate department of every works."—BUILDER.

"We accord the work unqualified praise. The information is given in a plain, straightforward manner, and bears throughout evidence of the intimate practical acquaintance of the author with every phase of commercial engineering."—MECHANICAL WORLD.

Engineering Construction.

PATTERN-MAKING: A Practical Treatise, embracing the Main Types of Engineering Construction, and including Gearing, both Hand and Machine made, Engine Work, Sheaves and Pulleys, Pipes and Columns, Screws, Machine Parts, Pumps and Cocks, the Moulding of Patterns in Loam and Greensand, &c., together with the methods of Estimating the weight of Castings; to which is added an Appendix of Tables for Workshop Reference. By A FOREMAN PATTERN MAKER. Second Edition, thoroughly Revised and much Enlarged. With upwards of 450 Illustrations. Crown 8vo, 7s. 6d. cloth.

[*Just published*.

"A well-written technical guide, evidently written by a man who understands and has practised what he has written about.... We cordially recommend it to engineering students, young journeymen, and others desirous of being initiated into the mysteries of pattern-making."—*Builder*.

"More than 450 illustrations help to explain the text, which is, however, always clear and explicit, thus rendering the work an excellent *vade mecum* for the apprentice who desires to become master of his trade."—*English Mechanic*.

Dictionary of Mechanical Engineering Terms.

LOCKWOOD'S DICTIONARY OF TERMS USED IN THE PRACTICE OF MECHANICAL ENGINEERING, embracing those current in the Drawing Office, Pattern Shop, Foundry, Fitting, Turning, Smith's and Boiler Shops, &c. &c. Comprising upwards of 6,000 Definitions. Edited by A FOREMAN PATTERN-MAKER, Author of "Pattern Making." Second Edition, Revised, with Additions. Crown 8vo, 7s. 6d. cloth.

"Just the sort of handy dictionary required by the various trades engaged in mechanical engineering. The practical engineering pupil will find the book of great value in his studies, and every foreman engineer and mechanic should have a copy."—*Building News*.

"Not merely a dictionary, but, to a certain extent, also a most valuable guide. It strikes us as a happy idea to combine with a definition of the phrase useful information on the subject of which it treats."—*Machinery Market*.

Mill Gearing.

TOOTHED GEARING: A Practical Handbook for Offices and Workshops. By A Foreman Pattern Maker, Author of "Pattern Making," "Lockwood's Dictionary of Mechanical Engineering Terms," &c. With 184 Illustrations. Crown 8vo, 6s. cloth.

[Just published.

Summary of Contents.

Chap. I. Principles.—II. Formation of Tooth Profiles.—III. Proportions of Teeth.—IV. Methods of Making Tooth Forms.—V. Involute Teeth.—VI. Some Special Tooth Forms.—VII. Bevel Wheels.—VIII. Screw Gears.—IX. Worm Gears.—X. Helical Wheels.—XI. Skew Bevels.—XII. Variable and other Gears.—XIII. Diametrical Pitch.—XIV. The Odontograph.—XV. Pattern Gears.—XVI. Machine Moulding Gears.—XVII. Machine Cut Gears.—XVIII. Proportion of Wheels.

"We must give the book our unqualified praise for its thoroughness of treatment, and we can heartily recommend it to all interested as the most practical book on the subject yet written.—*Mechanical World.*

Stone-working Machinery.

STONE-WORKING MACHINERY, and the Rapid and Economical Conversion of Stone. With Hints on the Arrangement and Management of Stone Works. By M. Powis Bale, M.I.M.E. With Illusts. Crown 8vo, 9s.

"The book should be in the hands of every mason or student of stone-work."—*Colliery Guardian.*

"A capital handbook for all who manipulate stone for building or ornamental purposes."—*Machinery Market.*

Pump Construction and Management.

PUMPS AND PUMPING: A Handbook for Pump Users. Being Notes on Selection, Construction and Management. By M. Powis Bale, M.I.M.E., Author of "Woodworking Machinery," "Saw Mills," &c. Second Edition, Revised. Crown 8vo, 2s. 6d. cloth.

"The matter is set forth as concisely as possible. In fact, condensation rather than diffuseness has been the author's aim throughout; yet he does not seem to have omitted anything likely to be of use."—*Journal of Gas Lighting.*

"Thoroughly practical and simply and clearly written."—*Glasgow Herald.*

Milling Machinery, etc.

MILLING MACHINES AND PROCESSES: A Practical Treatise on Shaping Metals by Rotary Cutters, including Information on Making and Grinding the Cutters. By Paul N. Hasluck, Author of "Lathe-work," "Handybooks for Handicrafts," &c. With upwards of 300 Engravings, including numerous Drawings by the Author. Large crown 8vo, 352 pages, 12s. 6d. cloth.

"A new departure in engineering literature.... We can recommend this work to all interested in milling machines; it is what it professes to be—a practical treatise."—*Engineer.*

"A capital and reliable book, which will no doubt be of considerable service, both to those who are already acquainted with the process as well as to those who contemplate its adoption."—*Industries.*

Turning.

LATHE-WORK: A Practical Treatise on the Tools, Appliances, and Processes employed in the Art of Turning. By PAUL N. HASLUCK. Fourth Edition, Revised and Enlarged. Cr. 8vo, 5s. cloth.

"Written by a man who knows, not only how work ought to be done, but who also knows how to do it, and how to convey his knowledge to others. To all turners this book would be valuable."—*Engineering.*

"We can safely recommend the work to young engineers. To the amateur it will simply be invaluable. To the student it will convey a great deal of useful information."—*Engineer.*

Screw-Cutting.

SCREW THREADS: And Methods of Producing Them. With Numerous Tables, and complete directions for using Screw-Cutting Lathes. By PAUL N. HASLUCK, Author of "Lathe-Work," &c. With Seventy-four Illustrations. Third Edition, Revised and Enlarged. Waistcoat-pocket size, 1s. 6d. cloth.

"Full of useful information, hints and practical criticism. Taps, dies and screwing-tools generally are illustrated and their action described."—*Mechanical World.*

"It is a complete compendium of all the details of the screw-cutting lathe; in fact a *multum in parvo* on all the subjects it treats upon."—*Carpenter and Builder.*

Smith's Tables for Mechanics, etc.

TABLES, MEMORANDA, AND CALCULATED RESULTS, FOR MECHANICS, ENGINEERS, ARCHITECTS, BUILDERS, etc. Selected and Arranged by FRANCIS SMITH. Fifth Edition, thoroughly Revised and Enlarged, with a New Section of ELECTRICAL TABLES, FORMULÆ, and MEMORANDA. Waistcoat-pocket size, 1s. 6d. limp leather.

"It would, perhaps, be as difficult to make a small pocket-book selection of notes and formulæ to suit ALL engineers as it would be to make a universal medicine; but Mr. Smith's waistcoat-pocket collection may be looked upon as a successful attempt."—*Engineer.*

"The best example we have ever seen of 270 pages of useful matter packed into the dimensions of a card-case."—*Building News.*

"A veritable pocket treasury of knowledge."—*Iron.*

French-English Glossary for Engineers, etc.

A POCKET GLOSSARY of TECHNICAL TERMS: ENGLISH-FRENCH, FRENCH-ENGLISH; with Tables suitable for the Architectural, Engineering, Manufacturing and Nautical Professions. By JOHN JAMES FLETCHER, Engineer and Surveyor. Second Edition, Revised and Enlarged, 200 pp. Waistcoat-pocket size, 1s. 6d. limp leather.

"It is a very great advantage for readers and correspondents in France and England to have so large a number of the words relating to engineering and manufacturers collected in a Liliputian volume. The little book will be useful both to students and travellers."—*Architect.*

"The glossary of terms is very complete, and many of the tables are new and well arranged. We cordially commend the book."—*Mechanical World.*

Year-Book of Engineering Formulæ, &c.

THE ENGINEER'S YEAR-BOOK FOR 1894. Comprising Formulæ, Rules, Tables, Data and Memoranda in Civil, Mechanical, Electrical, Marine and Mine Engineering. By H. R. KEMPE, A.M. Inst.C.E., M.I.E.E., Technical Officer of the Engineer-in-Chief's Office. General Post Office, London, Author of "A Handbook of Electrical Testing," "The Electrical Engineer's Pocket-Book," &c. With 700 Illustrations, specially Engraved for the work. Crown 8vo, 600 pages, 8s. leather.

[*Just published.*

"Represents an enormous quantity of work, and forms a desirable book of reference."—*The Engineer.*

"The book is distinctly in advance of most similar publications in this country."—*Engineering.*

"This valuable and well-designed book of reference meets the demands of all descriptions of engineers."—*Saturday Review.*

"Teems with up-to-date information in every branch of engineering and construction."—*Building News.*

"The needs of the engineering profession could hardly be supplied in a more admirable, complete and convenient form. To say that it more than sustains all comparisons is praise of the highest sort, and that may justly be said of it."—*Mining Journal.*

"There is certainly room for the new comer, which supplies explanations and directions, as well as formulæ and tables. It deserves to become one of the most successful of the technical annuals."—*Architect.*

"Brings together with great skill all the technical information which an engineer has to use day by day. It is in every way admirably equipped, and is sure to prove successful."—*Scotsman.*

"The up-to-dateness of Mr. Kempe's compilation is a quality that will not be lost on the busy people for whom the work is intended."—*Glasgow Herald.*

Portable Engines.

THE PORTABLE ENGINE; ITS CONSTRUCTION AND MANAGEMENT. A Practical Manual for Owners and Users of Steam Engines generally. By WILLIAM DYSON WANSBROUGH. With 90 Illustrations. Crown 8vo, 3s. 6d. cloth.

"This is a work of value to those who use steam machinery.... Should be read by everyone who has a steam engine, on a farm or elsewhere."—*Mark Lane Express.*

"We cordially commend this work to buyers and owners of steam engines, and to those who have to do with their construction or use."—*Timber Trades Journal.*

"Such a general knowledge of the steam engine as Mr. Wansbrough furnishes to the reader should be acquired by all intelligent owners and others who use the steam engine."—*Building News.*

"An excellent text-book of this useful form of engine. 'The Hints to Purchasers' contain a good deal of commonsense and practical wisdom."—*English Mechanic.*

Iron and Steel.

"IRON AND STEEL": A Work for the Forge, Foundry, Factory, and Office. Containing ready, useful, and trustworthy Information for Iron-masters and their Stock-takers; Managers of Bar, Rail, Plate, and Sheet Rolling Mills: Iron and Metal Founders; Iron Ship and Bridge Builders; Mechanical, Mining, and Consulting Engineers; Architects, Contractors, Builders, and Professional Draughtsmen. By CHARLES HOARE, Author of "The Slide Rule," &c. Eighth Edition, Revised throughout and considerably Enlarged. 32mo. 6s. leather.

"For comprehensiveness the book has not its equal."—*Iron.*

"One of the best of the pocket books."—*English Mechanic.*

"We cordially recommend this book to those engaged in considering the details of all kinds of iron and steel works."—*Naval Science.*

Elementary Mechanics.

CONDENSED MECHANICS. A Selection of Formulæ, Rules, Tables, and Data for the Use of Engineering Students, Science Classes, &c. In Accordance with the Requirements of the Science and Art Department. By W. G. CRAWFORD HUGHES, A.M.I.C.E. Crown 8vo, 2s. 6d. cloth.

"The book is well fitted for those who are either confronted with practical problems in their work, or are preparing for examination and wish to refresh their knowledge by going through their formulæ again."—*Marine Engineer.*

"It is well arranged, and meets the wants of those for whom it is intended."—*Railway News.*

Steam.

THE SAFE USE OF STEAM. Containing Rules for Unprofessional Steam-users. By an ENGINEER. Sixth Edition. Sewed, 6d.

"If steam-users would but learn this little book by heart, boiler explosions would become sensations by their rarity." — *English Mechanic.*

Warming.

HEATING BY HOT WATER: with Information and Suggestions on the best Methods of Heating Public, Private and Horticultural Buildings. By WALTER JONES. Second Edition. With 96 Illustrations. Crown 8vo, 2s. 6d. net.

"We confidently recommend all interested in heating by hot water to secure a copy of this valuable little treatise." — *The Plumber and Decorator.*

THE POPULAR WORKS OF MICHAEL REYNOLDS
("THE ENGINE DRIVER'S FRIEND").

Locomotive-Engine Driving.

LOCOMOTIVE-ENGINE DRIVING: A Practical Manual for Engineers in charge of Locomotive Engines. By MICHAEL REYNOLDS, Member of the Society of Engineers, formerly Locomotive Inspector L. B. and S. C. R. Ninth Edition. Including a KEY TO THE LOCOMOTIVE ENGINE. With Illustrations and Portrait of Author. Crown 8vo. 4s. 6d. cloth.

"Mr. Reynolds has supplied a want, and has supplied it well. We can confidently recommend the book, not only to the practical driver, but to everyone who takes an interest in the performance of locomotive engines." — *The Engineer.*

"Mr. Reynolds has opened a new chapter in the literature of the day. This admirable practical treatise, of the practical utility of which we have to speak in terms of warm commendation." — *Athenæum.*

"Evidently the work of one who knows his subject thoroughly." — *Railway Service Gazette.*

"Were the cautions and rules given in the book to become part of the every-day working of our engine-drivers, we might have fewer distressing accidents to deplore." — *Scotsman.*

Stationary Engine Driving.

STATIONARY ENGINE DRIVING: A Practical Manual for Engineers in charge of Stationary Engines. By MICHAEL REYNOLDS. Fifth Edition, Enlarged. With Plates and Woodcuts. Crown 8vo, 4s. 6d. cloth.

"The author is thoroughly acquainted with his subjects, and his advice on the various points treated is clear and practical.... He has produced a manual which is an exceedingly useful one for the class for whom it is specially intended." — *Engineering.*

"Our author leaves no stone unturned. He is determined that his readers shall not only know something about the stationary engine, but all about it." — *Engineer.*

"An engineman who has mastered the contents of Mr.Reynolds's book will require but little actual experience with boilers and engines before he can be trusted to look after them." — *English Mechanic.*

The Engineer, Fireman, and Engine-Boy.

THE MODEL LOCOMOTIVE ENGINEER, FIREMAN, and ENGINE-BOY. Comprising a Historical Notice of the Pioneer Locomotive Engines and their Inventors. By MICHAEL REYNOLDS. With numerous Illustrations and a fine Portrait of George Stephenson. Crown 8vo, 4s. 6d. cloth.

"From the technical knowledge of the author it will appeal to the railway man of to-day more forcibly than anything written by Dr. Smiles.... The volume contains information of a technical kind, and facts that every driver should be familiar with." — *English Mechanic.*

"We should be glad to see this book in the possession of everyone in the kingdom who has ever laid, or is to lay, hands on a locomotive engine." — *Iron.*

Continuous Railway Brakes.

CONTINUOUS RAILWAY BRAKES: *A Practical Treatise on the several Systems in Use in the United Kingdom; their Construction and Performance.* With copious Illustrations and numerous Tables. By MICHAEL REYNOLDS. Large crown 8vo, 9s. cloth.

"A popular explanation of the different brakes. It will be of great assistance in forming public opinion, and will be studied with benefit by those who take an interest in the brake."—*English Mechanic.*

"Written with sufficient technical detail to enable the principle and relative connection of the various parts of each particular brake to be readily grasped."—*Mechanical World.*

Engine-Driving Life.

ENGINE-DRIVING LIFE: *Stirring Adventures and Incidents in the Lives of Locomotive-Engine Drivers.* By MICHAEL REYNOLDS. Third and Cheaper Edition. Crown 8vo, 1s. 6d. cloth.

[*Just published.*

"From first to last perfectly fascinating. Wilkie Collins's most thrilling conceptions are thrown into the shade by true incidents, endless in their variety, related in every page."—*North British Mail.*

"Anyone who wishes to get a real insight into railway life cannot do better than read 'Engine-Driving Life' for himself; and if he once take it up he will find that the author's enthusiasm and real love of the engine-driving profession will carry him on till he has read every page."—*Saturday Review.*

Pocket Companion for Enginemen.

THE ENGINEMAN'S POCKET COMPANION AND PRACTICAL EDUCATOR FOR ENGINEMEN, BOILER ATTENDANTS, AND MECHANICS. By MICHAEL REYNOLDS. With Forty-five Illustrations and numerous Diagrams. Third Edition, Revised. Royal 18mo, 3s. 6d., strongly bound for pocket wear.

"This admirable work is well suited to accomplish its object, being the honest workmanship of a competent engineer."—*Glasgow Herald.*

"A most meritorious work, giving in a succinct and practical form all the information an engine-minder desirous of mastering the scientific principles of his daily calling would require."—*The Miller.*

"A boon to those who are striving to become efficient mechanics."—*Daily Chronicle.*

CIVIL ENGINEERING, SURVEYING, etc.

MR. HUMBER'S VALUABLE ENGINEERING BOOKS.

The Water Supply of Cities and Towns.

A COMPREHENSIVE TREATISE on the WATER-SUPPLY OF CITIES AND TOWNS. By WILLIAM HUMBER, A-M.Inst.C.E., and M. Inst. M.E., Author of "Cast and Wrought Iron Bridge Construction," &c. &c. Illustrated with 50 Double Plates, 1 Single Plate, Coloured Frontispiece, and upwards of 250 Woodcuts, and containing 400 pages of Text. Imp. 4to, £6 6s. elegantly and substantially half-bound in morocco.

List of Contents.

I. Historical Sketch of some of the means that have been adopted for the Supply of Water to Cities and Towns.—II. Water and the Foreign Matter usually associated with it.—III. Rainfall and Evaporation.—IV. Springs and the water-bearing formations of various districts.—V. Measurement and Estimation of the flow of Water.—VI. On the Selection of the Source of Supply.—VII. Wells.—VIII. Reservoirs.—IX. The Purification of Water.—X. Pumps.—XI. Pumping Machinery.—XII. Conduits.—XIII. Distribution of Water.—XIV. Meters, Service Pipes, and House Fittings.—XV. The Law and Economy of Water Works.—XVI. Constant and Intermittent Supply.—XVII. Description of Plates.—Appendices, giving Tables of Rates of Supply, Velocities, &c. &c., together with Specifications of several Works illustrated, among which will be found: Aberdeen, Bideford, Canterbury, Dundee, Halifax, Lambeth, Rotherham, Dublin, and others.

"The most systematic and valuable work upon water supply hitherto produced in English, or in any other language.... Mr. Humber's work is characterised almost throughout by an exhaustiveness much more distinctive of French and German than of English technical treatises."—*Engineer.*

"We can congratulate Mr. Humber on having been able to give so large an amount of Information on a subject so important as the water supply of cities and towns. The plates, fifty in number, are mostly drawings of executed works, and alone would have commanded the attention of every engineer whose practice may lie in this branch of the profession."—*Builder.*

Cast and Wrought Iron Bridge Construction.

A COMPLETE AND PRACTICAL TREATISE ON CAST AND WROUGHT IRON BRIDGE CONSTRUCTION, including Iron Foundations. In Three Parts—Theoretical, Practical, and Descriptive. By WILLIAM HUMBER, A.M.Inst.C.E., and M.Inst.M.E. Third Edition, Revised and much improved, with 115 Double Plates (20 of which now first appear in this edition), and numerous Additions to the Text. In Two Vols., imp. 4to, £6 16s. 6d. half-bound in morocco.

"A very valuable contribution to the standard literature of civil engineering. In addition to elevations, plans and sections, large scale details are given which very much enhance the instructive worth of those illustrations."—*Civil Engineer and Architect's Journal.*

"Mr. Humber's stately volumes, lately issued—in which the most important bridges erected during the last five years, under the direction of the late Mr. Brunel, Sir W. Cubitt, Mr. Hawkshaw, Mr. Page, Mr. Fowler, Mr. Hemans, and others among our most eminent engineers, are drawn and specified in great detail."—*Engineer.*

Strains, Calculation of.

A HANDY BOOK FOR THE CALCULATION OF STRAINS IN GIRDERS AND SIMILAR STRUCTURES, AND THEIR STRENGTH. Consisting of Formulæ and Corresponding Diagrams, with numerous details for Practical Application, &c. By WILLIAM HUMBER, A-M.Inst.C.E., &c. Fifth Edition. Crown 8vo, nearly 100 Woodcuts and 3 Plates, 7s. 6d. cloth.

"The formulae are neatly expressed, and the diagrams good."—*Athenæum.*

"We heartily commend this really *handy* book to our engineer and architect readers."—*English Mechanic.*

Barlow's Strength of Materials, enlarged by Humber.

A TREATISE ON THE STRENGTH OF MATERIALS: with Rules for Application in Architecture, the Construction of Suspension Bridges, Railways, &c. By PETER BARLOW, F.R.S. A New Edition, Revised by his Sons, P. W. BARLOW, F.R.S., and W. H. BARLOW, F.R.S.; to which are added, Experiments by HODGKINSON, FAIRBAIRN, and KIRKALDY; and Formulæ for Calculating Girders, &c. Arranged and Edited by WM. HUMBER, A-M.Inst.C.E. Demy 8vo, 400 pp., with 19 large Plates and numerous Woodcuts, 18s. cloth.

"Valuable alike to the student, tyro, and the experienced practitioner, it will always rank future, as it has hitherto done, as the standard treatise on that particular subject."—*Engineer.*

"There is no greater authority than Barlow."—*Building News.*

"As a scientific work of the first class, it deserves a foremost place on the bookshelves of every civil engineer and practical mechanic."—*English Mechanic.*

MR. HUMBER'S GREAT WORK ON MODERN ENGINEERING.

Complete in Four Volumes, imperial 4to, price £12 12s., half-morocco. Each Volume sold separately as follows:—

A RECORD OF THE PROGRESS OF MODERN ENGINEERING. FIRST SERIES. Comprising Civil, Mechanical, Marine, Hydraulic, Railway, Bridge, and other Engineering Works, &c. By WILLIAM HUMBER, A-M.Inst.C.E., &c. Imp. 4to, with 36 Double Plates, drawn to a large scale, Photographic Portrait of John Hawkshaw, C.E., F.R.S., &c., and copious descriptive Letterpress, Specifications, &c., £3 3s. half-morocco.

List of the Plates and Diagrams.

Victoria Station and Roof, L. B. & S. C. R. (3 plates); Southport Pier (2 plates); Victoria Station and Roof, L. C. & D. and G. W. R. (6 plates); Roof of Cremorne Music Hall; Bridge over G. N. Railway; Roof of Station, Dutch Rhenish Rail (2 plates); Bridge over the Thames, West London Extension Railway (5 plates); Armour Plates: Suspension Bridge, Thames (4 plates); The Allen Engine; Suspension Bridge, Avon (3 plates); Underground Railway (3 plates).

"Handsomely lithographed and printed. It will find favour with many who desire to preserve in a permanent form copies of the plans and specifications prepared for the guidance of the contractors for many important engineering works."—*Engineer.*

HUMBER'S PROGRESS OF MODERN ENGINEERING. SECOND SERIES. Imp. 4to, with 36 Double Plates, Photographic Portrait of Robert Stephenson, C.E., M.P., F.R.S., &c., and copious descriptive Letterpress, Specifications, &c., £3 3s. half-morocco.

List of the Plates and Diagrams.

Birkenhead Docks, Low Water Basin (15 plates); Charing Cross Station Roof, C. C. Railway (3 plates); Digswell Viaduct, Great Northern Railway; Robbery Wood Viaduct, Great Northern Railway; Iron Permanent Way; Clydach Viaduct, Merthyr, Tredegar, and Abergavenny Railway; Ebbw Viaduct, Merthyr, Tredegar, and Abergavenny Railway; College Wood Viaduct, Cornwall Railway; Dublin Winter Palace Roof (3 plates); Bridge over the Thames, L. C. & D. Railway (6 plates); Albert Harbour, Greenock (4 plates).

"Mr. Humber has done the profession good and true service, by the fine collection of examples he has here brought before the profession and the public."—*Practical Mechanic's Journal.*

HUMBER'S PROGRESS OF MODERN ENGINEERING. THIRD SERIES. Imp. 4to, with 40 Double Plates, Photographic Portrait of J. R. M'Clean, late Pres. Inst. C.E., and copious descriptive Letterpress, Specifications, &c., £3 3s. half-morocco.

List of the Plates and Diagrams.

MAIN DRAINAGE, METROPOLIS.—*North Side.*— Map showing Interception of Sewers; Middle Level Sewer (2 plates); Outfall Sewer, Bridge over River Lea (3 plates); Outfall Sewer, Bridge over Marsh Lane, North Woolwich Railway, and Bow and Barking Railway Junction; Outfall Sewer, Bridge over Bow and Barking Railway (3 plates); Outfall Sewer, Bridge over East London Waterworks' Feeder (2 plates); Outfall Sewer, Reservoir (2 plates); Outfall Sewer, Tumbling Bay and Outlet; Outfall Sewer, Penstocks. *South Side.*— Outfall Sewer, Bermondsey Branch (2 plates); Outfall Sewer, Reservoir and Outlet (4 plates); Outfall Sewer, Filth Hoist; Sections of Sewers (North and South Sides).

THAMES EMBANKMENT.—Section of River Wall; Steamboat Pier, Westminster (2 plates); Landing Stairs between Charing Cross and Waterloo Bridges; York Gate (2 plates); Overflow and Outlet at Savoy Street Sewer (3 plates); Steamboat Pier, Waterloo Bridge (3 plates); Junction of Sewers, Plans and Sections; Gullies, Plans and Sections; Rolling Stock; Granite and Iron Forts.

"The drawings have a constantly increasing value, and whoever desires to possess clear representations of the two great works carried out by our Metropolitan Board will obtain Mr. Humber's volume."—*Engineer.*

HUMBER'S PROGRESS OF MODERN ENGINEERING. FOURTH SERIES. Imp. 4to, with 36 Double Plates, Photographic Portrait of John Fowler, late Pres. Inst. C.E., and copious descriptive Letterpress, Specifications, &c., £3 3s. half-morocco.

List of the Plates and Diagrams.

Abbey Mills Pumping Station, Main Drainage, Metropolis (4 plates); Barrow Docks (5 plates); Manquis Viaduct, Santiago and Valparaiso Railway (2 plates); Adams Locomotive, St. Helen's Canal Railway (2 plates); Cannon Street Station Roof, Charing Cross Railway (3 plates); Road Bridge over the River Moka (2 plates); Telegraphic Apparatus for Mesopotamia; Viaduct over the River Wye, Midland Railway (3 plates); St. Germans Viaduct, Cornwall Railway (2 plates); Wrought-Iron Cylinder for Diving Bell; Millwall Docks (6 plates); Milroy's Patent Excavator; Metropolitan District Railway (6 plates); Harbours, Ports, and Breakwaters (3 plates).

"We gladly welcome another year's issue of this valuable publication from the able pen of Mr. Humber. The accuracy and general excellence of this work are well known, while its usefulness in giving the measurements and details of some of the latest examples of engineering, as carried out by the most eminent men in the profession, cannot be too highly prized."—*Artisan.*

Statics, Graphic and Analytic.

GRAPHIC AND ANALYTIC STATICS, in their Practical Application to the Treatment of Stresses in Roofs, Solid Girders, Lattice, Bowstring and Suspension Bridges, Braced Iron Arches and Piers, and other Frameworks. By R. HUDSON GRAHAM, C.E. Containing Diagrams and Plates to Scale. With numerous Examples, many taken from existing Structures. Specially arranged for Class-work in Colleges and Universities. Second Edition, Revised and Enlarged. 8vo, 16s. cloth.

"Mr. Graham's book will find a place wherever graphic and analytic statics are used or studied."—*Engineer.*

"The work is excellent from a practical point of view, and has evidently been prepared with much care. The directions for working are ample, and are illustrated by an abundance of well-selected examples. It is an excellent text-book for the practical draughtsman."—*Athenæum.*

Practical Mathematics.

MATHEMATICS FOR PRACTICAL MEN: Being a Commonplace Book of Pure and Mixed Mathematics. Designed chiefly for the use of Civil Engineers, Architects and Surveyors. By OLINTHUS GREGORY, L.L.D., F.R.A.S., Enlarged by HENRY LAW, C.E. 4th Edition, carefully Revised by J. R. YOUNG, formerly Professor of Mathematics, Belfast College. With 13 Plates. 8vo, £1 1s. cloth.

"The engineer or architect will here find ready to his hand rules for solving nearly every mathematical difficulty that may arise in his practice. The rules are in all cases explained by means of examples, in which every step of the process is clearly worked out."—*Builder.*

"One of the most serviceable books for practical mechanics.... It is an instructive book for the student, and a text-book for him who, having once mastered the subjects it treats of, needs occasionally to refresh his memory upon them."—*Building News.*

Hydraulic Tables.

HYDRAULIC TABLES, CO-EFFICIENTS, and FORMULÆ for finding the Discharge of Water from Orifices, Notches, Weirs, Pipes, and Rivers. With New Formulæ, Tables, and General Information on Rainfall, Catchment-Basins, Drainage, Sewerage, Water Supply for Towns and Mill Power. By JOHN NEVILLE, Civil Engineer, M.R.I.A. Third Ed., carefully Revised, with considerable Additions. Numerous Illusts. Cr. 8vo, 14s. cloth.

"Alike valuable to students and engineers in practice; its study will prevent the annoyance of avoidable failures, and assist them to select the readiest means of successfully carrying out any given work connected with hydraulic engineering."—*Mining Journal.*

"It is, of all English books on the subject, the one nearest to completeness.... From the good arrangement of the matter, the clear explanations, and abundance of formulæ, the carefully calculated tables, and, above all, the thorough acquaintance with both theory and construction, which is displayed from first to last, the book will be found to be an acquisition."—*Architect.*

Hydraulics.

HYDRAULIC MANUAL. Consisting of Working Tables and Explanatory Text. Intended as a Guide in Hydraulic Calculations and Field Operations. By LOWIS D'A. JACKSON, Author of "Aid to Survey Practice," "Modern Metrology," &c. Fourth Edition, Enlarged. Large cr. 8vo, 16s. cl.

"The author has had a wide experience in hydraulic engineering and has been a careful observer of the facts which have come under his notice, and from the great mass of material at his command he has constructed a manual which may be accepted as a trustworthy guide to this branch of the engineer's profession. We can heartily recommend this volume to all who desire to be acquainted with the latest development of this important subject."—*Engineering.*

"The standard-work in this department of mechanics."—*Scotsman.*

"The most useful feature of this work is its freedom from what is superannuated, and its thorough adoption of recent experiments; the text is, in fact, in great part a short account of the great modern experiments."—*Nature.*

Drainage.

ON THE DRAINAGE OF LANDS, TOWNS, AND BUILDINGS. By G. D. DEMPSEY, C.E., Author of "The Practical Railway Engineer," &c. Revised, with large Additions on RECENT PRACTICE IN DRAINAGE ENGINEERING, by D. KINNEAR CLARK, M.Inst.C.E. Author of "Tramways: Their Construction and Working," "A Manual of Rules, Tables, and Data for Mechanical Engineers," &c. Second Edition, Corrected. Fcap. 8vo, 5s. cloth.

"The new matter added to Mr. Dempsey's excellent work is characterised by the comprehensive grasp and accuracy of detail for which the name of Mr. D. K. Clark is a sufficient voucher."—*Athenæum*.

"As a work on recent practice in drainage engineering, the book is to be commended to all who are making that branch of engineering science their special study."—*Iron*.

"A comprehensive manual on drainage engineering, and a useful introduction to the student."—*Building News*.

Water Storage, Conveyance, and Utilisation.

WATER ENGINEERING: A Practical Treatise on the Measurement, Storage, Conveyance, and Utilisation of Water for the Supply of Towns, for Mill Power, and for other Purposes. By CHARLES SLAGG, Water and Drainage Engineer, A.M.Inst.C.E., Author of "Sanitary Work in the Smaller Towns, and in Villages," &c. With numerous Illusts. Cr. 8vo. 7s. 6d. cloth.

"As a small practical treatise on the water supply of towns, and on some applications of water-power, the work is in many respects excellent."—*Engineering*.

"The author has collated the results deduced from the experiments of the most eminent authorities, and has presented them in a compact and practical form, accompanied by very clear and detailed explanations.... The application of water as a motive power is treated very carefully and exhaustively."—*Builder*.

"For anyone who desires to begin the study of hydraulics with a consideration of the practical applications of the science there is no better guide."—*Architect*.

River Engineering.

RIVER BARS: The Causes of their Formation, and their Treatment by "Induced Tidal Scour;" with a Description of the Successful Reduction by this Method of the Bar at Dublin. By I. J. MANN, Assist. Eng. to the Dublin Port and Docks Board. Royal 8vo, 7s. 6d. cloth.

"We recommend all interested in harbour works—and, indeed, those concerned in the improvements of rivers generally—to read Mr. Mann's interesting work on the treatment of river bars."—*Engineer*.

Trusses.

TRUSSES OF WOOD AND IRON. Practical Applications of Science in Determining the Stresses, Breaking Weights, Safe Loads, Scantlings, and Details of Construction, with Complete Working Drawings. By WILLIAM GRIFFITHS, Surveyor, Assistant Master, Tranmere School of Science and Art. Oblong 8vo, 4s. 6d. cloth.

"This handy little book enters so minutely into every detail connected with the construction of roof trusses, that no student need be ignorant of these matters."—*Practical Engineer*.

Railway Working.

SAFE RAILWAY WORKING. A Treatise on Railway Accidents: Their Cause and Prevention; with a Description of Modern Appliances and Systems. By CLEMENT E. STRETTON, C.E., Vice-President and Consulting Engineer, Amalgamated Society of Railway Servants. With Illustrations and Coloured Plates. Third Edition, Enlarged. Crown 8vo, 3s. 6d. cloth.

"A book for the engineer, the directors, the managers; and, in short, all who wish for information on railway matters will find a perfect encyclopædia in 'Safe Railway Working.'"—*Railway Review*.

"We commend the remarks on railway signalling to all railway managers, especially where a uniform code and practice is advocated."—*Herepath's Railway Journal*.

"The author may be congratulated on having collected, in a very convenient form, much valuable information on the principal questions affecting the safe working of railways."—*Railway Engineer.*

Oblique Bridges.

A PRACTICAL AND THEORETICAL ESSAY ON OBLIQUE BRIDGES. With 13 large Plates. By the late GEORGE WATSON BUCK, M.I.C.E. Third Edition, revised by his Son, J. H. WATSON BUCK, M.I.C.E.; and with the addition of Description to Diagrams for Facilitating the Construction of Oblique Bridges, by W. H. BARLOW, M.I.C.E. Royal 8vo, 12s. cloth.

"The standard text-book for all engineers regarding skew arches is Mr. Buck's treatise, and it would be impossible to consult a better."—*Engineer.*

"Mr. Buck's treatise is recognised as a standard text-book, and his treatment has divested the subject of many of the intricacies supposed to belong to it. As a guide to the engineer and architect, on a confessedly difficult subject, Mr. Buck's work is unsurpassed."—*Building News.*

Tunnel Shafts.

THE CONSTRUCTION OF LARGE TUNNEL SHAFTS: A Practical and Theoretical Essay. By J. H. WATSON BUCK, M.Inst.C.E., Resident Engineer, London and North-Western Railway. Illustrated with Folding Plates. Royal 8vo, 12s. cloth.

"Many of the methods given are of extreme practical value to the mason; and the observations on the form of arch, the rules for ordering the stone, and the construction of the templates will be found of considerable use. We commend the book to the engineering profession."—*Building News.*

"Will be regarded by civil engineers as of the utmost value, and calculated to save much time and obviate many mistakes."—*Colliery Guardian.*

Student's Text-Book on Surveying.

PRACTICAL SURVEYING: A Text-Book for Students preparing for Examination or for Survey-work in the Colonies. By GEORGE W. USILL, A.M.I.C.E., Author of "The Statistics of the Water Supply of Great Britain." With Four Lithographic Plates and upwards of 330 Illustrations. Third Edition, Revised and Enlarged. Including Tables of Natural Sines, Tangents, Secants, &c. Crown 8vo, 7s. 6d. cloth; or, on THIN PAPER, bound in limp leather, gilt edges, rounded corners, for pocket use, 12s. 6d.

"The best forms of instruments are described as to their construction, uses and modes of employment, and there are innumerable hints on work and equipment such as the author, in his experience as surveyor, draughtsman, and teacher, has found necessary, and which the student in his inexperience will find most serviceable."—*Engineer.*

"The latest treatise in the English language on surveying, and we have no hesitation in saying that the student will find it a better guide than any of its predecessors.... Deserves to be recognised as the first book which should be put in the hands of a pupil of Civil Engineering, and every gentleman of education who sets out for the Colonies would find it well to have a copy."—*Architect.*

Survey Practice.

AID TO SURVEY PRACTICE, for Reference in Surveying, Levelling, and Setting-out; and in Route Surveys of Travellers by Land and Sea. With Tables, Illustrations, and Records. By LOWIS D'A. JACKSON, A.M.I.C.E., Author of "Hydraulic Manual," "Modern Metrology," &c. Second Edition, Enlarged. Large crown 8vo, 12s. 6d. cloth.

"A valuable *vade-mecum* for the surveyor. We can recommend this book as containing an admirable supplement to the teaching of the accomplished surveyor."—*Athenæum.*

"As a text-book we should advise all surveyors to place it in their libraries, and study well the matured instructions afforded in its pages."—*Colliery Guardian.*

"The author brings to his work a fortunate union of theory and practical experience which, aided by a clear and lucid style of writing, renders the book a very useful one."—*Builder.*

Surveying, Land and Marine.

LAND AND MARINE SURVEYING, in Reference to the Preparation of

Plans for Roads and Railways; Canals, Rivers, Towns' Water Supplies; Docks and Harbours. With Description and Use of Surveying Instruments. By W. D. HASKOLL, C.E., Author of "Bridge and Viaduct Construction," &c. Second Edition, Revised, with Additions. Large cr. 8vo, 9s. cl.

"This book must prove of great value to the student. We have no hesitation in recommending it, feeling assured that it will more than repay a careful study."—*Mechanical World.*

"A most useful and well arranged book. We can strongly recommend it as a carefully-written and valuable text-book. It enjoys a well-deserved repute among surveyors."—*Builder.*

"This volume cannot fail to prove of the utmost practical utility. It may be safely recommended to all students who aspire to become clean and expert surveyors."—*Mining Journal.*

Field-Book for Engineers.

THE ENGINEER'S, MINING SURVEYOR'S, AND CONTRACTOR'S FIELD-BOOK. Consisting of a Series of Tables, with Rules, Explanations of Systems, and use of Theodolite for Traverse Surveying and Plotting the Work with minute accuracy by means of Straight Edge and Set Square only; Levelling with the Theodolite, Casting-out and Reducing Levels to Datum, and Plotting Sections in the ordinary manner; setting-out Curves with the Theodolite by Tangential Angles and Multiples, with Right and Left-hand Readings of the Instrument: Setting-out Curves without Theodolite, on the System of Tangential Angles by sets of Tangents and Offsets; and Earthwork Tables to 80 feet deep, calculated for every 6 inches in depth. By W. D. HASKOLL, C.E. Fourth Edition. Crown 8vo, 12s. cloth.

"The book is very handy; the separate tables of sines and tangents to every minute will make it useful for many other purposes, the genuine traverse tables existing all the same."—*Athenæum.*

"Every person engaged in engineering field operations will estimate the importance of such a work and the amount of valuable time which will be saved by reference to a set of reliable tables prepared with the accuracy and fulness of those given in this volume."—*Railway News.*

Levelling.

A TREATISE ON THE PRINCIPLES AND PRACTICE OF LEVELLING. Showing its Application to purposes of Railway and Civil Engineering, in the Construction of Roads; with Mr. TELFORD'S Rules for the same. By FREDERICK W. SIMMS, F.G.S., M.Inst.C.E. Seventh Edition, with the addition of Law's Practical Examples for Setting-out Railway Curves, and TRAUTWINE'S Field Practice of Laying-out Circular Curves. With 7 Plates and numerous Woodcuts. 8vo, 8s. 6d. cloth. ⁂ TRAUTWINE on Curves may be had separate, 5s.

"The text-book on levelling in most of our engineering schools and colleges."—*Engineer.*

"The publishers have rendered a substantial service to the profession, especially to the younger members, by bringing out the present edition of Mr. Simms's useful work.—*Engineering.*

Trigonometrical Surveying.

AN OUTLINE OF THE METHOD OF CONDUCTING A TRIGONOMETRICAL SURVEY, for the Formation of Geographical and Topographical Maps and Plans, Military Reconnaissance, Levelling, &c., with Useful Problems, Formulæ, and Tables. By Lieut.-General FROME, R.E. Fourth Edition, Revised and partly Re-written by Major-General Sir CHARLES WARREN, G.C.M.G., R.E. With 19 Plates and 115 Woodcuts. Royal 8vo, 16s. cloth.

"The simple fact that a fourth edition has been called for is the best testimony to its merits. No words of praise from us can strengthen the position so well and so steadily maintained by this work. Sir Charles Warren has revised the entire work, and made such additions as were necessary to bring every portion of the contents up to the present date."—*Broad Arrow.*

Field Fortification.

A TREATISE ON FIELD FORTIFICATION, THE ATTACK OF FORTRESSES, MILITARY MINING, AND RECONNOITRING. By Colonel I. S. MACAULAY, late Professor of Fortification in the R.M.A., Woolwich. Sixth Edition. Crown 8vo, with separate Atlas of 12 Plates, 12s. cloth.

Tunnelling.

PRACTICAL TUNNELLING. Explaining in detail the Setting-out of the works, Shaft-sinking and Heading-driving, Ranging the Lines and Levelling underground, Sub-Excavating, Timbering, and the Construction of the Brickwork of Tunnels, with the amount of Labour required for, and the Cost of, the various portions of the work. By FREDERICK W. SIMMS, F.G.S., M.Inst.C.E. Third Edition, Revised and Extended by D. KINNEAR CLARK, M.Inst.C.E. Imperial 8vo, with 21 Folding Plates and numerous Wood Engravings, 30s. cloth.

"The estimation in which Mr. Simms's book on tunnelling has been held for over thirty years cannot be more truly expressed than in the words of the late Prof. Rankine: — 'The best source of information on the subject of tunnels is Mr. F. W. Simms's work on Practical Tunnelling.'" — *Architect.*

"It has been regarded from the first as a text-book of the subject.... Mr. Clark has added immensely to the value of the book." — *Engineer.*

Tramways and their Working.

TRAMWAYS: THEIR CONSTRUCTION AND WORKING. Embracing a Comprehensive History of the System; with an exhaustive Analysis of the various Modes of Traction, including Horse-Power, Steam, Cable Traction, Electric Traction, &c.; a Description of the Varieties of Rolling Stock; and ample Details of Cost and Working Expenses. New Edition, Thoroughly Revised, and Including the Progress recently made in Tramway Construction, &c. &c. By D. KINNEAR CLARK, M.Inst.C.E. With numerous Illustrations and Folding Plates. In One Volume, 8vo, 700 pages, price about 25s.

[*Nearly ready.*]

"All interested in tramways must refer to it, as all railway engineers have turned to the author's work 'Railway Machinery.'" — *Engineer.*

"An exhaustive and practical work on tramways, in which the history of this kind of locomotion, and a description and cost of the various modes of laying tramways, are to be found." — *Building News.*

"The best form of rails, the best mode of construction, and the best mechanical appliances are so fairly indicated in the work under review, that any engineer about to construct a tramway will be enabled at once to obtain the practical information which will be of most service to him." — *Athenæum.*

Curves, Tables for Setting-out.

TABLES OF TANGENTIAL ANGLES AND MULTIPLES *for Setting-out Curves from 5 to 200 Radius.* By ALEXANDER BEAZELEY, M.Inst.C.E. Fourth Edition. Printed on 48 Cards, and sold in a cloth box, waistcoat-pocket size, 3s. 6d.

"Each table is printed on a small card, which, being placed on the theodolite, leaves the hands free to manipulate the instrument—no small advantage as regards the rapidity of work." — *Engineer.*

"Very handy; a man may know that all his day's work must fall on two of these cards, which he puts into his own card-case, and leaves the rest behind." — *Athenæum.*

Earthwork.

EARTHWORK TABLES. Showing the Contents in Cubic Yards of Embankments, Cuttings, &c., of Heights or Depths up to an average of 80 feet. By JOSEPH BROADBENT, C.E., and FRANCIS CAMPIN, C.E. Crown 8vo, 5s. cloth.

"The way in which accuracy is attained, by a simple division of each cross section into three elements, two in which are constant and one variable, is ingenious."—*Athenæum*.

Heat, Expansion by.

EXPANSION OF STRUCTURES BY HEAT. By JOHN KEILY, C.E., late of the Indian Public Works and Victorian Railway Departments. Crown 8vo, 3s. 6d. cloth.

SUMMARY OF CONTENTS.

Section I. FORMULAS AND DATA.
Section II. METAL BARS.
Section III. SIMPLE FRAMES.
Section IV. COMPLEX FRAMES AND PLATES.
Section V. THERMAL CONDUCTIVITY.
Section VI. MECHANICAL FORCE OF HEAT.
Section VII. WORK OF EXPANSION AND CONTRACTION.
Section VIII. SUSPENSION BRIDGES.
Section IX. MASONRY STRUCTURES.

"The aim the author has set before him, viz., to show the effects of heat upon metallic and other structures, is a laudable one, for this is a branch of physics upon which the engineer or architect can find but little reliable and comprehensive data in books."—*Builder.*

"Whoever is concerned to know the effect of changes of temperature on such structures as suspension bridges and the like, could not do better than consult Mr. Keily's valuable and handy exposition of the geometrical principles involved in these changes."—*Scotsman.*

Earthwork, Measurement of.

A MANUAL ON EARTHWORK. By ALEX. J. S. GRAHAM, C.E. With numerous Diagrams. Second Edition. 18mo, 2s. 6d. cloth.

"A great amount of practical information, very admirably arranged, and available for rough estimates, as well as for the more exact calculations required in the engineer's and contractor's offices."—*Artizan.*

Strains in Ironwork.

THE STRAINS ON STRUCTURES OF IRONWORK: with Practical Remarks on Iron Construction. By F. W. SHEILDS, M.Inst.C.E. Second Edition, with 5 Plates. Royal 8vo, 5s. cloth.

"The student cannot find a better little book on this subject."—*Engineer.*

Cast Iron and other Metals, Strength of.

A PRACTICAL ESSAY ON THE STRENGTH OF CAST IRON AND OTHER METALS. By THOMAS TREDGOLD, C.E. Fifth Edition, including Hodgkinson's Experimental Researches. 8vo, 12s. cloth.

Oblique Arches.

A PRACTICAL TREATISE ON THE CONSTRUCTION OF OBLIQUE ARCHES. By JOHN HART. Third Edition, with Plates. Imperial 8vo, 8s. cloth.

Girders, Strength of.

GRAPHIC TABLE FOR FACILITATING THE COMPUTATION OF THE WEIGHTS OF WROUGHT IRON AND STEEL GIRDERS, etc., for Parliamentary and other Estimates. By J. H. WATSON BUCK, M.Inst.C.E. On a Sheet, 2s. 6d.

Water Supply and Water-Works.

A PRACTICAL TREATISE ON THE WATER SUPPLY OF TOWNS AND THE CONSTRUCTION OF WATER-WORKS. By W. K. BURTON, A.M.Inst.C E., Professor of Sanitary Engineering in the Imperial University, Tokyo, Japan, and Consulting Engineer to the Tokyo Water-Works. With an Appendix on **Water-Works in Countries subject to Earthquakes**, by JOHN MILNE, F.R.S., Professor of Mining in the Imperial University of Japan. With numerous Plates and Illusts.

[*In the press.*

MARINE ENGINEERING, SHIPBUILDING, NAVIGATION, etc.

Pocket-Book for Naval Architects and Shipbuilders.

THE NAVAL ARCHITECT'S AND SHIPBUILDER'S POCKET-BOOK of Formulæ, Rules, and Tables, and MARINE ENGINEER'S AND SURVEYOR'S Handy Book of Reference. By CLEMENT MACKROW, Member of the Institution of Naval Architects, Naval Draughtsman. Fifth Edition, Revised and Enlarged to 700 pages, with upwards of 300 Illustrations. Fcap., 12s. 6d. strongly bound in leather.

SUMMARY OF CONTENTS.

SIGNS AND SYMBOLS, DECIMAL FRACTIONS.—TRIGONOMETRY.—PRACTICAL GEOMETRY.—MENSURATION.—CENTRES AND MOMENTS OF FIGURES.—MOMENTS OF INERTIA AND RADII OF GYRATION.—ALGEBRAICAL EXPRESSIONS FOR SIMPSON'S RULES.—MECHANICAL PRINCIPLES.—CENTRE OF GRAVITY.—LAWS OF MOTION.—DISPLACEMENT, CENTRE OF BUOYANCY.—CENTRE OF GRAVITY OF SHIP'S HULL.—STABILITY CURVES AND METACENTRES.—SEA AND SHALLOW-WATER WAVES.—ROLLING OF SHIPS.—PROPULSION AND RESISTANCE OF VESSELS.—SPEED TRIALS.—SAILING, CENTRE OF EFFORT.—DISTANCES DOWN RIVERS, COAST LINES.—STEERING AND RUDDERS OF VESSELS.—LAUNCHING CALCULATIONS AND VELOCITIES.—WEIGHT OF MATERIAL AND GEAR.—GUN PARTICULARS AND WEIGHT.—STANDARD GAUGES.—RIVETED JOINTS AND RIVETING.—STRENGTH AND TESTS OF MATERIALS.—BINDING AND SHEARING STRESSES, ETC.—STRENGTH OF SHAFTING, PILLARS, WHEELS, ETC.—HYDRAULIC DATA, ETC.—CONIC SECTIONS, CATENARIAN CURVES.—MECHANICAL POWERS, WORK.—BOARD OF TRADE REGULATIONS FOR BOILERS AND ENGINES.—BOARD OF TRADE REGULATIONS FOR SHIPS.—LLOYD'S RULES FOR BOILERS.—LLOYD'S WEIGHT OF CHAINS.—LLOYD'S SCANTLINGS FOR SHIPS.—DATA OF ENGINES AND VESSELS.—SHIPS' FITTINGS AND TESTS.—SEASONING PRESERVING TIMBER.—MEASUREMENT OF TIMBER.—ALLOYS, PAINTS, VARNISHES.—DATA FOR STOWAGE.—ADMIRALTY TRANSPORT REGULATIONS.—RULES FOR HORSEPOWER, SCREW PROPELLERS, ETC.—PERCENTAGES FOR BUTT STRAPS, ETC.—PARTICULARS OF YACHTS.—MASTING AND RIGGING VESSELS.—DISTANCES OF FOREIGN PORTS.—TONNAGE TABLES.—VOCABULARY OF FRENCH AND ENGLISH TERMS.—ENGLISH WEIGHTS AND MEASURES.—FOREIGN WEIGHTS AND MEASURES.—DECIMAL EQUIVALENTS.—FOREIGN MONEY.—DISCOUNT AND WAGE TABLES.—USEFUL NUMBERS AND READY RECKONERS—TABLES OF CIRCULAR MEASURES.—TABLES OF AREAS OF AND CIRCUMFERENCES OF CIRCLES.—TABLES OF AREAS OF SEGMENTS OF CIRCLES.—TABLES OF SQUARES AND CUBES AND ROOTS OF NUMBERS.—TABLES OF LOGARITHMS OF NUMBERS.—TABLES OF HYPERBOLIC LOGARITHMS.—TABLES OF NATURAL SINES, TANGENTS, ETC.—TABLES OF LOGARITHMIC SINES, TANGENTS, ETC.

"In these days of advanced knowledge a work like this is of the greatest value. It contains a vast amount of information. We unhesitatingly say that it is the most valuable compilation for its specific purpose that has ever been printed. No naval architect, engineer, surveyor, or seaman, wood or iron shipbuilder, can afford to be without this work."—*Nautical Magazine.*

"Should be used by all who are engaged in the construction or designs of vessels.... Will be found to contain the most useful tables and formulæ required by shipbuilders, carefully collected from the best authorities, and put together in a popular and simple form."—*Engineer.*

"The professional shipbuilder has now, in a convenient and accessible form, reliable data for solving many of the numerous problems that present themselves in the course of his work."—*Iron.*

"There is no doubt that a pocket-book of this description must be a necessity in the shipbuilding trade.... The volume contains a mass of useful information clearly expressed and presented in a handy form."—*Marine Engineer.*

Marine Engineering.

MARINE ENGINES AND STEAM VESSELS (A Treatise on). By ROBERT MURRAY, C.E. Eighth Edition, thoroughly Revised, with considerable Additions by the Author and by GEORGE CARLISLE, C.E., Senior Surveyor to the Board of Trade at Liverpool. 12mo, 5s. cloth boards.

"Well adapted to give the young steamship engineer or marine engine and boiler maker a general introduction into his practical work."—*Mechanical World.*

"We feel sure that this thoroughly revised edition will continue to be as popular in the future as it has been in the past, as, for its size, it contains more useful information than any similar treatise."—*Industries.*

"As a compendious and useful guide to engineers of our mercantile and royal naval services, we should say it cannot be surpassed."—*Building News.*

"The information given is both sound and sensible, and well qualified to direct young sea-going hands on the straight road to the extra chief's certificate.... Most useful to surveyors, inspectors, draughtsmen, and young engineers."—*Glasgow Herald.*

Pocket-Book for Marine Engineers.

A POCKET-BOOK OF USEFUL TABLES AND FORMULÆ FOR MARINE ENGINEERS. By FRANK PROCTOR, A.I.N.A, Third Edition. Royal 32mo, leather, gilt edges, with strap, 4s.

"We recommend it to our readers as going far to supply a long-felt want."—*Naval Science.*

"A most useful companion to all marine engineers."—*United Service Gazette.*

Introduction to Marine Engineering.

ELEMENTARY ENGINEERING: A Manual for Young Marine Engineers and Apprentices. In the Form of Questions and Answers on Metals, Alloys, Strength of Materials, Construction and Management of Marine Engines and Boilers, Geometry, &c. &c. With an Appendix of Useful Tables. By JOHN SHERREN BREWER, Government Marine Surveyor, Hong-Kong. Second Edition, Revised. Small crown 8vo, 2s. cloth.

"Contains much valuable information for the class for whom it is intended, especially in the chapters on the management of boilers and engines."—*Nautical Magazine.*

"A useful introduction to the more elaborate text-books."—*Scotsman.*

"To a student who has the requisite desire and resolve to attain a thorough knowledge, Mr. Brewer offers decidedly useful help."—*Athenæum.*

Navigation.

PRACTICAL NAVIGATION. Consisting of THE SAILOR'S SEA-BOOK, by JAMES GREENWOOD and W. H. ROSSER; together with the requisite Mathematical and Nautical Tables for the Working of the Problems, by HENRY LAW, C.E., and Professor J. R. YOUNG. Illustrated. 12mo, 7s. strongly half-bound.

Drawing for Marine Engineers.

LOCKIE'S MARINE ENGINEER'S DRAWING-BOOK. Adapted to the Requirements of the Board of Trade Examinations. By JOHN LOCKIE, C.E. With 22 Plates, Drawn to Scale. Royal 3vo, 3s. 6d. cloth.

"The student who learns from these drawings will have nothing to unlearn."—*Engineer.*

"The examples chosen are essentially practical, and are such as should prove of service to engineers generally, while admirably fulfilling their specific purpose."—*Mechanical World.*

Sailmaking.

THE ART AND SCIENCE OF SAILMAKING. By SAMUEL B. SADLER.

Practical Sailmaker, late in the employment of Messrs. Ratsey and Lapthorne, of Cowes and Gosport. With Plates and other Illustrations. Small 4to, 12s. 6d. cloth.

SUMMARY OF CONTENTS.

CHAP. I. THE MATERIALS USED AND THEIR RELATION TO SAILS.—II. ON THE CENTRE OF EFFORT.—III. ON MEASURING.—IV. ON DRAWING.—V. ON THE NUMBER OF CLOTHS REQUIRED.—VI. ON ALLOWANCES.—VII. ON CALCULATION OF GORES.—VIII. ON CUTTING OUT.—IX. ON ROPING.—X. ON DIAGONAL-CUT SAILS.—XI. CONCLUDING REMARKS.

"This work is very ably written, and is illustrated by diagrams and carefully worked calculations. The work should be in the hands of every sailmaker, whether employer or employed, as it cannot fail to assist them in the pursuit of their important avocations."—*Isle of Wight Herald.*

"This extremely practical work gives a complete education in all the branches of the manufacture, cutting out, roping, seaming, and goring. It is copiously illustrated, and will form a first-rate text-book and guide."—*Portsmouth Times.*

"The author of this work has rendered a distinct service to all interested in the art of sailmaking. The subject of which he treats is a congenial one. Mr. Sadler is a practical sailmaker, and has devoted years of careful observation and study to the subject; and the results of the experience thus gained he has set forth in the volume before us."—*Steamship.*

Chain Cables.

CHAIN CABLES AND CHAINS. Comprising Sizes and Curves of Links, Studs, &c., Iron for Cables and Chains, Chain Cable and Chain Making, Forming and Welding Links, Strength of Cables and Chains, Certificates for Cables, Marking Cables, Prices of Chain Cables and Chains, Historical Notes, Acts of Parliament, Statutory Tests, Charges for Testing, List of Manufacturers of Cables, &c. &c. By THOMAS W. TRAILL, F.E.R.N., M. Inst. C.E., Engineer Surveyor in Chief, Board of Trade, Inspector of Chain Cable and Anchor Proving Establishments, and General Superintendent, Lloyd's Committee on Proving Establishments. With numerous Tables, Illustrations and Lithographic Drawings. Folio, £2 2s. cloth, bevelled boards.

"It contains a vast amount of valuable Information. Nothing seems to be wanting to make it a complete and standard work of reference on the subject."—*Nautical Magazine.*

MINING AND METALLURGY.

Mining Machinery.

MACHINERY FOR METALLIFEROUS MINES: A Practical Treatise for Mining Engineers, Metallurgists, and Managers of Mines. By E. HENRY DAVIES, M.E., F.G.S. Crown 8vo, 580 pp., with upwards of 300 Illustrations, 12s. 6d. cloth.

[Just published.

"Mr. Davies, in this handsome volume, has done the advanced student and the manager of mines good service. Almost every kind of machinery in actual use is carefully described, and the woodcuts and plates are good."—*Athenæum.*

"From cover to cover the work exhibits all the same characteristics which excite the confidence and attract the attention of the student as he peruses the first page. The work may safely be recommended. By its publication the literature connected with the industry will be enriched, and the reputation of its author enhanced."—*Mining Journal.*

"Mr. Davies has endeavoured to bring before his readers the best of everything in modern mining appliances. His work carries internal evidence of the author's impartiality, and this constitutes one of the great merits of the book. Throughout his work the criticisms are based on his own or other reliable experience."—*Iron and Steel Trades' Journal.*

"The work deals with nearly every class of machinery or apparatus likely to be met with or required in connection with metalliferous mining, and is one which we have every confidence in recommending."—*Practical Engineer.*

Metalliferous Minerals and Mining.

A TREATISE ON METALLIFEROUS MINERALS AND MINING. By D. C. DAVIES, F.G.S., Mining Engineer, &c., Author of "A Treatise on Slate and Slate Quarrying." Fifth Edition, thoroughly Revised and much Enlarged, by his Son, E. HENRY DAVIES, M.E., F.G.S. With about 150 Illustrations. Crown 8vo, 12s. 6d. cloth.

"Neither the practical miner nor the general reader interested in mines can have a better book for his companion and his guide."—*Mining Journal.*

"We are doing our readers a service in calling their attention to this valuable work."—*Mining World.*

"A book that will not only be useful to the geologist, the practical miner, and the metallurgist, but also very interesting to the general public."—*Iron.*

"As a history of the present state of mining throughout the world this book has a real value, end it supplies an actual want."—*Athenæum.*

Earthy Minerals and Mining.

A TREATISE ON EARTHY & OTHER MINERALS AND MINING. By D. C. DAVIES, F.G.S., Author of "Metalliferous Minerals," &c. Third Edition, revised and Enlarged, by his Son, E. HENRY DAVIES, M.E., F.G.S. With about 100 Illustrations. Crown 8vo, 12s. 6d. cloth.

"We do not remember to have met with any English work on mining matters that contains the same amount of information packed in equally convenient form."—*Academy.*

"We should be inclined to rank it as among the very best of the handy technical and trades manuals which have recently appeared."—*British Quarterly Review.*

Metalliferous Mining in the United Kingdom.

BRITISH MINING: A Treatise on the History, Discovery, Practical Development, and Future Prospects of Metalliferous Mines in the United Kingdom. By ROBERT HUNT, F.R.S., Editor of "Ure's Dictionary of Arts, Manufactures, and Mines," &c. Upwards of 950 pp., with 230 Illustrations. Second Edition, Revised. Super-royal 8vo, £2 2s. cloth.

"One of the most valuable works of reference of modern times. Mr. Hunt, as Keeper of Mining Records of the United Kingdom, has had opportunities for such a task not enjoyed by anyone else, and has evidently made the most of them.... The language and style adopted are good, and the treatment of the various subjects laborious, conscientious, and scientific."—*Engineering.*

"The book is, in fact, a treasure-house of statistical information on mining subjects, and we know of no other work embodying so great a mass of matter of this kind. Were this the only merit of Mr. Hunt's volume, it would be sufficient to render it indispensable in the library of everyone interested in the development of the mining and metallurgical industries of this Country.—*Athenæum.*

"A mass of Information not elsewhere available, and of the greatest value to those who may be interested in our great mineral industries."—*Engineer.*

Underground Pumping Machinery.

MINE DRAINAGE. Being a Complete and Practical Treatise on Direct-Acting Underground Steam Pumping Machinery, with a Description of a large number of the best known Engines, their General Utility and the Special Sphere of their Action, the Mode of their Application, and their merits compared with other forms of Pumping Machinery. By STEPHEN MICHELL. 8vo, 15s. cloth.

"Will be highly esteemed by colliery owners and lessees, mining engineers, and students generally who require to be acquainted with the best means of securing the drainage of mines. It is a most valuable work, and stands almost alone in the literature of steam pumping machinery."—*Colliery Guardian.*

"Much valuable Information is given, so that the book is thoroughly worthy of an extensive circulation amongst practical men and purchasers of machinery."'—*Mining Journal.*

Prospecting for Gold and other Metals.

THE PROSPECTOR'S HANDBOOK: A Guide for the Prospector and

Traveller in Search of Metal-Bearing or other Valuable Minerals. By J. W. ANDERSON, M.A. (Camb.), F.R.G.S., Author of "Fiji and New Caledonia." Fifth Edition, thoroughly Revised and Enlarged. Small crown 8vo, 3s. 6d. cloth.

"Will supply a much felt want, especially among Colonists, in whose way are so often thrown many mineralogical specimens the value of which it is difficult to determine."—*Engineer.*

"How to find commercial minerals, and how to identify them when they are found, are the leading points to which attention is directed. The author has managed to pack as much practical detail into his pages as would supply material for a book three times its size."—*Mining Journal.*

Mining Notes and Formulæ.

NOTES AND FORMULÆ FOR MINING STUDENTS. By JOHN HERMAN MERIVALE, M.A., Certificated Colliery Manager, Professor of Mining in the Durham College of Science, Newcastle-upon-Tyne. Third Edition, Revised and Enlarged. Small crown 8vo, 2s. 6d. cloth.

"Invaluable to anyone who is working up for an examination on mining subjects."—*Iron and Coal Trades Review.*

"The author has done his work in an exceedingly creditable manner, and has produced a book that will be of service to students, and those who are practically engaged in mining operations."—*Engineer.*

Handybook for Miners.

THE MINER'S HANDBOOK: A Handybook of Reference on the Subjects of Mineral Deposits, Mining Operations, Ore Dressing, &c. For the Use of Students and others interested in Mining matters. Compiled by JOHN MILNE, F.R.S., Professor of Mining in the Imperial University of Japan. Square 18mo, 7s. 6d. cloth.

[*Just published.*]

"Professor Milne's handbook is sure to be received with favour by all connected with mining, and will be extremely popular among students."—*Athenæum.*

Miners' and Metallurgists' Pocket-Book.

A POCKET-BOOK FOR MINERS AND METALLURGISTS. Comprising Rules, Formulæ, Tables, and Notes, for Use in Field and Office Work. By F. DANVERS POWER, F.G.S., M.E. Fcap. 8vo, 9s. leather, gilt edges.

"This excellent book is an admirable example of its kind, and ought to find a large sale amongst English-speaking prospectors and mining engineers."—*Engineering.*

"Miners and metallurgists will find in this work a useful vade-mecum containing a mass of rules, formulæ, tables, and various other information, the necessity for reference to which occurs in. their daily duties."—*Iron.*

Mineral Surveying and Valuing.

THE MINERAL SURVEYOR AND VALUER'S COMPLETE GUIDE, comprising a Treatise on Improved Mining Surveying and the Valuation of Mining Properties, with New Traverse Tables. By WM. LINTERN. Third Edition, Enlarged. 12mo, 4s. cloth.

"Mr. Lintern's book forms a valuable and thoroughly trustworthy guide."—*Iron and Coal Trades Review.*

Asbestos and its Uses.

ASBESTOS: *Its Properties, Occurrence, and Uses.* With some Account of the Mines of Italy and Canada. By ROBERT H. JONES. With Eight Collotype Plates and other Illustrations. Crown 8vo, 12s. 6d. cloth.

"An interesting and invaluable work."—*Colliery Guardian.*

Explosives.

A HANDBOOK ON MODERN EXPLOSIVES. Being a Practical Treatise on the Manufacture and Application of Dynamite, Gun-Cotton, Nitro-Glycerine, and other Explosive Compounds. Including the Manufacture of Collodion-Cotton. By M. EISSLER, Mining Engineer and Metallurgical Chemist, Author of "The Metallurgy of Gold," "The Metallurgy of Silver," &c. With about 100 Illusts. Crown 8vo, 10s. 6d. cloth.

"Useful not only to the miner, but also to officers of both services to whom blasting and the use of explosives generally may at any time become a necessary auxiliary."—*Nature*.

"A veritable mine of information on the subject of explosives employed for military, mining, and blasting purposes."—*Army and Navy Gazette*.

Colliery Management.

THE COLLIERY MANAGER'S HANDBOOK: A Comprehensive Treatise on the Laying-out and Working of Collieries, Designed as a Book of Reference for Colliery Managers, and for the Use of Coal-Mining Students preparing for First-class Certificates. By CALEB PAMELY, Mining Engineer and Surveyor; Member of the North of England Institute of Mining and Mechanical Engineers; and Member of the South Wales Institute of Mining Engineers. With nearly 500 Plans, Diagrams, and other Illustrations. Second Edition, Revised, with Additions. Medium 8vo, about 700 pages. Price £1 5s. strongly bound.

SUMMARY OF CONTENTS.

GEOLOGY.—SEARCH FOR COAL.—MINERAL LEASES AND OTHER HOLDINGS.—SHAFT SINKING.—FITTING UP THE SHAFT AND SURFACE ARRANGEMENTS.—STEAM BOILERS AND THEIR FITTINGS.—TIMBERING AND WALLING.—NARROW WORK AND METHODS OF WORKING.—UNDERGROUND CONVEYANCE.—DRAINAGE.—THE GASES MET WITH IN MINES; VENTILATION.—ON THE FRICTION OF AIR IN MINES.—THE PRIESTMAN OIL ENGINE; PETROLEUM AND NATURAL GAS.—SURVEYING AND PLANNING.—SAFETY LAMPS AND FIRE-DAMP DETECTORS.—SUNDRY AND INCIDENTAL OPERATIONS AND APPLIANCES.—COLLIERY EXPLOSIONS.—MISCELLANEOUS QUESTIONS & ANSWERS.

Appendix: SUMMARY OF REPORT OF H.M. COMMISSIONERS ON ACCIDENTS IN MINES.

⁂ OPINIONS OF THE PRESS.

"Mr. Pamely has not only given us a comprehensive reference book of a very high order, suitable to the requirements of mining-engineers and colliery managers, but at the same time has provided mining students with a class-book that is as interesting as it is instructive."—*Colliery Manager*.

"Mr. Pamely's work is eminently suited to the purpose for which it is intended—being clear, interesting, exhaustive, rich in detail, and up to date, giving descriptions of the very latest machines in every department.... A mining engineer could scarcely go wrong who followed this work."—*Colliery Guardian*.

"This is the most complete 'all round' work on coal-mining published in the English language.... No library of coal-mining books is complete without it."—*Colliery Engineer* (Scranton, Pa., U.S.A.).

"Mr. Pamely's work is in all respects worthy of our admiration. No person in any responsible position connected with mines should be without a copy."—*Westminster Review*.

Coal and Iron.

THE COAL AND IRON INDUSTRIES OF THE UNITED KINGDOM. Comprising a Description of the Coal Fields, and of the Principal Seams of Coal, with Returns of their Produce and its Distribution, and Analyses of Special Varieties. Also an Account of the occurrence of Iron Ores in Veins or Seams; Analyses of each Variety; and a History of the Rise and Progress of Pig Iron Manufacture. By RICHARD MEADE, Assistant Keeper of Mining Records. With Maps. 8vo, £1 8s. cloth.

"The book is one which must find a place on the shelves of all interested in coal and iron production, and in the iron, steel and other metallurgical industries."—*Engineer*.

"Of this book we may unreservedly say that it is the best of its class which we have ever met.... A book of reference which no one engaged in the iron or coal trades should omit from his library."—*Iron and Coal Trades Review.*

Coal Mining.

COAL AND COAL MINING: A Rudimentary Treatise on. By the late Sir WARINGTON W. SMYTH, M.A., F.R.S., &c., Chief Inspector of the Mines of the Crown. Seventh Edition, Revised and Enlarged. With numerous Illustrations. 12mo, 4s. cloth boards.

"As an outline is given of every known coal-field in this and other countries, as well as of the principal methods of working, the book will doubtless interest a very large number of readers."—*Mining Journal.*

Subterraneous Surveying.

SUBTERRANEOUS SURVEYING, Elementary and Practical Treatise on, with and without the Magnetic Needle. By THOMAS FENWICK, Surveyor of Mines, and THOMAS BAKER, C.E. Illust. 12mo, 3s. cloth boards.

Granite Quarrying.

GRANITES AND OUR GRANITE INDUSTRIES. By GEORGE F. HARRIS, F.G.S., Membre de la Société Belge de Géologie, Lecturer on Economic Geology at the Birkbeck Institution, &c. With Illustrations. Crown 8vo, 2s. 6d. cloth.

"A clearly and well-written manual on the granite industry."—*Scotsman.*

"An interesting work, which will be deservedly esteemed."—*Colliery Guardian.*

"An exceedingly interesting and valuable monograph on a subject which has hitherto received unaccountably little attention in the shape of systematic literary treatment."—*Scottish Leader.*

Gold, Metallurgy of.

THE METALLURGY OF GOLD: A Practical Treatise on the Metallurgical Treatment of Gold-bearing Ores. Including the Processes of Concentration and Chlorination, and the Assaying, Melting, and Refining of Gold. By M. EISSLER, Mining Engineer and Metallurgical Chemist, formerly Assistant Assayer of the U. S. Mint, San Francisco. Third Edition, Revised and greatly Enlarged. With 187 Illustrations. Crown 8vo, 12s. 6d. cloth.

"This book thoroughly deserves its title of a 'Practical Treatise.' The whole process of gold milling, from the breaking of the quartz to the assay of the bullion, is described in clear and orderly narrative and with much, but not too much, fulness of detail."—*Saturday Review.*

"The work is a storehouse of information and valuable data, and we strongly recommend it to all professional men engaged in the gold-mining industry."—*Mining Journal.*

Silver, Metallurgy of.

THE METALLURGY OF SILVER: A Practical Treatise on the Amalgamation, Roasting, and Lixiviation of Silver Ores. Including the Assaying, Melting and Refining, of Silver Bullion. By M. EISSLER, Author of "The Metallurgy of Gold," &c. Second Edition, Enlarged. With 150 Illustrations. Crown 8vo, 10s. 6d. cloth.

"A practical treatise, and a technical work which we are convinced will supply a long-felt want amongst practical men, and at the same time be of value to students and others indirectly connected with the industries."—*Mining Journal.*

"From first to last the book is thoroughly sound and reliable."—*Colliery Guardian.*

"For chemists, practical miners, assayers, and investors alike, we do not know of any work on the subject so handy and yet so comprehensive."—*Glasgow Herald.*

Lead, Metallurgy of.

THE METALLURGY OF ARGENTIFEROUS LEAD: A Practical Treatise on the Smelting of Silver-Lead Ores and the Refining of Lead Bullion.

Including Reports on various Smelting Establishments and Descriptions of Modern Smelting Furnaces and Plants in Europe and America. By M. EISSLER, M.E., Author of "The Metallurgy of Gold," &c. Crown 8vo, 400 pp., with 183 Illustrations, 12s. 6d. cloth.

"The numerous metallurgical processes, which are fully and extensively treated of, embrace all the stages experienced in the passage of the lead from the various natural states to its issue from the refinery as an article of commerce." — *Practical Engineer.*

"The present volume fully maintains the reputation of the author. Those who wish to obtain a thorough insight into the present state of this industry cannot do better than read this volume, and all mining engineers cannot fail to find many useful hints and suggestions in it." — *Industries.*

"It is most carefully written and illustrated with capital drawings and diagrams. In fact, it is the work of an expert for experts, by whom it will be prized as an indispensable text-book." — *Bristol Mercury.*

Iron, Metallurgy of.

METALLURGY OF IRON. Containing History of Iron Manufacture, Methods of Assay, and Analyses of Iron Ores, Processes of Manufacture of Iron and Steel, &c. By H. BAUERMAN, F.G.S., A.R.S.M. With numerous Illustrations. Sixth Edition, Revised and Enlarged. 12mo, 5s. 6d. cloth.

"Carefully written, it has the merit of brevity and conciseness, as to less important points, while all material matters are very fully and thoroughly entered into." — *Standard.*

Iron Mining.

THE IRON ORES OF GREAT BRITAIN AND IRELAND: Their Mode of Occurrence, Age, and Origin, and the Methods of Searching for and Working them, with a Notice of some of the Iron Ores of Spain. By J. D. KENDALL, F.G.S., Mining Engineer. With Plates and Illustrations. Crown 8vo, 16s. cloth.

"The author has a thorough practical knowledge of his subject, and has supplemented a careful study of the available literature by unpublished information derived from his own observations. The result is a very useful volume which cannot fail to be of value to all interested in the iron industry of the country." — *Industries.*

"Constitutes a systematic and careful account of our present knowledge of the origin and occurrence of the iron ores of Great Britain, and embraces a description of the means employed in reaching and working these ores." — *Iron.*

"Mr. Kendall is a great authority on this subject and writes from personal observation." — *Colliery Guardian.*

"Mr. Kendall's book is thoroughly well done. In it there are the outlines of the history of ore mining in every centre and there is everything that we want to know as to the character of the ores of each district, their commercial value and the cost of working them." — *Iron and Steel Trades Journal.*

ELECTRICITY, ELECTRICAL ENGINEERING, etc.

Electrical Engineering.

THE ELECTRICAL ENGINEER'S POCKET-BOOK OF MODERN RULES, FORMULÆ, TABLES, AND DATA. By H. R. KEMPE, M.Inst.E.E., A.M.Inst.C.E., Technical Officer, Postal Telegraphs, Author of "A Handbook of Electrical Testing," &c. Second Edition, thoroughly Revised, with Additions. With numerous Illustrations. Royal 32mo, oblong, 5s. leather.

"There is very little in the shape of formulæ or data which the electrician is likely to want in a hurry which cannot be found in its pages." — *Practical Engineer.*

"A very useful book of reference for daily use in practical electrical engineering and its various applications to the industries of the present day." — *Iron.*

"It is the best book of its kind." — *Electrical Engineer.*

"Well arranged and compact. The 'Electrical Engineer's Pocket-Book' is a good one." — *Electrician.*

"Strongly recommended to those engaged in the various electrical industries." — *Electrical Review.*

Electric Lighting.

ELECTRIC LIGHT FITTING: A Handbook for Working Electrical Engineers, embodying Practical Notes on Installation Management. By JOHN W. URQUHART, Electrician, Author of "Electric Light," &c. With numerous Illustrations. Second Edition, Revised, with Additional Chapters. Crown 8vo, 5s. cloth.

"This volume deals with what may be termed the mechanics of electric lighting, and is addressed to men who are already engaged in the work or are training for it. The work traverses a great deal of ground, and may be read as a sequel to the same author's useful work on 'Electric Light.'"—*Electrician*.

"This is an attempt to state in the simplest language the precautions which should be adopted in installing the electric light, and to give information, for the guidance of those who have to run the plant when installed. The book is well worth the perusal of the workmen for whom it is written."—*Electrical Review*.

"We have read this book with a good deal of pleasure. We believe that the book will be of use to practical workmen, who will not be alarmed by finding mathematical formulæ which they are unable to understand."—*Electrical Plant*.

"Eminently practical and useful.... Ought to be in the hands of everyone in charge of an electric light plant."—*Electrical Engineer*.

"Mr. Urquhart has succeeded in producing a really capital book, which we have no hesitation in recommending to the notice of working electricians and electrical engineers."—*Mechanical World*.

Electric Light.

ELECTRIC LIGHT: Its Production and Use. Embodying Plain Directions for the Treatment of Dynamo-Electric Machines, Batteries, Accumulators, and Electric Lamps. By J. W. URQUHART, C.E., Author of "Electric Light Fitting," "Electroplating," &c. Fifth Edition, carefully Revised, with Large Additions and 145 Illustrations. Crown 8vo, 7s. 6d. cloth.

"The whole ground of electric lighting is more or less covered and explained in a very clear and concise manner."—*Electrical Review*.

"Contains a good deal of very interesting information, especially in the parts where the author gives dimensions and working costs."—*Electrical Engineer*.

"A miniature *vade-mecum* of the salient facts connected with the science of electric lighting."—*Electrician*.

"You cannot for your purpose have a better book than 'Electric Light,' by Urquhart."—*Engineer*.

"The book is by far the best that we have yet met with on the subject."—*Athenæum*.

Construction of Dynamos.

DYNAMO CONSTRUCTION: A Practical Handbook for the Use of Engineer Constructors and Electricians-in-Charge. Embracing Framework Building, Field Magnet and Armature Winding and Grouping, Compounding, &c. With Examples of leading English, American, and Continental Dynamos and Motors. By J. W. URQUHART, Author of "Electric Light," "Electric Light Fitting," &c. With upwards of 100 Illustrations. Crown 8vo, 7s. 6d. cloth.

"Mr. Urquhart's book is the first one which deals with these matters in such a way that the engineering student can understand them. The book is very readable, and the author leads his readers up to difficult subjects by reasonably simple tests."—*Engineering Review*.

"The author deals with his subject in a style so popular as to make his volume a handbook of great practical value to engineer contractors and electricians in charge of lighting installations."—*Scotsman*.

"'Dynamo Construction' more than sustains the high character of the author's previous publications. It is sure to be widely read by the large and rapidly-increasing number of practical electricians."—*Glasgow Herald*.

"A book for which a demand has long existed."—*Mechanical World*.

A New Dictionary of Electricity.

THE STANDARD ELECTRICAL DICTIONARY. A Popular Dictionary of Words and Terms Used in the Practice of Electrical Engineering. Containing upwards of 3,000 Definitions. By T. O'Connor Sloane, A.M., Ph.D., Author of "The Arithmetic of Electricity," &c. Crown 8vo, 630 pp., 350 Illustrations, 7s. 6d. cloth.

[*Just published.*]

"The work has many attractive features in it, and is beyond doubt, a well put together and useful publication. The amount of ground covered may be gathered from the fact that in the index about 5,000 references will be found. The inclusion of such comparatively modern words as 'impedence,' 'reluctance,' &c., shows that the author has desired to be up to date, and indeed there are other indications of carefulness of compilation. The work is one which does the author great credit and it should prove of great value, especially to students."—*Electrical Review.*

"We have found the book very complete and reliable, and can, therefore, commend it heartily."—*Mechanical World.*

"Very complete and contains a large amount of useful information."—*Industries.*

"An encyclopædia of electrical science in the compass of a dictionary. The information given is sound and clear. The book is well printed, well illustrated, and well up to date, and may be confidently recommended."—*Builder.*

"We hail the appearance of this little work as one which will meet a want that has been keenly felt for some time.... The author is to be congratulated on the excellent manner in which he has accomplished his task."—*Practical Engineer.*

"The volume is excellently printed and illustrated, and should form part of the library of every one who is directly or indirectly connected with electrical matters."—*Hardware Trade Journal.*

Electric Lighting of Ships.

ELECTRIC SHIP-LIGHTING: A Handbook on the Practical Fitting and Running of Ship's Electrical Plant. For the Use of Shipowners and Builders, Marine Electricians, and Sea-going Engineers-in-Charge. By J. W. Urquhart, C.E., Author of "Electric Light," &c. With 88 Illustrations. Crown 8vo, 7s. 6d. cloth.

"The subject of ship electric lighting is one of vast importance in these days, and Mr. Urquhart is to be highly complimented for placing such a valuable work at the service of the practical marine electrician."—*The Steamship.*

"Distinctly a book which of its kind stands almost alone, and for which there should be a demand."—*Electrical Review.*

Electric Lighting.

THE ELEMENTARY PRINCIPLES OF ELECTRIC LIGHTING. By Alan A. Campbell Swinton, Associate I.E.E. Third Edition, Enlarged and Revised. With 16 Illustrations. Crown 8vo, 1s. 6d. cloth.

"Anyone who desires a short and thoroughly clear exposition of the elementary principles of electric lighting cannot do better than read this little work."—*Bradford Observer.*

Dynamic Electricity.

THE ELEMENTS OF DYNAMIC ELECTRICITY AND MAGNETISM. By Philip Atkinson, A.M., Ph.D., Author of "Elements of Static Electricity," "The Elements of Electric Lighting," &c. &c. Crown 8vo, 417 pp., with 120 Illustrations, 10s. 6d. cloth.

Electric Motors, &c.

THE ELECTRIC TRANSFORMATION OF POWER and its Application by the Electric Motor, including Electric Railway Construction. By P. Atkinson, A.M., Ph.D., Author of "The Elements of Electric Lighting," &c. With 96 Illustrations. Crown 8vo, 7s. 6d. cloth.

Dynamo Construction.

HOW TO MAKE A DYNAMO: A Practical Treatise for Amateurs. Containing numerous Illustrations and Detailed Instructions for Constructing a Small Dynamo, to Produce the Electric Light. By ALFRED CROFTS. Fourth Edition, Revised and Enlarged. Crown 8vo, 2s. cloth.

"The instructions given in this unpretentious little book are sufficiently clear and explicit to enable any amateur mechanic possessed of average skill and the usual tools to be found in an amateur's workshop, to build a practical dynamo machine."—*Electrician.*

Text Book of Electricity.

THE STUDENT'S TEXT-BOOK OF ELECTRICITY. By HENRY M. NOAD, Ph.D., F.R.S. New Edition, carefully Revised. With Introduction and Additional Chapters, by W. H. PREECE, M.I.C.E. Crown 8vo, 12s. 6d. cloth.

Electricity.

A MANUAL OF ELECTRICITY: Including Galvanism, Magnetism, Dia-Magnetism, Electro-Dynamics. By HENRY M. NOAD, Ph.D., F.R.S. Fourth Edition (1859). 8vo, £1 4s. cloth.

ARCHITECTURE, BUILDING, etc.

Building Construction.

PRACTICAL BUILDING CONSTRUCTION: A Handbook for Students Preparing for Examinations, and a Book of Reference for Persons Engaged in Building. By JOHN PARNELL ALLEN, Surveyor, Lecturer on Building Construction at the Durham College of Science, Newcastle-on-Tyne. Medium 8vo, 450 pages, with 1,000 Illustrations. 12s. 6d. cloth.

[*Just published.*

"This volume is one of the most complete expositions of building construction we have seen. It contains all that is necessary to prepare students for the various examinations in building construction."—*Building News.*

"The author depends nearly as much on his diagrams as on his type. The pages suggest the hand of a man of experience in building operations—and the volume must be a blessing to many teachers as well as to students."—*The Architect.*

"This volume promises to be the recognised handbook in all advanced classes where building construction is taught from a practical point of view. We strongly commend the book to the notice of all teachers of building construction."—*Technical World.*

"The work is sure to prove a formidable rival to great and small competitors alike, and bids fair to take a permanent place as a favourite students' text-book. The large number of illustrations deserve particular mention for the great merit they possess for purposes of reference, in exactly corresponding to convenient scales."—*Jour. Inst. Brit. Archts.*

Concrete.

CONCRETE: ITS NATURE AND USES. A Book for Architects, Builders, Contractors, and Clerks of Works. By GEORGE L. SUTCLIFFE, A.R.I.B.A. 350 pages, with numerous Illustrations. Crown 8vo, 7s. 6d. cloth.

[*Just published.*

"The author treats a difficult subject in a lucid manner. The manual fills a long-felt gap. It is careful and exhaustive; equally useful as a student's guide and a architect's book of reference."—*Journal of Royal Institution of British Architects.*

"There is room for this new book, which will probably be for some time the standard work on the subject for a builder's purpose."—*Glasgow Herald.*

"A thoroughly useful and comprehensive work."—*British Architect.*

Mechanics for Architects.

THE MECHANICS OF ARCHITECTURE: A Treatise on Applied Mechanics, especially Adapted to the Use of Architects. By E. W. TARN,

M.A., Author of "The Science of Building," &c. Second Edition, Enlarged. Illust. with 125 Diagrams. Cr. 8vo, 7s. 6d. cloth.

[Just published.

"The book is a very useful and helpful manual of architectural mechanics, and really contains sufficient to enable a careful and painstaking student to grasp the principles bearing upon the majority of building problems.... Mr. Tarn has added, by this volume, to the debt of gratitude which is owing to him by architectural students for the many valuable works which he has produced for their use."—*The Builder.*

"The mechanics in the volume are really mechanics, and are harmoniously wrought in with the distinctive professional manner proper to the subject. The diagrams and type are commendably clear."—*The Schoolmaster.*

The New Builder's Price Book, 1894.

LOCKWOOD'S BUILDER'S PRICE BOOK FOR 1894. A Comprehensive Handbook of the Latest Prices and Data for Builders, Architects, Engineers, and Contractors. *Re-constructed, Re-written, and Greatly Enlarged.* By FRANCIS T. W. MILLER. 700 closely-printed pages, crown 8vo, 4s. cloth.

"This book is a very useful one, and should find a place in every English office connected with the building and engineering professions."—*Industries.*

"An excellent book of reference."—*Architect.*

"In its new and revised form this Price Book is what a work of this kind should be—comprehensive, reliable, well arranged, legible, and well bound."—*British Architect.*

Designing Buildings.

THE DESIGN OF BUILDINGS: Being Elementary Notes on the Planning, Sanitation and Ornamentive Formation of Structures, based on Modern Practice. Illustrated with Nine Folding Plates. By W. WOODLEY, Assistant Master, Metropolitan Drawing Classes, &c. Demy 8vo, 6s. cloth.

[Just published.

Sir Wm. Chambers's Treatise on Civil Architecture.

THE DECORATIVE PART OF CIVIL ARCHITECTURE. By Sir WILLIAM CHAMBERS, F.R.S. With Portrait, Illustrations, Notes, and an Examination of Grecian Architecture, by JOSEPH GWILT, F.S.A. Revised and Edited by W. H. LEEDS. 66 Plates, 4to, 21s. cloth.

Villa Architecture.

A HANDY BOOK OF VILLA ARCHITECTURE: Being a Series of Designs for Villa Residences in various Styles. With Outline Specifications and Estimates. By C. WICKES, Architect, Author of "The Spires and Towers of England," &c. 61 Plates, 4to, £1 11s. 6d. half-morocco.

"The whole of the designs bear evidence of their being the work of an artistic architect, and they will prove very valuable and suggestive."—*Building News.*

Text-Book for Architects.

THE ARCHITECT'S GUIDE: Being a Text-Book of Useful Information for Architects, Engineers, Surveyors, Contractors, Clerks of Works, &c. &c. By FREDERICK ROGERS, Architect. Third Edition. Crown 8vo, 3s. 6d. cloth.

"As a text-book of useful information for architects, engineers, surveyors, &c., it would be hard to find a handier or more complete little volume."—*Standard.*

Taylor and Cresy's Rome.

THE ARCHITECTURAL ANTIQUITIES OF ROME. By the late G. L. TAYLOR, Esq., F.R.I.B.A., and EDWARD CRESY, Esq. New Edition, thoroughly Revised by the Rev. ALEXANDER TAYLOR, M.A. (son of the late G. L. Taylor, Esq.), Fellow of Queen's College, Oxford, and Chaplain of Gray's

Inn. Large folio, with 130 Plates, £3 3s. half-bound.

<small>"Taylor and Cresy's work has from its first publication been ranked among those professional books which cannot be bettered."—*Architect*.</small>

Linear Perspective.

ARCHITECTURAL PERSPECTIVE: The whole Course and Operations of the Draughtsman in Drawing a Large House in Linear Perspective. Illustrated by 39 Folding Plates. By F. O. FERGUSON. 8vo, 3s. 6d. boards.

<small>"It is the most intelligible of the treatises on this ill-treated subject that I have met with."—E. INGRESS BELL, Esq., in the *R.I.B.A. Journal*.</small>

Architectural Drawing.

PRACTICAL RULES ON DRAWING, *for the Operative Builder and Young Student in Architecture.* By GEORGE PYNE. With 14 Plates, 4to, 7s. 6d. boards.

Vitruvius' Architecture.

THE ARCHITECTURE *of MARCUS VITRUVIUS POLLIO.* Translated by JOSEPH GWILT, F.S.A., F.R.A.S. New Edition, Revised by the Translator. With 23 Plates. Fcap. 8vo, 5s. cloth.

Designing, Measuring, and Valuing.

THE STUDENT'S GUIDE to the PRACTICE of MEASURING AND VALUING ARTIFICERS' WORK. Containing Directions for taking Dimensions, Abstracting the same, and bringing the Quantities into Bill, with Tables of Constants for Valuation of Labour, and for the Calculation of Areas and Solidities. Originally edited by EDWARD DOBSON, Architect. With Additions by E. WYNDHAM TARN, M.A. Sixth Edition. With 8 Plates and 63 Woodcuts. Crown 8vo, 7s. 6d. cloth.

<small>"This edition will be found the most complete treatise on the principles of measuring and valuing artificers' work that has yet been published."—*Building News*.</small>

Pocket Estimator and Technical Guide.

THE POCKET TECHNICAL GUIDE, MEASURER, AND ESTIMATOR FOR BUILDERS AND SURVEYORS. Containing Technical Directions for Measuring Work in all the Building Trades, Complete Specifications for Houses, Roads, and Drains, and an easy Method of Estimating the parts of a Building collectively. By A. C. BEATON. Sixth Edit. Waistcoat-pocket size, 1s. 6d. leather, gilt edges.

<small>"No builder, architect, surveyor, or valuer should be without his 'Beaton.'"—*Building News*.</small>

Donaldson on Specifications.

THE HANDBOOK OF SPECIFICATIONS; or, Practical Guide to the Architect, Engineer, Surveyor, and Builder, in drawing up Specifications and Contracts for Works and Constructions. Illustrated by Precedents of Buildings actually executed by eminent Architects and Engineers. By Professor T. L. DONALDSON, P.R.I.B.A., &c. New Edition. 8vo, with upwards of 1,000 pages of Text, and 33 Plates. £1 11s. 6d. cloth.

<small>"Valuable as a record, and more valuable still as a book of precedents.... Suffice it to say that Donaldson's 'Handbook of Specifications' must be bought by all architects."—*Builder*.</small>

Bartholomew and Rogers' Specifications.

SPECIFICATIONS FOR PRACTICAL ARCHITECTURE. A Guide to the Architect, Engineer, Surveyor, and Builder. With an Essay on the Structure and Science of Modern Buildings. Upon the Basis of the Work by ALFRED BARTHOLOMEW, thoroughly Revised, Corrected, and greatly added to by FREDERICK ROGERS, Architect. Third Edition, Revised, with Additions. With numerous Illustrations. Medium 8vo, 15s. cloth.

"The collection of specifications prepared by Mr. Rogers on the basis of Bartholomew's work is too well known to need any recommendation from us. It is one of the books with which every young architect must be equipped." —*Architect.*

Construction.

THE SCIENCE OF BUILDING: An Elementary Treatise on the Principles of Construction. By E. WYNDHAM TARN, M.A., Architect. Third Edition, Revised and Enlarged. With 59 Engravings. Fcap. 8vo, 4s. cl.

"A very valuable book, which we strongly recommend to all students." —*Builder.*

House Building and Repairing.

THE HOUSE-OWNER'S ESTIMATOR; or, What will it Cost to Build, Alter, or Repair? A Price Book for Unprofessional People, as well as the Architectural Surveyor and Builder. By JAMES D. SIMON. Edited by FRANCIS T. W. MILLER, A.R.I.B.A. Fourth Edition. Crown 8vo, 3s. 6d. cloth.

"In two years it will repay its cost a hundred times over." —*Field.*

Cottages and Villas.

COUNTRY AND SUBURBAN COTTAGES AND VILLAS: How to Plan and Build Them. Containing 33 Plates, with Introduction, General Explanations, and Description of each Plate. By JAMES W. BOGUE, Architect, Author of "Domestic Architecture," &c. 4to, 10s. 6d. cloth.

Building; Civil and Ecclesiastical.

A BOOK ON BUILDING, Civil and Ecclesiastical, including Church Restoration; with the Theory of Domes and the Great Pyramid, &c. By Sir EDMUND BECKETT, Bart., LL.D., F.R.A.S. Second Edition. Fcap. 8vo, 5s. cloth.

"A book which is always amusing and nearly always instructive." —*Times.*

Sanitary Houses, etc.

THE SANITARY ARRANGEMENTS OF DWELLING-HOUSES. By A. J. WALLIS TAYLER, A.M. Inst. C.E. Crown 8vo, with numerous Illustrations. Price about 3s. cloth.

[Nearly ready.

Ventilation of Buildings.

VENTILATION. A Text Book to the Practice of the Art of Ventilating Buildings. By W. P. BUCHAN, R.P. 12mo, 4s. cloth.

"Contains a great amount of useful practical information, as thoroughly interesting as it is technically reliable." —*British Architect.*

The Art of Plumbing.

PLUMBING. A Text Book to the Practice of the Art or Craft of the Plumber. By WILLIAM PATON BUCHAN, R.P. Sixth Edition, Enlarged. 12mo, 4s. cloth.

"A text-book which may be safely put in the hands of every young plumber." —*Builder.*

Geometry for the Architect, Engineer, etc.

PRACTICAL GEOMETRY, for the Architect, Engineer, and Mechanic. Giving Rules for the Delineation and Application of various Geometrical Lines, Figures and Curves. By E. W. TARN, M.A., Architect. 8vo, 9s. cloth.

"No book with the same objects in view has ever been published in which the clearness of the rules laid down and the illustrative diagrams have been so satisfactory."—*Scotsman*.

The Science of Geometry.
THE GEOMETRY OF COMPASSES; or, Problems Resolved by the mere Description of Circles, and the use of Coloured Diagrams and Symbols. By OLIVER BYRNE. Coloured Plates. Crown 8vo, 3s. 6d. cloth.

CARPENTRY, TIMBER, etc.

Tredgold's Carpentry, Revised & Enlarged by Tarn.
THE ELEMENTARY PRINCIPLES OF CARPENTRY. A Treatise on the Pressure and Equilibrium of Timber Framing, the Resistance of Timber, and the Construction of Floors, Arches, Bridges, Roofs, Uniting Iron and Stone with Timber, &c. To which is added an Essay on the Nature and Properties of Timber, &c., with Descriptions of the kinds of Wood used in Building; also numerous Tables of the Scantlings of Timber for different purposes, the Specific Gravities of Materials, &c. By THOMAS TREDGOLD, C.E. With an Appendix of Specimens of Various Roofs of Iron and Stone, Illustrated. Seventh Edition, thoroughly revised and considerably enlarged by E. WYNDHAM TARN, M.A., Author of "The Science of Building," &c. With 61 Plates, Portrait of the Author, and several Woodcuts. In One large Vol., 4to, price £1 5s. cloth.

"Ought to be in every architect's and every builder's library."—*Builder*.

"A work whose monumental excellence must commend it wherever skilful carpentry is concerned. The author's principles are rather confirmed than impaired by time. The additional plates are of great intrinsic value."—*Building News*.

Woodworking Machinery.
WOODWORKING MACHINERY: Its Rise, Progress, and Construction. With Hints on the Management of Saw Mills and the Economical Conversion of Timber. Illustrated with Examples of Recent Designs by leading English, French, and American Engineers. By M. POWIS BALE, A.M.Inst.C.E., M.I.M.E. Second Edition, Revised, with large Additions. Large crown 8vo, 440 pp., 9s. cloth.

[Just published.

"Mr. Bale is evidently an expert on the subject and he has collected so much information that his book is all-sufficient for builders and others engaged in the conversion of timber."—*Architect*.

"The most comprehensive compendium of wood-working machinery we have seen. The author is a thorough master of his subject."—*Building News*.

Saw Mills.
SAW MILLS: Their Arrangement and Management, and the Economical Conversion of Timber. (A Companion Volume to "Woodworking Machinery.") By M. POWIS BALE. Crown 8vo, 10s. 6d. cloth.

"The *administration* of a large sawing establishment is discussed, and the subject examined from a financial standpoint. Hence the size, shape, order, and disposition of saw-mills and the like are gone into in detail, and the course of the timber is traced from its reception to its delivery in its converted state. We could not desire a more complete or practical treatise."—*Builder*.

Nicholson's Carpentry.
THE CARPENTER'S NEW GUIDE; or, Book of Lines for Carpenters; comprising all the Elementary Principles essential for acquiring a knowledge of Carpentry. Founded on the late PETER NICHOLSON'S Standard Work. New Edition, Revised by A. ASHPITEL, F.S.A. With Practical Rules

on Drawing, by G. Pyne. With 74 Plates, 4to, £1 1s. cloth.

Handrailing and Stairbuilding.

A PRACTICAL TREATISE ON HANDRAILING: Showing New and Simple Methods for Finding the Pitch of the Plank, Drawing the Moulds, Bevelling, Jointing-up, and Squaring the Wreath. By George Collings. Second Edition, Revised and Enlarged, to which is added A Treatise on Stairbuilding. 12mo, 2s. 6d. cloth limp.

"Will be found of practical utility in the execution of this difficult branch of joinery."—*Builder.*

"Almost every difficult phase of this somewhat intricate branch of joinery is elucidated by the aid of plates and explanatory letterpress."—*Furniture Gazette.*

Circular Work.

CIRCULAR WORK IN CARPENTRY AND JOINERY: A Practical Treatise on Circular Work of Single and Double Curvature. By George Collings. With Diagrams. Second Edit, 12mo, 2s. 6d. cloth limp.

"An excellent example of what a book of this kind should be. Cheap in price, clear in definition and practical in the examples selected."—*Builder.*

Handrailing.

HANDRAILING COMPLETE IN EIGHT LESSONS. On the Square-Cut System. By J. S. Goldthorp, Teacher of Geometry and Building Construction at the Halifax Mechanic's Institute. With Eight Plates and over 150 Practical Exercises. 4to, 3s. 6d. cloth.

"Likely to be of considerable value to joiners and others who take a pride in good work. We heartily commend it to teachers and students."—*Timber Trades Journal.*

Timber Merchant's Companion.

THE TIMBER MERCHANT'S AND BUILDER'S COMPANION. Containing New and Copious Tables of the Reduced Weight and Measurement of Deals and Battens, of all sizes, from One to a Thousand Pieces, and the relative Price that each size bears per Lineal Foot to any given Price per Petersburg Standard Hundred; the Price per Cube Foot of Square Timber to any given Price per Load of 50 Feet; the proportionate Value of Deals and Battens by the Standard, to Square Timber by the Load of 50 Feet; the readiest mode of ascertaining the Price of Scantling per Lineal Foot of any size, to any given Figure per Cube Foot, &c. &c. By William Dowsing. Fourth Edition, Revised and Corrected. Cr. 8vo, 3s. cl.

"Everything is as concise and clear as it can possibly be made. There can be no doubt that every timber merchant and builder ought to possess it."—*Hull Advertiser.*

"We are glad to see a fourth edition of these admirable tables, which for correctness and simplicity of arrangement leave nothing to be desired."—*Timber Trades Journal.*

Practical Timber Merchant.

THE PRACTICAL TIMBER MERCHANT. Being a Guide for the use of Building Contractors, Surveyors, Builders, &c., comprising useful Tables for all purposes connected with the Timber Trade, Marks of Wood, Essay on the Strength of Timber, Remarks on the Growth of Timber, &c. By W. Richardson. Fcap. 8vo, 3s. 6d. cloth.

"This handy manual contains much valuable information for the use of timber merchants, builders, foresters, and all others connected with the growth, sale, and manufacture of timber."—*Journal of Forestry.*

Timber Freight Book.

THE TIMBER MERCHANT'S, SAW MILLER'S, AND IMPORTER'S FREIGHT BOOK AND ASSISTANT. Comprising Rules, Tables, and Memoranda relating to the Timber Trade. By William Richardson, Timber

Broker; together with a Chapter on "SPEEDS OF SAW MILL MACHINERY," by M. POWIS BALE, M.I.M.E., &c. 12mo, 3s. 6d. cl. boards.

"A very useful manual of rules, tables, and memoranda relating to the timber trade. We recommend it as a compendium of calculation to all timber measurers and merchants, and as supplying a real want in the trade."—*Building News.*

Packing-Case Makers, Tables for.

PACKING-CASE TABLES; showing the number of Superficial Feet in Boxes or Packing-Cases, from six inches square and upwards. By W. RICHARDSON, Timber Broker. Third Edition. Oblong 4to, 3s. 6d. cl.

"Invaluable labour-saving tables."—*Ironmonger.*
"Will save much labour and calculation."—*Grocer.*

Superficial Measurement.

THE TRADESMAN'S GUIDE TO SUPERFICIAL MEASUREMENT. Tables calculated from 1 to 200 inches in length, by 1 to 108 inches in breadth. For the use of Architects, Surveyors, Engineers, Timber Merchants, Builders, &c. By JAMES HAWKINGS. Fourth Edition. Fcap., 3s. 6d. cloth.

"A useful collection of tables to facilitate rapid calculation of surfaces. The exact area of any surface of which the limits have been ascertained can be instantly determined. The book will be found of the greatest utility to all engaged in building operations."—*Scotsman.*

"These tables will be found of great assistance to all who require to make calculations in superficial measurement."—*English Mechanic.*

Forestry.

THE ELEMENTS OF FORESTRY. Designed to afford Information concerning the Planting and Care of Forest Trees for Ornament or Profit, with Suggestions upon the Creation and Care of Woodlands. By F. B. HOUGH. Large crown 8vo, 10s. cloth.

Timber Importer's Guide.

THE TIMBER IMPORTER'S, TIMBER MERCHANT'S, AND BUILDER'S STANDARD GUIDE. By RICHARD E. GRANDY. Comprising an Analysis of Deal Standards, Home and Foreign, with Comparative Values and Tabular Arrangements for fixing Net Landed Cost on Baltic and North American Deals, including all intermediate Expenses, Freight, Insurance, &c. &c. Together with copious Information for the Retailer and Builder. Third Edition, Revised. 12mo, 2s. cloth limp.

"Everything it pretends to be: built up gradually, it leads one from a forest to a treenail, and throws in, as a makeweight, a host of material concerning bricks, columns, cisterns, &c."—*English Mechanic.*

DECORATIVE ARTS, etc.

Woods and Marbles (Imitation of).

SCHOOL OF PAINTING FOR THE IMITATION OF WOODS AND MARBLES, as Taught and Practised by A. R. VAN DER BURG and P. VAN DER BURG, Directors of the Rotterdam Painting Institution. Royal folio, 18½ by 12½ in., Illustrated with 24 full-size Coloured Plates; also 12 plain Plates, comprising 154 Figures. Second and Cheaper Edition. Price £1 11s. 6d.

List of Plates.

1. Various Tools required for Wood Painting—2, 3. Walnut: Preliminary Stages of Graining and Finished Specimen—4. Tools used for Marble Painting and Method of Manipulation—5, 6. St. Remi Marble: Earlier Operations and Finished Specimen—7. Methods of Sketching different Grains, Knots, &c.—8, 9. Ash: Preliminary Stages and Finished Specimen—10. Methods of Sketching Marble Grains—11, 12. Breche Marble: Preliminary Stages of Working and Finished Specimen—13. Maple: Methods of Producing the different Grains—14, 15. Bird's-eye Maple: Preliminary Stages and Finished Specimen—16. Methods of Sketching the different Species of White Marble—17, 18. White Marble: Preliminary Stages of Process and Finished Specimen—19. Mahogany: Specimens of various Grains and Methods of Manipulation—20, 21. Mahogany: Earlier Stages and Finished Specimen—22, 23, 24. Sienna Marble: Varieties of Grain, Preliminary Stages and Finished Specimen—25, 26, 27. Juniper Wood: Methods of producing Grain, &c.: Preliminary Stages and Finished Specimen—28, 29, 30. Vert de Mer Marble: Varieties of Grain and Methods of Working Unfinished and Finished Specimens—31, 32, 33. Oak: Varieties of Grain, Tools Employed, and Methods of Manipulation, Preliminary Stages and Finished Specimen—34, 35, 36. Waulsort Marble: Varieties of Grain, Unfinished and Finished Specimens.

"Those who desire to attain skill in the art of painting woods and marbles will find advantage in consulting this book.... Some of the Working Men's Clubs should give their young men the opportunity to study it."—*Builder.*

"A comprehensive guide to the art. The explanations of the processes, the manipulation and management of the colours, and the beautifully executed plates will not be the least valuable to the student who aims at making his work a faithful transcript of nature."—*Building News.*

Wall Paper.

WALL PAPER DECORATION. By ARTHUR SEYMOUR JENNINGS, Author of "Practical Paper Hanging." With numerous Illustrations. Demy 8vo.

[*In preparation.*

House Decoration.

ELEMENTARY DECORATION. A Guide to the Simpler Forms of Everyday Art. Together with PRACTICAL HOUSE DECORATION. By JAMES W. FACEY. With numerous Illustrations. In One Vol., 5s. strongly half-bound.

House Painting, Graining, etc.

HOUSE PAINTING, GRAINING, MARBLING, AND SIGN WRITING, A Practical Manual of. By ELLIS A. DAVIDSON. Sixth Edition. With Coloured Plates and Wood Engravings. 12mo, 6s. cloth boards.

"A mass of information, of use to the amateur and of value to the practical man."—*English Mechanic.*

Decorators, Receipts for.

THE DECORATOR'S ASSISTANT: A Modern Guide to Decorative Artists and Amateurs, Painters, Writers, Gilders, &c. Containing upwards of 600 Receipts, Rules and Instructions; with a variety of Information for General Work connected with every Class of Interior and Exterior Decorations, &c. Fifth Edition, Revised. 152 pp., crown 8vo, 1s. in wrapper.

"Full of receipts of value to decorators, painters, gilders. &c. The book contains the gist of larger treatises on colour and technical processes. It would be difficult to meet with a work so full of varied information on the painter's art."—*Building News.*

Moyr Smith on Interior Decoration.

ORNAMENTAL INTERIORS, ANCIENT AND MODERN. By J. MOYR SMITH. Super-royal 8vo, with 32 full-page Plates and numerous smaller Illustrations, handsomely bound in cloth, gilt top, price 18s.

"The book is well illustrated and handsomely got up, and contains some true criticism and a good many good examples of decorative treatment."—*The Builder.*

British and Foreign Marbles.

MARBLE DECORATION and the Terminology of British and Foreign Marbles. A Handbook for Students. By GEORGE H. BLAGROVE, Author of "Shoring and its Application," &c. With 28 Illustrations. Crown 8vo, 3s. 6d. cloth.

"This most useful and much wanted handbook should be in the hands of every architect and builder."—*Building World*.

"A carefully and usefully written treatise; the work is essentially practical."—*Scotsman*.

Marble Working, etc.

MARBLE AND MARBLE WORKERS: A Handbook for Architects, Artists, Masons, and Students. By ARTHUR LEE, Author of "A Visit to Carrara," "The Working of Marble," &c. Small crown 8vo, 2s. cloth.

"A really valuable addition to the technical literature of architects and masons."—*Building News*.

DELAMOTTE'S WORKS ON ILLUMINATION AND ALPHABETS.

A PRIMER OF THE ART OF ILLUMINATION, *for the Use of Beginners*: with a Rudimentary Treatise on the Art, Practical Directions for its Exercise, and Examples taken from Illuminated MSS., printed in Gold and Colours. By F. DELAMOTTE. New and Cheaper Edition. Small 4to, 6s. ornamental boards.

"The examples of ancient MSS. recommended to the student, which, with much good sense, the author chooses from collections accessible to all, are selected with judgment and knowledge, as well as taste."—*Athenæum*.

ORNAMENTAL ALPHABETS, Ancient and Mediæval, *from the Eighth Century, with Numerals;* including Gothic, Church-Text, large and small, German, Italian, Arabesque, Initials for Illumination, Monograms, Crosses, &c. &c., for the use of Architectural and Engineering Draughtsmen, Missal Painters, Masons, Decorative Painters, Lithographers, Engravers, Carvers, &c. &c. Collected and Engraved by F. DELAMOTTE, and printed in Colours. New and Cheaper Edition. Royal 8vo, oblong, 2s. 6d. ornamental boards.

"For those who insert enamelled sentences round gilded chalices, who blazon shop legends over shop-doors, who letter church walls with pithy sentences from the Decalogue, this book will be useful."—*Athenæum*.

EXAMPLES OF MODERN ALPHABETS, *Plain and Ornamental;* including German, Old English, Saxon, Italic, Perspective, Greek, Hebrew, Court Hand, Engrossing, Tuscan, Riband, Gothic, Rustic, and Arabesque; with several Original Designs, and an Analysis of the Roman and Old English Alphabets, large and small, and Numerals, for the use of Draughtsmen, Surveyors, Masons, Decorative Painters, Lithographers, Engravers, Carvers, &c. Collected and Engraved by F. DELAMOTTE, and printed in Colours. New and Cheaper Edition. Royal 8vo, oblong, 2s. 6d. ornamental boards.

"There is comprised in it every possible shape into which the letters of the alphabet and numerals can be formed, and the talent which has been expended in the conception of the various plain and ornamental letters is wonderful." — *Standard*.

MEDIÆVAL ALPHABETS AND INITIALS FOR ILLUMINATORS. By F. G. DELAMOTTE. Containing 21 Plates and Illuminated Title, printed in Gold and Colours. With an Introduction by J. WILLIS BROOKS. Fourth and Cheaper Edition. Small 4to, 4s. ornamental boards.

"A volume in which the letters of the alphabet come forth glorified in gilding and all the colours of the prism interwoven and intertwined and interm ingled." — *Sun*.

THE EMBROIDERER'S BOOK OF DESIGN. Containing Initials, Emblems, Cyphers, Monograms, Ornamental Borders, Ecclesiastical Devices, Mediæval and Modern Alphabets, and National Emblems. Collected by F. DELAMOTTE, and printed in Colours. Oblong royal 8vo, 1s. 6d. ornamental wrapper.

"The book will be of great assistance to ladies and young children who are endowed with the art of plying the needle in this most ornamental and useful pretty work." — *East Anglian Times*.

Wood Carving.

INSTRUCTIONS IN WOOD-CARVING, *for Amateurs*: with Hints on Design. By A LADY. With Ten Plates. New and Cheaper Edition. Crown 8vo, 2s. in emblematic wrapper.

"The handicraft of the wood-carver, so well as a book can impart it, may be learnt from 'A Lady's' publication." — *Athenæum*.

NATURAL SCIENCE, etc.

The Heavens and their Origin.

THE VISIBLE UNIVERSE: Chapters on the Origin and Construction of the Heavens. By J. E. GORE, F.R.A.S., Author of "Star Groups," &c. Illustrated by 6 Stellar Photographs and 12 Plates. Demy 8vo, 16s. cloth, gilt top.

"A valuable and lucid summary of recent astronomical theory, rendered more valuable and attractive by a series of stellar photographs and other illustrations." — *The Times*.

"In presenting a clear and concise account of the present state of our knowledge, Mr. Gore has made a valuable addition to the literature of the subject." — *Nature*.

"One of the finest works on astronomical science that has recently appeared in our language. In spirit and in method it is scientific from cover to cover, but the style is so clear and attractive that it will be as acceptable and as readable to those who make no scientific pretensions as to those who devote themselves specially to matters astronomical." — *Leeds Mercury*.

"As interesting as a novel, and instructive withal; the text being made still more luminous by stellar photographs and other illustrations.... A most valuable book." — *Manchester Examiner*.

The Constellations.

STAR GROUPS: A Student's Guide to the Constellations. By J. ELLARD GORE, F.R.A.S., M.R.I.A., &c., Author of "The Visible Universe," "The Scenery of the Heavens." With 30 Maps. Small 4to, 5s. cloth, silvered.

"A knowledge of the principal constellations visible in our latitudes may be easily acquired from the thirty maps and accompanying text contained in this work." — *Nature*.

"The volume contains thirty maps showing stars of the sixth magnitude — the usual naked-eye limit — and each is accompanied by a brief commentary, adapted to facilitate recognition and bring to notice objects of special interest. For the purpose of a preliminary survey of the midnight pomp of the heavens, nothing could be better than a set of delineations averaging scarcely twenty square inches in area, and including nothing that cannot at once be identified." — *Saturday Review*.

"A very compact and handy guide to the constellations." — *Athenæum*.

Astronomical Terms.

AN ASTRONOMICAL GLOSSARY; or, Dictionary of Terms used in Astronomy. With Tables of Data and Lists of Remarkable and Interesting Celestial Objects. By J. ELLARD GORE, F.R.A.S., Author of "The Visible Universe," &c. Small crown 8vo, 2s. 6d. cloth.

"A very useful little work for beginners in astronomy, and not to be despised by more advanced students."—*The Times*.

"A very handy book ... the utility of which is much increased by its valuable tables of astronomical data."—*The Athenæum*.

"Astronomers of all kinds will be glad to have it for reference."—*Guardian*.

The Microscope.

THE MICROSCOPE: Its Construction and Management, including Technique, Photo-micrography, and the Past and Future of the Microscope. By Dr. HENRI VAN HEURCK, Director of the Antwerp Botannical Gardens. English Edition, Re-Edited and Augmented by the Author from the Fourth French Edition, and Translated by WYNNE E. BAXTER, F.R.M.S., F.G.S., &c. About 400 pages, with Three Plates and upwards of 250 Woodcuts. Imp. 8vo, 18s. cloth gilt.

"A translation of a well-known work, at once popular and comprehensive."—*Times*.
"The translation is as felicitious as it is accurate."—*Nature*.

Astronomy.

ASTRONOMY. By the late Rev. ROBERT MAIN, M.A., F.R.S. Third Edition, Revised, by WM. THYNNE LYNN, B.A., F.R.A.S., formerly of the Royal Observatory, Greenwich, 12mo, 2s. cloth limp.

"A sound and simple treatise, and a capital book for beginners."—*Knowledge*.
"Accurately brought down to the requirements of the present time."—*Educational Times*.

Recent and Fossil Shells.

A MANUAL OF THE MOLLUSCA: Being a Treatise on Recent and Fossil Shells. By S. P. WOODWARD, A.L.S., F.G.S., late Assistant Palæontologist in the British Museum. With an Appendix on *Recent and Fossil Conchological Discoveries*, by RALPH TATE, A.L.S., F.G.S. Illustrated by A. N. WATERHOUSE and JOSEPH WILSON LOWRY. With 23 Plates and upwards of 300 Woodcuts. Reprint of Fourth Ed., 1880. Cr. 8vo, 7s. 6d. cl.

"A most valuable storehouse of conchological and geological information."—*Science Gossip*.

Geology and Genesis.

THE TWIN RECORDS OF CREATION; or, Geology and Genesis: their Perfect Harmony and Wonderful Concord. By GEORGE W. VICTOR LE VAUX. Fcap. 8vo, 5s. cloth.

"A valuable contribution to the evidences of Revelation, and disposes very conclusively of the arguments of those who would set God's Works against God's Word. No real difficulty is shirked and no sophistry is left unexposed."—*The Rock*.

DR. LARDNER'S COURSE OF NATURAL PHILOSOPHY.

THE HANDBOOK OF MECHANICS. Enlarged and almost Rewritten by BENJAMIN LOEWY, F.R.A.S. With 378 Illustrations. Post 8vo, 6s. cloth.

"The perspicuity of the original has been retained, and chapters which had become obsolete have been replaced by others of more modern character. The explanations throughout are studiously popular, and care has been taken to show the application of the various branches of physics to the industrial arts, and to the practical business of life."—*Mining Journal*.

"Mr. Loewy has carefully revised the book, and brought it up to modern requirements."—*Nature*.

"Natural philosophy has had few exponents more able or better skilled in the art of popularising the subject than Dr. Lardner; and Mr. Loewy is doing good service in fitting this treatise, and the others of the series, for use at the present time."—*Scotsman*.

THE HANDBOOK OF HYDROSTATICS AND PNEUMATICS. New Edition, Revised and Enlarged, by BENJAMIN LOEWY, F.R.A.S. With 236 Illustrations. Post 8vo, 5s. cloth.

"For those 'who desire to attain an accurate knowledge of physical science without the profound methods of mathematical investigation,' this work is not merely intended, but well adapted."—*Chemical News.*

"The volume before us has been carefully edited, augmented to nearly twice the bulk of the former edition, and all the most recent matter has been added.... It is a valuable text-book."—*Nature.*

"Candidates for pass examinations will find it, we think, specially suited to their requirements."—*English Mechanic.*

THE HANDBOOK OF HEAT. Edited and almost entirely Rewritten by BENJAMIN LOEWY, F.R.A.S., &c. 117 Illusts. Post 8vo, 6s. cloth.

"The style is always clear and precise, and conveys instruction without leaving any cloudiness or lurking doubts behind."—*Engineering.*

"A most exhaustive book on the subject on which it treats, and is so arranged that it can be understood by all who desire to attain an accurate knowledge of physical science.... Mr. Loewy has included all the latest discoveries in the varied laws and effects of heat."—*Standard.*

"A complete and handy text-book for the use of students and general readers."—*English Mechanic.*

THE HANDBOOK OF OPTICS. By DIONYSIUS LARDNER, D.C.L., formerly Professor of Natural Philosophy and Astronomy in University College, London. New Edition. Edited by T. OLVER HARDING, B.A. Lond., of University College, London. With 298 Illustrations. Small 8vo, 448 pages, 5s. cloth.

"Written by one of the ablest English scientific writers, beautifully and elaborately illustrated."—*Mechanic's Magazine.*

THE HANDBOOK OF ELECTRICITY, MAGNETISM, AND ACOUSTICS. By Dr. LARDNER. Ninth Thousand. Edit. by GEORGE CAREY FOSTER, B.A., F.C.S. With 400 Illustrations. Small 8vo, 5s. cloth.

"The book could not have been entrusted to anyone better calculated to preserve the terse and lucid style of Lardner, while correcting his errors and bringing up his work to the present state of scientific knowledge."—*Popular Science Review.*

THE HANDBOOK OF ASTRONOMY. Forming a Companion to the "Handbook of Natural Philosophy." By DIONYSIUS LARDNER, D.C.L., formerly Professor of Natural Philosophy and Astronomy in University College, London. Fourth Edition, Revised and Edited by EDWIN DUNKIN, F.R.A.S., Royal Observatory, Greenwich. With 38 Plates and upwards of 100 Woodcuts. In One Vol., small 8vo, 550 pages, 9s. 6d. cloth.

"Probably no other book contains the same amount of information in so compendious and well-arranged a form—certainly none at the price at which this is offered to the public."—*Athenæum.*

"We can do no other than pronounce this work a most valuable manual of astronomy, and we strongly recommend it to all who wish to acquire a general—but at the same time correct—acquaintance with this sublime science."—*Quarterly Journal of Science.*

"One of the most deservedly popular books on the subject.... We would recommend not only the student of the elementary principles of the science, but he who aims at mastering the higher and mathematical branches of astronomy, not to be without this work beside him."—*Practical Magazine.*

Geology.

RUDIMENTARY TREATISE ON GEOLOGY, PHYSICAL AND HISTORICAL. Consisting of "Physical Geology," which sets forth the leading Principles of the Science; and "Historical Geology," which treats of the Mineral and Organic Conditions of the Earth at each successive epoch, especial reference being made to the British Series of Rocks. By RALPH TATE, A.L.S., F.G.S., &c. With 250 Illustrations. 12mo, 5s. cl. bds.

"The fulness of the matter has elevated the book into a manual. Its information is exhaustive and well arranged."—*School Board Chronicle.*

DR. LARDNER'S MUSEUM OF SCIENCE AND ART.

THE MUSEUM OF SCIENCE AND ART. Edited by DIONYSIUS LARDNER, D.C.L., formerly Professor of Natural Philosophy and Astronomy in University College, London. With upwards of 1,200 Engravings on Wood. In 6 Double Volumes, £1 1s. in a new and elegant cloth binding; or handsomely bound in half-morocco, 31s. 6d.

⁂ OPINIONS OF THE PRESS.

"This series, besides affording popular but sound instruction on scientific subjects, with which the humblest man in the country ought to be acquainted, also undertakes that teaching of 'Common Things' which every well-wisher of his kind is anxious to promote. Many thousand copies of this serviceable publication have been printed, in the belief and hope that the desire for instruction and improvement widely prevails; and we have no fear that such enlightened faith will meet with disappointment."—*Times.*

"A cheap and interesting publication, alike informing and attractive. The papers combine subjects of importance and great scientific knowledge, considerable inductive powers, and a popular style of treatment."—*Spectator.*

"The 'Museum of Science and Art' is the most valuable contribution that has ever been made to the Scientific Instruction of every class of society."—Sir DAVID BREWSTER, in the *North British Review.*

"Whether we consider the liberality and beauty of the illustrations, the charm of the writing, or the durable interest of the matter, we must express our belief that there is hardly to be found among the new books one that would be welcomed by people of so many ages and classes as a valuable present."—*Examiner.*

⁂ *Separate books formed from the above, suitable for Workmen's Libraries, Science Classes, etc.*

Common Things Explained. Containing Air, Earth, Fire, Water, Time, Man, the Eye, Locomotion, Colour, Clocks and Watches, &c. 233 Illustrations, cloth gilt, 5s.

The Microscope. Containing Optical Images, Magnifying Glasses, Origin and Description of the Microscope, Microscopic Objects, the Solar Microscope, Microscopic Drawing and Engraving, &c. 147 Illustrations, cloth gilt, 2s.

Popular Geology. Containing Earthquakes and Volcanoes, the Crust of the Earth, &c. 201 Illustrations, cloth gilt, 2s. 6d.

Popular Physics. Containing Magnitude and Minuteness, the Atmosphere, Meteoric Stones, Popular Fallacies, Weather Prognostics, the Thermometer, the Barometer, Sound, &c. 85 Illustrations, cloth gilt, 2s. 6d.

Steam, and its Uses. Including the Steam Engine, the Locomotive, and Steam Navigation. 89 Illustrations, cloth gilt, 2s.

Popular Astronomy. Containing How to observe the Heavens—The Earth, Sun, Moon, Planets, Light, Comets, Eclipses, Astronomical Influences, &c. 182 Illustrations, cloth gilt, 4s. 6d.

The Bee and White Ants: Their Manners and Habits. With Illustrations of Animal Instinct and Intelligence. 135 Illustrations, cloth gilt, 2s.

The Electric Telegraph Popularised. To render intelligible to all who can Read, irrespective of any previous Scientific Acquirements, the various forms of Telegraphy in Actual Operation. 100 Illustrations, cloth gilt, 1s. 6d.

Dr. Lardner's School Handbooks.

NATURAL PHILOSOPHY FOR SCHOOLS. By Dr. LARDNER. 328 Illustrations. Sixth Edition. One Vol., 3s. 6d. cloth.

"A very convenient class-book for junior students in private schools. It is intended to convey in clear and precise terms, general notions of all the principal divisions of Physical Science."—*British Quarterly Review.*

ANIMAL PHYSIOLOGY FOR SCHOOLS. By Dr. LARDNER. With 190 Illustrations. Second Edition. One Vol., 3s. 6d. cloth.

"Clearly written, well arranged, and excellently illustrated."—*Gardener's Chronicle.*

Lardner and Bright on the Electric Telegraph.

THE ELECTRIC TELEGRAPH. By Dr. LARDNER. Revised and Re-written by E. B. BRIGHT, F.R.A.S. 140 Illustrations. Small 8vo, 2s. 6d. cloth.

"One of the most readable books extant on the Electric Telegraph."—*English Mechanic.*

CHEMICAL MANUFACTURES, CHEMISTRY.

Chemistry for Engineers, etc.

ENGINEERING CHEMISTRY; A Practical Treatise for the Use of Analytical Chemists, Engineers, Iron Masters, Iron Founders, Students, and others. Comprising Methods of Analysis and Valuation of the Principal Materials used in Engineering Work, with numerous Analyses, Examples, and Suggestions. By H. JOSHUA PHILLIPS, F.I.C., F.C.S. formerly Analytical and Consulting Chemist to the Great Eastern Railway. Second Edition, Revised and Enlarged. Crown 8vo, 400 pp., with Illustrations, 10s. 6d. cloth.

[*Just published.*

"In this work the author has rendered no small service to a numerous body of practical men.... The analytical methods may be pronounced most satisfactory, being as accurate as the despatch required of engineering chemists permits."—*Chemical News.*

"Those in search of a handy treatise on the subject of analytical chemistry as applied to the every-day requirements of workshop practice will find this volume of great assistance."—*Iron.*

"The first attempt to bring forward a Chemistry specially written for the use of engineers, and we have no hesitation whatever in saying that it should at once be in the possession of every railway engineer."—*The Railway Engineer.*

"The book will be very useful to those who require a handy and concise *resume* of approved methods of analysing and valuing metals, oils, fuels, &c. It is, in fact, a work for chemists, a guide to the routine of the engineering laboratory.... The book is full of good things. As a handbook of technical analysis, it is very welcome."—*Builder.*

"Considering the extensive ground which such a subject as Engineering Chemistry covers, the work is complete, and recommends itself to both the practising analyst and the analytical student."—*Chemical Trade Journal.*

"The analytical methods given are, as a whole, such as are likely to give rapid and trustworthy results in experienced hands. There is much excellent descriptive matter in the work, the chapter on 'Oils and Lubrication' being specially noticeable in this respect."—*Engineer.*

Alkali Trade, Manufacture of Sulphuric Acid, etc.

A MANUAL OF THE ALKALI TRADE, including the Manufacture of Sulphuric Acid, Sulphate of Soda, and Bleaching Powder. By JOHN LOMAS, Alkali Manufacturer, Newcastle-upon-Tyne and London. With 232 Illustrations and Working Drawings, and containing 390 pages of Text. Second Edition, with Additions. Super-royal 8vo, £1 10s. cloth.

"This book is written by a manufacturer for manufacturers. The working details of the most approved forms of apparatus are given, and these are accompanied by no less than 232 wood engravings, all of which may be used for the purposes of construction. Every step in the manufacture is very fully described in this manual, and each improvement explained."—*Athenæum.*

"We find not merely a sound and luminous explanation of the chemical principles of the trade, but a notice of numerous matters which have a most important bearing on the successful conduct of alkali works, but which are generally overlooked by even experienced technological authors."—*Chemical Review.*

The Blowpipe.

THE BLOWPIPE IN CHEMISTRY, MINERALOGY, AND GEOLOGY. Containing all known Methods of Anhydrous Analysis, many Working Examples, and Instructions for Making Apparatus. By Lieut.-Colonel W. A. ROSS, R.A., F.G.S. With 120 Illustrations. Second Edition, Revised and Enlarged. Crown 8vo, 5s. cloth.

"The student who goes through the course of experimentation here laid down will gain a better insight into inorganic chemistry and mineralogy than if he had 'got up' any of the best text-books, and passed any number of examinations in their contents."—*Chemical News.*

Commercial Chemical Analysis.

THE COMMERCIAL HANDBOOK OF CHEMICAL ANALYSIS; or, Practical Instructions for the determination of the Intrinsic or Commercial Value of Substances used in Manufactures, in Trades, and in the Arts. By A. NORMANDY, Editor of Rose's "Treatise on Chemical Analysis." New Edition, to a great extent Re-written by HENRY M. NOAD, Ph.D., F.R.S. With numerous Illustrations. Crown 8vo, 12s. 6d. cloth.

"We strongly recommend this book to our readers as a guide, alike indispensable to the housewife as to the pharmaceutical practitioner."—*Medical Times.*

"Essential to the analysts appointed under the new Act. The most recent results are given and the work is well edited and carefully written."—*Nature.*

Dye-Wares and Colours.

THE MANUAL OF COLOURS AND DYE-WARES: Their Properties, Applications, Valuations, Impurities, and Sophistications. For the use of Dyers, Printers, Drysalters, Brokers, &c. By J. W. SLATER. Second Edition, Revised and greatly Enlarged. Crown 8vo, 7s. 6d. cloth.

"A complete encyclopædia of the *materia tinctoria.* The information given respecting each article is full and precise, and the methods of determining the value of articles such as these, so liable to sophistication, are given with clearness, and are practical as well as valuable."—*Chemist and Druggist.*

"There is no other work which covers precisely the same ground. To students preparing for examinations in dyeing and printing it will prove exceedingly useful."—*Chemical News.*

Modern Brewing and Malting.

A HANDYBOOK FOR BREWERS: Being a Practical Guide to the Art of Brewing and Malting. Embracing the Conclusions of Modern Research which bear upon the Practice of Brewing. By HERBERT EDWARDS WRIGHT, M.A., Author of "A Handbook for Young Brewers." Crown 8vo, 530 pp., 12s. 6d. cloth.

"May be consulted with advantage by the student who is preparing himself for examinational tests, while the scientific brewer will find in it a *resume* of all the most important discoveries of modern times. The work is written throughout in a clear and concise manner, and the author takes great care to discriminate between vague theories and well-established facts."—*Brewers' Journal.*

"We have great pleasure in recommending this handybook, and have no hesitation in saying that it is one of the best—if not the best—which has yet been written on the subject of beer-brewing in this country, and it should have a place on the shelves of every brewer's library."—*The Brewer's Guardian.*

"Although the requirements of the student are primarily considered, an acquaintance of half-an-hour's duration cannot fail to impress the practical brewer with the sense of having found a trustworthy guide and practical counsellor in brewery matters."—*Chemical Trade Journal.*

Analysis and Valuation of Fuels.

FUELS: SOLID, LIQUID, AND GASEOUS, Their Analysis and Valuation. For the Use of Chemists and Engineers. By H. J. PHILLIPS, F.C.S., formerly Analytical and Consulting Chemist to the Great Eastern Railway. Second Edition, Revised and Enlarged. Crown 8vo, 5s. cloth.

"Ought to have its place in the laboratory of every metallurgical establishment, and wherever fuel is used on a large scale."—*Chemical News.*

"Cannot fail to be of wide interest, especially at the present time."—*Railway News.*

Pigments.

THE ARTIST'S MANUAL OF PIGMENTS. Showing their Composition, Conditions of Permanency, Non-Permanency, and Adulterations; Effects in Combination with Each Other and with Vehicles; and the most Reliable Tests of Purity Together with the Science and Art Department's Examination Questions on Painting. By H. C. STANDAGE. Second Edition.

Crown 8vo, 2s. 6d. cloth.

"This work is indeed *multum-in-parvo*, and we can, with good conscience, recommend it to all who come in contact with pigments, whether as makers, dealers or users." — *Chemical Review.*

Gauging. Tables and Rules for Revenue Officers, Brewers, etc.

A POCKET BOOK OF MENSURATION AND GAUGING: Containing Tables, Rules and Memoranda for Revenue Officers, Brewers, Spirit Merchants, &c. By J. B. MANT (Inland Revenue). Second Edition, Revised. 18mo, 4s. leather.

"This handy and useful book is adapted to the requirements of the Inland Revenue Department, and will be a favourite book of reference. The range of subjects is comprehensive, and the arrangement simple and clear." — *Civilian.*

"Should be in the hands of every practical brewer." — *Brewers' Journal.*

INDUSTRIAL ARTS, TRADES, AND MANUFACTURES.

Cotton Spinning.

COTTON MANUFACTURE: A Practical Manual. Embracing the various operations of Cotton Manufacture, Dyeing, &c. For the Use of Operatives, Overlookers, and Manufacturers. By JOHN LISTER, Technical Instructor, Pendleton. With numerous Illustrations. Demy 8vo, 7s. 6d. cloth.

[Just published.

Flour Manufacture, Milling, etc.

FLOUR MANUFACTURE: A Treatise on Milling Science and Practice. By FRIEDRICH KICK, Imperial Regierungsrath, Professor of Mechanical Technology in the Imperial German Polytechnic Institute, Prague. Translated from the Second Enlarged and Revised Edition with Supplement. By H. H. P. POWLES, Assoc. Memb. Institution of Civil Engineers. Nearly 400 pp. Illustrated with 28 Folding Plates, and 167 Woodcuts. Royal 8vo, 25s. cloth.

"This valuable work is, and will remain, the standard authority on the science of milling.... The miller who has read and digested this work will have laid the foundation, so to speak, of a successful career; he will have acquired a number of general principles which he can proceed to apply. In this handsome volume we at last have the accepted text-book of modern milling in good, sound English, which has little, if any, trace of the German idiom." — *The Miller.*

"The appearance of this celebrated work in English is very opportune, and British millers will, we are sure, not be slow in availing themselves of its pages." — *Millers' Gazette.*

Agglutinants.

CEMENTS, PASTES, GLUES AND GUMS: A Practical Guide to the Manufacture and Application of the various Agglutinants required in the Building, Metal-Working, Wood-Working and Leather-Working Trades, and for Workshop, Laboratory or Office Use. With upwards of 900 Recipes and Formulæ. By H. C. STANDAGE, Chemist. Crown 8vo, 2s. 6d. cloth.

[Just published.

"We have pleasure in speaking favourably of this volume. So far as we have had experience, which is not inconsiderable, this manual is trustworthy." — *Athenæum.*

"As a revelation of what are considered trade secrets, this book will arouse an amount of curiosity among the large number of industries it touches." — *Daily Chronicle.*

"In this goodly collection of receipts it would be strange if a cement for any purpose cannot be found."—*Oil and Colourman's Journal.*

Soap-making.

THE ART OF SOAP-MAKING: A Practical Handbook of the Manufacture of Hard and Soft Soaps, Toilet Soaps, etc. Including many New Processes, and a Chapter on the Recovery of Glycerine from Waste Leys. By ALEXANDER WATT. Fourth Edition, Enlarged. Crown 8vo, 7s. 6d. cloth.

"The work will prove very useful, not merely to the technological student, but to the practical soap-boiler who wishes to understand the theory of his art."—*Chemical News.*

"A thoroughly practical treatise on an art which has almost no literature in our language. We congratulate the author on the success of his endeavour to fill a void in English technical literature."—*Nature.*

Paper Making.

PRACTICAL PAPER-MAKING: A Manual for Paper-makers and Owners and Managers of Paper-Mills. With Tables, Calculations, &c. By G. CLAPPERTON, Paper-maker. With Illustrations of Fibres from Micro-Photographs. Crown 8vo, 5s. cloth.

[*Just published.*

"The author caters for the requirements of responsible mill hands, apprentices, &c., whilst his manual will be found of great service to students of technology, as well as to veteran paper makers and mill owners. The illustrations form an excellent feature."—*Paper Trade Review.*

"We recommend everybody interested in the trade to get a copy of this thoroughly practical book."—*Paper Making.*

Paper Making.

THE ART OF PAPER MAKING: A Practical Handbook of the Manufacture of Paper from Rags, Esparto, Straw, and other Fibrous Materials. Including the Manufacture of Pulp from Wood Fibre, with a Description of the Machinery and Appliances used. To which are added Details of Processes for Recovering Soda from Waste Liquors. By ALEXANDER WATT, Author of "The Art of Soap-Making" With Illusts. Crown 8vo, 7s. 6d. cloth.

"It may be regarded as the standard work on the subject. The book is full of valuable information. The 'Art of Paper-making,' is in every respect a model of a text-book, either for a technical class or for the private student."—*Paper and Printing Trades Journal.*

Leather Manufacture.

THE ART OF LEATHER MANUFACTURE. Being a Practical Handbook, in which the Operations of Tanning, Currying, and Leather Dressing are fully Described, and the Principles of Tanning Explained, and many Recent Processes Introduced; as also the Methods for the Estimation of Tannin, and a Description of the Arts of Glue Boiling, Gut Dressing, &c. By ALEXANDER WATT, Author of "Soap-Making," &c. Second Edition. Crown 8vo, 9s. cloth.

"A sound, comprehensive treatise on tanning and its accessories. It is an eminently valuable production, which redounds to the credit of both author and publishers."—*Chemical Review.*

Boot and Shoe Making.

THE ART OF BOOT AND SHOE-MAKING. A Practical Handbook, including Measurement, Last-Fitting, Cutting-Out, Closing, and Making, with a Description of the most approved Machinery employed. By JOHN B. LENO, late Editor of *St. Crispin*, and *The Boot and Shoe-Maker*. 12mo, 2s. cloth limp.

"This excellent treatise is by far the best work ever written. The chapter on clicking, which shows how waste may be prevented, will save fifty times the price of the book."—*Scottish Leather Trader.*

Dentistry Construction.

MECHANICAL DENTISTRY: *A Practical Treatise on the Construction of the various kinds of Artificial Dentures.* Comprising also Useful Formulæ, Tables, and Receipts for Gold Plate, Clasps, Solders, &c. &c. By CHARLES HUNTER. Third Edition. Crown 8vo, 3s. 6d. cloth.

"We can strongly recommend Mr. Hunter's treatise to all students preparing for the profession of dentistry, as well as to every mechanical dentist."—*Dublin Journal of Medical Science.*

Wood Engraving.

WOOD ENGRAVING: *A Practical and Easy Introduction to the Study of the Art.* By WILLIAM NORMAN BROWN. Second Edition. With numerous Illustrations. 12mo, 1s. 6d. cloth limp.

"The book is clear and complete, and will be useful to anyone wanting to understand the first elements of the beautiful art of wood engraving."—*Graphic.*

Horology.

A TREATISE ON MODERN HOROLOGY, in Theory and Practice. Translated from the French of CLAUDIUS SAUNIER, ex-Director of the School of Horology at Maçon, by JULIEN TRIPPLIN, F.R A.S., Besançon Watch Manufacturer, and EDWARD RIGG, M.A., Assayer in the Royal Mint. With 78 Woodcuts and 22 Coloured Copper Plates. Second Edition. Super-royal 8vo, £2 2s. cloth; £2 10s. half-calf.

"There is no horological work in the English language at all to be compared to this production of M. Saunier's for clearness and completeness. It is alike good as a guide for the student and as a reference for the experienced horolegist and skilled workman."—*Horological Journal.*

"The latest, the most complete, and the most reliable of those literary productions to which continental watchmakers are indebted for the mechanical superiority over their English brethren—in fact, the Book of Books, is M. Saunier's 'Treatise.'"—*Watchmaker, Jeweller and Silversmith.*

Watchmaking.

THE WATCHMAKER'S HANDBOOK. Intended as a Workshop Companion for those engaged in Watchmaking and the Allied Mechanical Arts. Translated from the French of CLAUDIUS SAUNIER, and considerably enlarged by JULIEN TRIPPLIN, F.R.A.S., Vice-President of the Horological Institute, and EDWARD RIGG, M.A., Assayer in the Royal Mint. With numerous Woodcuts and 14 Copper Plates. Third Edition. Crown 8vo, 9s. cloth.

"Each part is truly a treatise in itself. The arrangement is good and the language is clear and concise. It is an admirable guide for the young watchmaker."—*Engineering.*

"It is impossible to speak too highly of its excellence. It fulfils every requirement in a handbook intended for the use of a workman. Should be found in every workshop."—*Watch and Clockmaker.*

"This book contains an immense number of practical details bearing on the daily occupation of a watchmaker."—*Watchmaker and Metalworker* (Chicago).

Watches and Timekeepers.

A HISTORY OF WATCHES AND OTHER TIMEKEEPERS. By JAMES F. KENDAL, M.B.H.Inst. 1s. 6d. boards; or 2s. 6d. cloth gilt.

"Mr. Kendal's book, for its size, is the best which has yet appeared on this subject in the English language."—*Industries.*

"Open the book where you may, there is interesting matter in it concerning the ingenious devices of the ancient or modern horologer. The subject is treated in a liberal and entertaining spirit, as might be expected of a historian who is a master of the craft."—*Saturday Review.*

Electrolysis of Gold, Silver, Copper, etc.

ELECTRO-DEPOSITION: *A Practical Treatise on the Electrolysis of Gold, Silver, Copper, Nickel, and other Metals and Alloys.* With descriptions of Voltaic Batteries, Magneto and Dynamo-Electric Machines, Thermopiles, and of the Materials and Processes used in every Department of the Art, and several Chapters on Electro-Metallurgy. By ALEXANDER WATT, Author of "Electro-Metallurgy," &c. Third Edition, Revised. Crown 8vo, 9s. cloth.

"Eminently a book for the practical worker in electro-deposition. It contains practical descriptions of methods, processes and materials as actually pursued and used in the workshop."—ENGINEER.

Electro-Metallurgy.

ELECTRO-METALLURGY: *Practically Treated*. By ALEXANDER WATT, Author of "Electro-Deposition," &c. Ninth Edition, including the most recent Processes. 12mo, 4s. cloth boards.

"From this book both amateur and artisan may learn everything necessary for the successful prosecution of electroplating."—*Iron*.

Working in Gold.

THE JEWELLER'S ASSISTANT IN THE ART OF WORKING IN GOLD: A Practical Treatise for Masters and Workmen, Compiled from the Experience of Thirty Years' Workshop Practice. By GEORGE E. GEE, Author of "The Goldsmith's Handbook," &c. Cr. 8vo, 7s. 6d. cloth.

"This manual of technical education is apparently destined to be a valuable auxiliary to a handicraft which is certainty capable of great improvement."—*The Times*.

"Very useful in the workshop, as the knowledge is practical, having been acquired by long experience, and all the recipes and directions are guaranteed to be successful."—*Jeweller and Metalworker*.

Electroplating.

ELECTROPLATING: A Practical Handbook on the Deposition of Copper, Silver, Nickel, Gold, Aluminium, Brass, Platinum, &c. &c. With Descriptions of the Chemicals, Materials, Batteries, and Dynamo Machines used in the Art. By J. W. URQUHART, C.E., Author of "Electric Light," &c. Third Edition, Revised, with Additions. Numerous Illustrations. Crown 8vo, 5s. cloth.

"An excellent practical manual."—*Engineering.*
"An excellent work, giving the newest information."—*Horological Journal.*

Electrotyping.

ELECTROTYPING: The Reproduction and Multiplication of Printing Surfaces and Works of Art by the Electro-deposition of Metals. By J. W. URQUHART, C.E. Crown 8vo, 5s. cloth.

"The book is thoroughly practical. The reader is, therefore, conducted through the leading laws of electricity, then through the metals used by electrotypers, the apparatus, and the depositing processes, up to the final preparation of the work."—*Art Journal.*

Goldsmiths' Work.

THE GOLDSMITH'S HANDBOOK. By GEORGE E. GEE, Jeweller, &c. Third Edition, considerably Enlarged, 12mo, 3s. 6d. cl. bds.

"A good, sound educator, and will be generally accepted as an authority."—*Horological Journal.*

Silversmiths' Work.

THE SILVERSMITH'S HANDBOOK. By GEORGE E. GEE, Jeweller, &c. Second Edition, Revised, with numerous Illustrations. 12mo, 3s. 6d. cloth boards.

"The chief merit of the work is its practical character.... The workers in the trade will speedily discover its merits when they sit down to study it."—*English Mechanic.*

∴ *The above two-works together, strongly half-bound, price 7s.*

Bread and Biscuit Baking.

THE BREAD AND BISCUIT BAKER'S AND SUGAR-BOILER'S ASSISTANT. Including a large variety of Modern Recipes. With Remarks on the Art of Bread-making. By ROBERT WELLS, Practical Baker. Second Edition, with Additional Recipes. Crown 8vo, 2s. cloth.

"A large number of wrinkles for the ordinary cook, as well as the baker."—*Saturday Review.*

Confectionery for Hotels and Restaurants.

THE PASTRYCOOK AND CONFECTIONER'S GUIDE. For Hotels, Restaurants and the Trade in general, adapted also for Family Use. By ROBERT WELLS, Author of "The Bread and Biscuit Baker's and Sugar-Boiler's Assistant." Crown 8vo, 2s. cloth.

"We cannot speak too highly of this really excellent work. In these days of keen competition our readers cannot do better than purchase this book."—*Bakers' Times.*

Ornamental Confectionery.

ORNAMENTAL CONFECTIONERY: A Guide for Bakers. Confectioners and Pastrycooks; including a variety of Modern Recipes, and Remarks on Decorative and Coloured Work. With 129 Original Designs. By ROBERT WELLS, Practical Baker, Author of "The Bread and Biscuit Baker's and Sugar-Boiler's Assistant," &c. Crown 8vo, cloth gilt, 5s.

"A valuable work, practical, and should be in the hands of every baker and confectioner. The illustrative designs are alone worth treble the amount charged for the whole work."—*Bakers' Times.*

Flour Confectionery.

THE MODERN FLOUR CONFECTIONER. Wholesale and Retail. Containing a large Collection of Recipes for Cheap Cakes, Biscuits, &c. With Remarks on the Ingredients used in their Manufacture. To which are added Recipes for Dainties for the Working Man's Table. By R. WELLS, Author of "The Bread and Biscuit Baker," &c. Crown 8vo, 2s. cl.

"The work is of a decidedly practical character, and in every recipe regard is had to economical working."—*North British Daily Mail.*

Laundry Work.

LAUNDRY MANAGEMENT. A Handbook for Use in Private and Public Laundries, Including Descriptive Accounts of Modern Machinery and Appliances for Laundry Work. By the EDITOR of "The Laundry Journal." With numerous Illustrations. Second Edition. Crown 8vo, 2s. 6d. cloth.

"This book should certainly occupy an honoured place on the shelves of all housekeepers who wish to keep themselves *au courant* of the newest appliances and methods."—*The Queen.*

HANDYBOOKS FOR HANDICRAFTS.

By PAUL N. HASLUCK,

EDITOR OF "WORK" (NEW SERIES); AUTHOR OF "LATHEWORK," "MILLING MACHINES," &c.

Crown 8vo, 144 pages, cloth, price 1s. each.

☞ *These* HANDYBOOKS *have been written to supply information for* WORKMEN, STUDENTS, *and* AMATEURS *in the several Handicrafts, on the actual* PRACTICE *of the* WORKSHOP, *and are intended to convey in plain language* TECHNICAL KNOWLEDGE *of the several* CRAFTS. *In describing the processes employed, and the manipulation of material, workshop terms are used; workshop practice is fully explained; and the text is freely illustrated with drawings of modern tools, appliances, and processes.*

THE METAL TURNER'S HANDYBOOK. A Practical Manual for Workers at the Foot-Lathe. With over 100 Illustrations. Price 1s.

"The book will be of service alike to the amateur and the artisan turner. It displays thorough knowledge of the subject."—*Scotsman.*

THE WOOD TURNER'S HANDYBOOK. A Practical Manual for Workers at the Lathe. With over 100 Illustrations. Price 1s.

"We recommend the book to young turners and amateurs. A multitude of workmen have hitherto sought in vain for a manual of this special industry."—*Mechanical World.*

THE WATCH JOBBER'S HANDYBOOK. A Practical Manual on Cleaning, Repairing, and Adjusting. With upwards of 100 Illustrations. Price 1s.

"We strongly advise all young persons connected with the watch trade to acquire and study this inexpensive work."—*Clerkenwell Chronicle.*

THE PATTERN MAKER'S HANDYBOOK. A Practical Manual on the Construction of Patterns for Founders. With upwards of 100 Illustrations. Price 1s.

"A most valuable, if not indispensable, manual for the pattern maker."—*Knowledge.*

THE MECHANIC'S WORKSHOP HANDYBOOK. A Practical Manual on Mechanical Manipulation. Embracing Information on various Handicraft Processes, with Useful Notes and Miscellaneous Memoranda.

Comprising about 200 Subjects. Price 1s.

"A very clever and useful book, which should be found in every workshop; and it should certainly find a place in all technical schools." — *Saturday Review.*

THE MODEL ENGINEER'S HANDYBOOK. A Practical Manual on the Construction of Model Steam Engines. With upwards of 100 Illustrations. Price 1s.

"Mr. Hasluck has produced a very good little book." — *Builder.*

THE CLOCK JOBBER'S HANDYBOOK. A Practical Manual on Cleaning, Repairing, and Adjusting. With upwards of 100 Illustrations. Price 1s.

"It is of inestimable service to those commencing the trade." — *Coventry Standard.*

THE CABINET WORKER'S HANDYBOOK: A Practical Manual on the Tools, Materials, Appliances, and Processes employed in Cabinet Work. With upwards of 100 Illustrations. Price 1s.

"Mr. Hasluck's thoroughgoing little Handybook is amongst the most practical guides we have seen for beginners in cabinet-work." — *Saturday Review.*

THE WOODWORKER'S HANDYBOOK OF MANUAL INSTRUCTION. Embracing Information on the Tools, Materials, Appliances and Processes employed in Woodworking. With 104 Illustrations. Price 1s.

[*Just published.*

THE METALWORKER'S HANDYBOOK. With upwards of 100 Illustrations.

[*In preparation.*

∴ OPINIONS OF THE PRESS.

"Written by a man who knows, not only how work ought to be done, but how to do it, and how to convey his knowledge to others."—*Engineering.*

"Mr. Hasluck writes admirably, and gives complete instructions."—*Engineer.*

"Mr. Hasluck combines the experience of a practical teacher with the manipulative skill and scientific knowledge of processes of the trained mechanician, and the manuals are marvels of what can be produced at a popular price."—*Schoolmaster.*

"Helpful to workmen of all ages and degrees of experience."—*Daily Chronicle.*

"Practical, sensible, and remarkably cheap."—*Journal of Education.*

"Concise, clear and practical."—*Saturday Review.*

COMMERCE, COUNTING-HOUSE WORK, TABLES, etc.

Commercial Education.

LESSONS IN COMMERCE. By Professor R. GAMBARO, of the Royal High Commercial School at Genoa. Edited and Revised by JAMES GAULT, Professor of Commerce and Commercial Law in King's College, London. Crown 8vo, 3s. 6d. cloth.

"The publishers of this work have rendered considerable service to the cause of commercial education by the opportune production of this volume.... The work is peculiarly acceptable to English readers and an admirable addition to existing class-books. In a phrase, we think the work attains its object in furnishing a brief account of those laws and customs of British trade with which the commercial man interested therein should be familiar."—*Chamber of Commerce Journal.*

"An invaluable guide in the hands of those who are preparing for a commercial career."—*Counting House.*

Foreign Commercial Correspondence.

THE FOREIGN COMMERCIAL CORRESPONDENT: Being Aids to Commercial Correspondence in Five Languages—English, French, German, Italian, and Spanish. By CONRAD E. BAKER. Second Edition. Crown 8vo, 3s. 6d. cloth.

"Whoever wishes to correspond in all the languages mentioned by Mr. Baker cannot do better than study this work, the materials of which are excellent and conveniently arranged. They consist not of entire specimen letters but—what are far more useful—short passages, sentences, or phrases expressing the same general idea in various forms."—*Athenæum.*

"A careful examination has convinced us that it is unusually complete, well arranged, and reliable. The book is a thoroughly good one."—*Schoolmaster.*

Accounts for Manufacturers.

FACTORY ACCOUNTS: Their Principles and Practice. A Handbook for Accountants and Manufacturers, with Appendices on the Nomenclature of Machine Details; the Income Tax Acts; the Rating of Factories; Fire and Boiler Insurance; the Factory and Workshop Acts, &c., including also a Glossary of Terms and a large number of Specimen Rulings. By EMILE GARCKE and J. M. FELLS. Fourth Edition, Revised and Enlarged. Demy 8vo, 250 pages, 6s. strongly bound.

"A very interesting description of the requirements of Factory Accounts.... The principle of assimilating the Factory Accounts to the general commercial books is one which we thoroughly agree with."—*Accountants' Journal.*

"Characterised by extreme thoroughness. There are few owners of factories who would not derive great benefit from the perusal of this most admirable work."—*Local Government Chronicle.*

Intuitive Calculations.

THE COMPENDIOUS CALCULATOR; or, Easy and Concise Methods of Performing the various Arithmetical Operations required in Commercial and Business Transactions, together with Useful Tables. By DANIEL O'GORMAN. Corrected and Extended by Professor J. R. YOUNG. Twenty-seventh Edition, Revised by C. NORRIS. Fcap. 8vo, 2s. 6d. cloth limp; or, 3s. 6d. strongly half-bound in leather.

"It would be difficult to exaggerate the usefulness of a book like this to everyone engaged in commerce or manufacturing industry. It is crammed full of rules and formulæ for shortening and employing calculations."—*Knowledge.*

Modern Metrical Units and Systems.

MODERN METROLOGY: *A Manual of the Metrical Units and Systems of the Present Century.* With an Appendix containing a proposed English System. By LOWIS D'A. JACKSON, A.M.Inst.C.E., Author of "Aid to Survey Practice," &c. Large crown 8vo, 12s. 6d. cloth.

"We recommend the work to all interested in the practical reform of our weights and measures."—*Nature.*

The Metric System and the British Standards.

A SERIES OF METRIC TABLES, *in which the British Standard Measures and Weights are compared with those of the Metric System at present in Use on the Continent.* By C. H. DOWLING, C.E. 8vo, 10s. 6d. strongly bound.

"Mr. Dowling's Tables are well put together as a ready-reckoner for the conversion of one system into the other."—*Athenæum.*

Iron and Metal Trades' Calculator.

THE IRON AND METAL TRADES' COMPANION. For expeditiously ascertaining the Value of any Goods bought or sold by Weight, from 1s. per cwt. to 112s. per cwt., and from one farthing per pound to one shilling per pound. By THOMAS DOWNIE. 396 pp., 9s. leather.

"A most useful set of tables; nothing like them before existed."—*Building News.*

"Although specially adapted to the iron and metal trades, the tables will be found useful in every other business in which merchandise is bought and sold by weight."—*Railway News.*

Chadwick's Calculator for Numbers and Weights Combined.

THE NUMBER, WEIGHT, AND FRACTIONAL CALCULATOR. Containing upwards of 250,000 Separate Calculations, showing at a glance the value at 422 different rates, ranging from 1/125th of a Penny to 20s. each, or per cwt., and £20 per ton, of any number of articles consecutively, from 1 to 470.—Any number of cwts., qrs., and lbs., from 1 cwt. to 470 cwts.—Any number of tons, cwts., qrs., and lbs., from 1 to 1,000 tons. By WILLIAM CHADWICK, Public Accountant. Third Edition, Revised and Improved. 8vo, 18s., strongly bound for Office wear and tear.

☞ *Is adapted for the use of Accountants and Auditors, Railway Companies, Canal Companies, Shippers, Shipping Agents, General Carriers, etc. Ironfounders, Brassfounders, Metal Merchants, Iron Manufacturers, Ironmongers, Engineers, Machinists, Boiler Makers. Millwrights, Roofing, Bridge and Girder Makers, Colliery Proprietors, etc. Timber Merchants, Builders, Contractors, Architects, Surveyors, Auctioneers, Valuers, Brokers, Mill Owners and Manufacturers, Mill Furnishers, Merchants, and General Wholesale Tradesmen. Also for the Apportionment of Mileage Charges for Railway Traffic.*

⁂ OPINIONS OF THE PRESS.

"It is as easy of reference for any answer or any number of answers as a dictionary, and the references are even more quickly made. For making up accounts or estimates the book must prove invaluable to all who have any considerable quantity of calculations involving price and measure in any combination to do."—*Engineer.*

"The most complete and practical ready reckoner which it has been our fortune yet to see. It is difficult to imagine a trade or occupation in which it could not be of the greatest use, either in saving human labour or in checking work. The publishers have placed within the reach of every commercial man an invaluable and unfailing assistant."—*The Miller.*

"The most perfect work of the kind yet prepared."—*Glasgow Herald.*

Harben's Comprehensive Weight Calculator.

THE WEIGHT CALCULATOR. Being a Series of Tables upon a New and Comprehensive Plan, exhibiting at One Reference the exact Value of any Weight from 1 lb. to 15 tons, at 300 Progressive Rates, from 1d. to 168s. per cwt., and containing 186,000 Direct Answers, which, with their Combinations, consisting of a single addition (mostly to be performed at sight), will afford an aggregate of 10,266,000 Answers; the whole being calculated and designed to ensure correctness and promote despatch. By HENRY HARBEN, Accountant. Fourth Edition, carefully Corrected. Royal 8vo, £1 5s. strongly half-bound.

"A practical and useful work of reference for men of business generally; it is the best of the kind we have seen."—*Ironmonger.*

"Of priceless value to business men. It is a necessary book in all mercantile offices."—*Sheffield Independent.*

Harben's Comprehensive Discount Guide.

THE DISCOUNT GUIDE. Comprising several Series of Tables for the use of Merchants, Manufacturers, Ironmongers, and others, by which may be ascertained the exact Profit arising from any mode of using Discounts, either in the Purchase or Sale of Goods, and the method of either Altering a Rate of Discount or Advancing a Price, so as to produce, by one operation, a sum that will realise any required profit after allowing one or more Discounts: to which are added Tables of Profit or Advance from 1¼ to 90 per cent., Tables of Discount from 1¼ to 98¾ per cent., and Tables of Commission, &c., from ⅛ to 10 per cent. By HENRY HARBEN, Accountant, Author of "The Weight Calculator." New Edition, carefully Revised and Corrected. Demy 8vo, 544 pp., £1 5s. half-bound.

"A book such as this can only be appreciated by business men to whom the saving of time means saving of money. We have the high authority of Professor J. R. Young that the tables throughout the work are constructed upon strictly accurate principles. The work is a model of typographical clearness, and must prove of great value to merchants, manufacturers, and general traders."—*British Trade Journal.*

Iron Shipbuilders' and Merchants' Weight Tables.

IRON-PLATE WEIGHT TABLES: For Iron Shipbuilders, Engineers, and Iron Merchants. Containing the Calculated Weights of upwards of 150,000 different sizes of Iron Plates, from 1 foot by 6 in. by ¼ in. to 10 feet by 5 feet by 1 in. Worked out on the basis of 40 lbs. to the square foot of Iron of 1 inch in thickness. Carefully compiled and thoroughly Revised by H. BURLINSON and W. H. SIMPSON. Oblong 4to, 25s. half-bound.

"This work will be found of great utility. The authors have had much practical experience of what is wanting in making estimates: and the use of the book will save much time in making elaborate calculations."—*English Mechanic.*

AGRICULTURE, FARMING, GARDENING, etc.

Dr. Fream's New Edition of "The Standard Treatise on Agriculture."

THE COMPLETE GRAZIER, and FARMER'S and CATTLE-BREEDER'S ASSISTANT: A Compendium of Husbandry. Originally Written by WILLIAM YOUATT. Thirteenth Edition, entirely Re-written, considerably Enlarged, and brought up to the Present Requirements of Agricultural Practice, by WILLIAM FREAM, LL.D., Steven Lecturer in the University of Edinburgh, Author of "The Elements of Agriculture," &c. Royal 8vo, 1,100 pp., with over 450 Illustrations. £1 11s. 6d. strongly and handsomely bound.

EXTRACT FROM PUBLISHERS' ADVERTISEMENT.

"A treatise that made its original appearance in the first decade of the century, and that enters upon its Thirteenth Edition before the century has run its course, has undoubtedly established its position as a work of permanent value.... The phenomenal progress of the last dozen years in the Practice and Science of Farming has rendered it necessary, however, that the volume should be re-written, ... and for this undertaking the publishers were fortunate enough to secure the services of Dr. FREAM, whose high attainments in all matters pertaining to agriculture have been so emphatically recognised by the highest professional and official authorities. In carrying out his editorial duties, Dr. FREAM has been favoured with valuable contributions by Prof. J. WORTLEY AXE, Mr. E. BROWN, Dr. BERNARD DYER, Mr. W. J. MALDEN, Mr. R. H. REW, Prof. SHELDON, Mr. J. SINCLAIR, Mr. SANDERS SPENCER, and others.

"As regards the illustrations of the work, no pains have been spared to make them as representative and characteristic as possible, so as to be practically useful to the Farmer and Grazier."

SUMMARY OF CONTENTS.

BOOK I. ON THE VARIETIES, BREEDING, REARING, FATTENING, AND MANAGEMENT OF CATTLE.
BOOK II. ON THE ECONOMY AND MANAGEMENT OF THE DAIRY.
BOOK III. ON THE BREEDING, REARING, AND MANAGEMENT OF HORSES.
BOOK IV. ON THE BREEDING, REARING, AND FATTENING OF SHEEP.
BOOK V. ON THE BREEDING, REARING, AND FATTENING OF SWINE.
BOOK VI. ON THE DISEASES OF LIVE STOCK.
BOOK VII. ON THE BREEDING, REARING, AND MANAGEMENT OF POULTRY.
BOOK VIII. ON FARM OFFICES AND IMPLEMENTS OF HUSBANDRY.
BOOK IX. ON THE CULTURE AND MANAGEMENT OF GRASS LANDS.
BOOK X. ON THE CULTIVATION AND APPLICATION OF GRASSES, PULSE, AND ROOTS.
BOOK XI. ON MANURES AND THEIR APPLICATION TO GRASS LAND&CROPS.
BOOK XII. MONTHLY CALENDARS OF FARMWORK.

⁂ OPINIONS OF THE PRESS ON THE NEW EDITION.

"Dr. Fream is to be congratulated on the successful attempt he has made to give us a work which will at once become the standard classic of the farm practice of the country. We believe that it will be found that it has no compeer among the many works at present in existence.... The illustrations are admirable, while the frontispiece, which represents the well-known bull, New Year's Gift, bred by the Queen, is a work of art."—*The Times.*

"The book must be recognised as occupying the proud position of the most exhaustive work of reference in the English language on the subject with which it deals."—*Athenæum.*

"The most comprehensive guide to modern farm practice that exists in the English language to-day.... The book is one that ought to be on every farm and in the library of every landowner."—*Mark Lane Express.*

"In point of exhaustiveness and accuracy the work will certainly hold a pre-eminent and unique position among books dealing with scientific agricultural practice. It is, in fact, an agricultural library of itself."—*North British Agriculturist.*

"A compendium of authoratative and well-ordered knowledge on every conceivable branch of the work of the live stock farmer; probably without an equal in this or any other country."—*Yorkshire Post.*

"The best and brightest guide to the practice of husbandry, one that has no superior—no equal we might truly say—among the agricultural literature now before the public.... In every section in which we have tested it, the work has been found thoroughly up to date."—*Bell's Weekly Messenger.*

British Farm Live Stock.

FARM LIVE STOCK OF GREAT BRITAIN. By ROBERT WALLACE, F.L.S., F.R.S.E., &c., Professor of Agriculture and Rural Economy in the University of Edinburgh. Third Edition, thoroughly Revised and considerably Enlarged. With over 120 Phototypes of Prize Stock. Demy 8vo, 384 pp., with 79 Plates and Maps, 12s. 6d. cloth.

"A really complete work on the history, breeds, and management of the farm stock of Great Britain, and one which is likely to find its way to the shelves of every country gentleman's library."—*The Times.*

"The latest edition of 'Farm Live Stock of Great Britain' is a production to be proud of, and its issue not the least of the services which its author has rendered to agricultural science."—*Scottish Farmer.*

"The book is very attractive ... and we can scarcely imagine the existence of a farmer who would not like to have a copy of this beautiful work."—*Mark Lane Express.*

"A work which will long be regarded as a standard authority whenever a concise history and description of the breeds of live stock in the British Isles is required."—*Bell's Weekly Messenger.*

Dairy Farming.

BRITISH DAIRYING. A Handy Volume on the Work of the Dairy-Farm. For the Use of Technical Instruction Classes, Students in Agricultural Colleges, and the Working Dairy-Farmer. By Prof. J. P. SHELDON, late Special Commissioner of the Canadian Government, Author of "Dairy Farming," &c. With numerous Illustrations. Crown 8vo, 2s. 6d. cloth.

"May be confidently recommended as a useful text-book on dairy farming.—*Agricultural Gazette.*

"Probably the best half-crown manual on dairy work that has yet been produced.—*North British*

Agriculturist.

"It is the soundest little work we have yet seen on the subject."—*The Times.*

Dairy Manual.

MILK, CHEESE AND BUTTER: Their Composition, Character and the Processes of their Production. A Practical Manual for Students and Dairy Farmers. By JOHN OLIVER, late Principal of the Western Dairy Institute, Berkeley. Crown 8vo, 380 pages, with Coloured Test Sheets and numerous Illustrations, 7s. 6d. cloth.

[Just published.

Agricultural Facts and Figures.

NOTE-BOOK OF AGRICULTURAL FACTS AND FIGURES FOR FARMERS AND FARM STUDENTS. By PRIMROSE McCONNELL, B.Sc. Fifth Edition. Royal 32mo, roan, gilt edges, with band, 4s.

"Literally teems with information, and we can cordially recommend it to all connected with agriculture."—*North British Agriculturist.*

Small Farming.

SYSTEMATIC SMALL FARMING; or, The Lessons of my Farm. Being an Introduction to Modern Farm Practice for Small Farmers. By ROBERT SCOTT BURN, Author of "Outlines of Modern Farming," &c. With numerous Illustrations, crown 8vo, 6s. cloth.

"This is the completest book of its class we have seen, and one which every amateur farmer will read with pleasure and accept as a guide."—*Field.*

Modern Farming.

OUTLINES OF MODERN FARMING. By R. SCOTT BURN. Soils, Manures, and Crops—Farming and Farming Economy—Cattle, Sheep, and Horses—Management of Dairy, Pigs, and Poultry—Utilisation of Town-Sewage, Irrigation, &c. Sixth Edition. In One Vol., 1,250 pp., half-bound, profusely Illustrated, 12s.

"The aim of the author has been to make his work at once comprehensive and trustworthy, and he has succeeded to a degree which entitles him to much credit."—*Morning Advertiser.*

Agricultural Engineering.

FARM ENGINEERING, THE COMPLETE TEXT-BOOK OF. Comprising Draining and Embanking; Irrigation and Water Supply; Farm Roads, Fences, and Gates; Farm Buildings; Barn Implements and Machines; Field Implements and Machines; Agricultural Surveying, &c. By Prof. JOHN SCOTT, In One Vol., 1,150 pages, half-bound, with over 600 Illustrations, 12s.

"Written with great care, as well as with knowledge and ability. The author has done his work well; we have found him a very trustworthy guide wherever we have tested his statements. The volume will be of great value to agricultural students."—*Mark Lane Express.*

Agricultural Text-Book.

THE FIELDS OF GREAT BRITAIN: A Text-Book of Agriculture, adapted to the Syllabus of the Science and Art Department. For Elementary and Advanced Students. By HUGH CLEMENTS (Board of Trade). Second Edition, Revised, with Additions. 18mo, 2s. 6d. cloth.

"A most comprehensive volume, giving a mass of information."—*Agricultural Economist.*

"It is a long time since we have seen a book which has pleased us more, or which contains such a vast and useful fund of knowledge."—*Educational Times.*

Tables for Farmers, etc.

TABLES, MEMORANDA, AND CALCULATED RESULTS for *Farmers, Graziers, Agricultural Students, Surveyors, Land Agents, Auctioneers,* etc. With a New System of Farm Book-keeping. Selected and Arranged by SIDNEY FRANCIS. Third Edition, Revised. 272 pp., waistcoat-pocket size, 1s. 6d. limp leather.

"Weighing less than 1 oz., and occupying no more space than a match box, it contains a mass of facts and calculations which has never before, in such handy form, been obtainable. Every operation on the farm is dealt with. The work may be taken as thoroughly accurate, the whole of the tables having been revised by Dr. Fream. We cordially recommend it."—*Bell's Weekly Messenger.*

The Management of Bees.

BEES FOR PLEASURE AND PROFIT: A Guide to the Manipulation of Bees, the Production of Honey, and the General Management of the Apiary. By G. GORDON SAMSON. With numerous Illustrations. Crown 8vo, 1s. cloth.

"The intending bee-keeper will find exactly the kind of information required to enable him to make a successful start with his hives. The author is a thoroughly competent teacher, and his book may be commended."—*Morning Post.*

Farm and Estate Book-keeping.

BOOK-KEEPING FOR FARMERS & ESTATE OWNERS. A Practical Treatise, presenting, in Three Plans, a System adapted for all Classes of Farms. By JOHNSON M. WOODMAN, Chartered Accountant. Second Edition, Revised. Crown 8vo, 3s. 6d. cloth boards; or 2s. 6d. cloth limp.

"The volume is a capital study of a most important subject."—*Agricultural Gazette.*

"The young farmer, land agent, and surveyor will find Mr. Woodman's treatise more than repay its cost and study."—*Building News.*

Farm Account Book.

WOODMAN'S YEARLY FARM ACCOUNT BOOK. Giving a Weekly Labour Account and Diary, and showing the Income and Expenditure under each Department of Crops, Live Stock, Dairy, &c. &c. With Valuation, Profit and Loss Account, and Balance Sheet at the end of the Year. By JOHNSON M. WOODMAN, Chartered Accountant, Author of "Book-keeping for Farmers." Folio, 7s. 6d. half-bound.

"Contains every requisite form for keeping farm accounts readily and accurately."—*Agriculture.*

Early Fruits, Flowers, and Vegetables.

THE FORCING GARDEN; or, How to Grow Early Fruits, Flowers, and Vegetables. With Plans and Estimates for Building Glasshouses, Pits, and Frames. With Illustrations. By SAMUEL WOOD. Crown 8vo, 3s. 6d. cloth.

"A good book, and fairly fills a place that was in some degree vacant. The book is written with great care, and contains a great deal of valuable teaching."—*Gardeners' Magazine.*

Good Gardening.

A PLAIN GUIDE TO GOOD GARDENING; or, How to Grow Vegetables, Fruits, and Flowers. By S. WOOD. Fourth Edition, with considerable Additions, &c., and numerous Illustrations. Crown 8vo, 3s. 6d. cl.

"A very good book, and one to be highly recommended as a practical guide. The practical directions are excellent."—*Athenæum.*

"May be recommended to young gardeners, cottagers, and specially to amateurs, for the plain, simple, and trustworthy information it gives on common matters too often neglected."—*Gardeners' Chronicle.*

Gainful Gardening.

MULTUM-IN-PARVO GARDENING; or, How to make One Acre of Land produce £620 a-year by the Cultivation of Fruits and Vegetables; also,

How to Grow Flowers in Three Glass Houses, so as to realise £176 per annum clear Profit. By SAMUEL WOOD, Author of "Good Gardening," &c. Fifth and Cheaper Edition, Revised, with Additions. Crown 8vo, 1s. sewed.

"We are bound to recommend it as not only suited to the case of the amateur and gentleman's gardener, but to the market grower."—*Gardeners' Magazine*.

Gardening for Ladies.

THE LADIES' MULTUM-IN-PARVO FLOWER GARDEN, and Amateurs' Complete Guide. With Illusts. By S. WOOD. Cr. 8vo, 3s. 6d. cl.

"This volume contains a good deal of sound common sense instruction."—*Florist*.
"Full of shrewd hints and useful instructions, based on a lifetime of experience."—*Scotsman*.

Receipts for Gardeners.

GARDEN RECEIPTS. Edited by CHARLES W. QUIN. 12mo, 1s. 6d. cloth limp.

"A useful and handy book, containing a good deal of valuable information."—*Athenæum*.

Market Gardening.

MARKET AND KITCHEN GARDENING. By Contributors to "The Garden." Compiled by C. W. SHAW, late Editor of "Gardening Illustrated." 12mo, 3s. 6d. cloth boards.

"The most valuable compendium of kitchen and market-garden work published."—*Farmer*.

Cottage Gardening.

COTTAGE GARDENING; or, Flowers, Fruits, and Vegetables for Small Gardens. By E. HOBDAY. 12mo, 1s. 6d. cloth limp.

"Contains much useful information at a small charge."—*Glasgow Herald*.

AUCTIONEERING, VALUING, LAND SURVEYING ESTATE AGENCY, etc.

Auctioneer's Assistant.

THE APPRAISER, AUCTIONEER, BROKER, HOUSE AND ESTATE AGENT AND VALUER'S POCKET ASSISTANT, for the Valuation for Purchase, Sale, or Renewal of Leases, Annuities and Reversions, and of property generally; with Prices for Inventories, &c. By JOHN WHEELER, Valuer, &c. Sixth Edition, Re-written and greatly extended by C. NORRIS, Surveyor, Valuer, &c. Royal 32mo, 5s. cloth.

"A neat and concise book of reference, containing an admirable and clearly-arranged list of prices for inventories, and a very practical guide to determine the value of furniture, &c."—*Standard*.

"Contains a large quantity of varied and useful information as to the valuation for purchase, sale, or renewal of leases, annuities and reversions, and of property generally, with prices for inventories, and a guide to determine the value of interior fittings and other effects."—*Builder*.

Auctioneering.

AUCTIONEERS: THEIR DUTIES AND LIABILITIES. A Manual of Instruction and Counsel for the Young Auctioneer. By ROBERT SQUIBBS, Auctioneer. Second Edition, Revised and partly Re-written. Demy 8vo, 12s. 6d. cloth.

❖ OPINIONS OF THE PRESS.

"The standard text-book on the topics of which it treats." —*Athenæum.*

"The work is one of general excellent character, and gives much information in a compendious and satisfactory form." —*Builder.*

"May be recommended as giving a great deal of information on the law relating to auctioneers, in a very readable form." —*Law Journal.*

"Auctioneers may be congratulated on having so pleasing a writer to minister to their special needs." —*Solicitors' Journal.*

"Every auctioneer ought to possess a copy of this excellent work." —*Ironmonger.*

"Of great value to the profession.... We readily welcome this book from the fact that it treats the subject in a manner somewhat new to the profession." —*Estates Gazette.*

Inwood's Estate Tables.

TABLES FOR THE PURCHASING OF ESTATES, Freehold, Copyhold, or Leasehold; Annuities, Advowsons, etc., and for the Renewing of Leases held under Cathedral Churches, Colleges, or other Corporate bodies, for Terms of Years certain, and for Lives: also for Valuing Reversionary Estates, Deferred Annuities, Next Presentations, &c.; together with SMART'S Five Tables of Compound Interest, and an Extension of the same to Lower and Intermediate Rates. By W. INWOOD. 24th Edition, with considerable Additions, and new and valuable Tables of Logarithms for the more Difficult Computations of the Interest of Money, Discount, Annuities, &c., by M. FEDOR THOMAN, of the Société Crédit Mobilier of Paris. Crown 8vo, 8s. cloth.

"Those interested in the purchase and sale of estates, and in the adjustment of compensation cases, as well as in transactions in annuities, life insurances, &c., will find the present edition of eminent service." —*Engineering.*

"'Inwood's Tables' still maintain a most enviable reputation. The new issue has been enriched by large additional contributions by M. Fedor Thoman, whose carefully arranged Tables cannot fail to be of the utmost utility." —*Mining Journal.*

Agricultural Valuer's Assistant.

THE AGRICULTURAL VALUER'S ASSISTANT. A Practical Handbook on the Valuation of Landed Estates; including Rules and Data for Measuring and Estimating the Contents, Weights, and Values of Agricultural Produce and Timber, and the Values of Feeding Stuffs, Manures, and Labour; with Forms of Tenant-Right-Valuations, Lists of Local Agricultural Customs, Scales of Compensation under the Agricultural Holdings Act, &c. &c. By TOM BRIGHT, Agricultural Surveyor. Second Edition, much Enlarged. Crown 8vo, 5s. cloth.

"Full of tables and examples in connection with the valuation of tenant-right, estates, labour, contents, and weights of timber, and farm produce of all kinds." —*Agricultural Gazette.*

"An eminently practical handbook, full of practical tables and data of undoubted interest and value to surveyors and auctioneers in preparing valuations of all kinds." —*Farmer.*

Plantations and Underwoods.

POLE PLANTATIONS AND UNDERWOODS: A Practical Handbook on Estimating the Cost of Forming, Renovating, Improving, and Grubbing Plantations and Underwoods, their Valuation for Purposes of Transfer, Rental, Sale, or Assessment. By TOM BRIGHT, Author of "The Agricultural Valuer's Assistant," &c. Crown 8vo, 3s. 6d. cloth.

"To valuers, foresters and agents it will be a welcome aid." —*North British Agriculturist.*

"Well calculated to assist the valuer in the discharge of his duties, and of undoubted interest and use both to surveyors and auctioneers in preparing valuations of all kinds." —*Kent Herald.*

Hudson's Land Valuer's Pocket-Book.

THE LAND VALUER'S BEST ASSISTANT: Being Tables on a very much Improved Plan, for Calculating the Value of Estates. With Tables for reducing Scotch, Irish, and Provincial Customary Acres to Statute Measure, &c. By R. HUDSON, C.E. New Edition. Royal 32mo, leather, elastic band, 4s.

"Of incalculable value to the country gentleman and professional man."—*Farmers' Journal.*

Ewart's Land Improver's Pocket-Book.

THE LAND IMPROVER'S POCKET-BOOK OF FORMULÆ, TABLES, and MEMORANDA required in any Computation relating to the Permanent Improvement of Landed Property. By JOHN EWART, Land Surveyor and Agricultural Engineer. Second Edition, Revised. Royal 32mo, oblong, leather, gilt edges, with elastic band, 4s.

"A compendious and handy little volume."—*Spectator.*

Complete Agricultural Surveyor's Pocket-Book.

THE LAND VALUER'S AND LAND IMPROVER'S COMPLETE POCKET-BOOK. Being of the above Two Works bound together. Leather, with strap, 7s. 6d.

House Property.

HANDBOOK OF HOUSE PROPERTY. A Popular and Practical Guide to the Purchase, Mortgage, Tenancy, and Compulsory Sale of Houses and Land, including the Law of Dilapidations and Fixtures; with Examples of all kinds of Valuations, Useful Information on Building, and Suggestive Elucidations of Fine Art. By E. L. TARBUCK, Architect and Surveyor. Fifth Edition, Enlarged. 12mo, 5s. cloth.

"The advice is thoroughly practical."—*Law Journal.*

"For all who have dealings with house property, this is an indispensable guide."—*Decoration.*

"Carefully brought up to date, and much improved by the addition of a division on fine art.... A well-written and thoughtful work."—*Land Agent's Record.*

LAW AND MISCELLANEOUS.

Private Bill Legislation and Provisional Orders.

HANDBOOK FOR THE USE OF SOLICITORS AND ENGINEERS Engaged in Promoting Private Acts of Parliament and Provisional Orders, for the Authorization of Railways, Tramways, Works for the Supply of Gas and Water, and other undertakings of a like character. By L. LIVINGSTON MACASSEY, of the Middle Temple, Barrister-at-Law, M.Inst.C.E.; Author of "Hints on Water Supply." Demy 8vo, 950 pp., 25s. cl.

"The author's double experience as an engineer and barrister has enabled him to approach the subject alike from an engineering and legal point of view."—*Local Government Chronicle.*

Law of Patents.

PATENTS FOR INVENTIONS, AND HOW TO PROCURE THEM. Compiled for the Use of Inventors, Patentees and others. By G. G. M. HARDINGHAM, Assoc.Mem.Inst.C.E., &c. Demy 8vo, 1s. 6d. cloth.

Labour Disputes.

INDUSTRIAL CONCILIATION AND ARBITRATION: An Historical Sketch, with Practical Suggestions for the Settlement of Labour Disputes. By J. S. JEANS, Author of "Railway Problems," "England's Supremacy," &c. Crown 8vo, 200 pp., 2s. 6d. cloth.

[*Just published.*

Pocket-Book for Sanitary Officials.

THE HEALTH OFFICER'S POCKET-BOOK: A Guide to Sanitary Practice and Law. For Medical Officers of Health, Sanitary Inspectors, Members of Sanitary Authorities, &c. By EDWARD F. WILLOUGHBY, M.D. (Lond.), &c., Author of "Hygiene and Public Health." Fcap. 8vo, 7s. 6d. cloth, red edges, rounded corners.

[Just published.

"A mine of condensed information of a pertinent and useful kind on the various subjects of which it treats. The matter seems to have been carefully compiled and arranged for facility of reference, and it is well illustrated by diagrams and woodcuts. The different subjects are succinctly but fully and scientifically dealt with."—*The Lancet.*

"Ought to be welcome to those for whose use it is designed, since it practically boils down a reference library into a pocket volume.... It combines, with an uncommon degree of efficiency, the qualities of accuracy, conciseness and comprehensiveness."—*Scotsman.*

"An excellent publication, dealing with the scientific, technical and legal matters connected with the duties of medical officers of health and sanitary inspectors. The work is replete with information."—*Local Government Journal.*

A Complete Epitome of the Laws of this Country.

EVERY MAN'S OWN LAWYER: A Handy-Book of the Principles of Law and Equity. By A BARRISTER. Thirty-first Edition, carefully Revised, and including the Legislation of 1893. Comprising (amongst other Acts) the *Voluntary Conveyances Act,* 1893; the *Married Women's Property Act,* 1893; the *Trustee Act,* 1893; the *Savings Bank Act,* 1893; the *Barbed Wire Act,* 1893; the *Industrial and Provident Societies Act,* 1893; the *Hours of Labour of Railway Servants Act,* 1893; the *Fertiliser and Feeding Stuffs Act,* 1893, &c., as well as the *Betting and Loans (Infants) Act,* 1892; the *Gaming Act,* 1892; the *Shop Hours Act,* 1892; the *Conveyancing and Real Property Act,* 1892; the *Small Holdings Act,* 1892; and many other new Acts. Crown 8vo, 700 pp., price 6s. 8d. (saved at every consultation!), strongly bound in cloth.

[Just published.

⁂ *The Book will be found to comprise (amongst other matter)—*

THE RIGHTS AND WRONGS OF INDIVIDUALS—LANDLORD AND TENANT—VENDORS AND PURCHASERS—PARTNERS AND AGENTS—COMPANIES AND ASSOCIATIONS—MASTERS, SERVANTS, AND WORKMEN—LEASES AND MORTGAGES—LIBEL AND SLANDER—CONTRACTS AND AGREEMENTS—BONDS AND BILLS OF SALE—CHEQUES, BILLS, AND NOTES—RAILWAY AND SHIPPING LAW—BANKRUPTCY AND INSURANCE—BORROWERS, LENDERS, AND SURETIES—CRIMINAL LAW—PARLIAMENTARY ELECTIONS—COUNTY COUNCILS—MUNICIPAL CORPORATIONS—PARISH LAW, CHURCH-WARDENS, ETC.—PUBLIC HEALTH AND NUISANCES—COPYRIGHT AND PATENTS—TRADE MARKS AND DESIGNS—HUSBAND AND WIFE, DIVORCE, ETC.—TRUSTEES AND EXECUTORS—GUARDIAN AND WARD, INFANTS, ETC.—GAME LAWS AND SPORTING—HORSES, HORSE DEALING, AND DOGS—INN-KEEPERS, LICENSING, ETC.—FORMS OF WILLS, AGREEMENTS ETC. ETC.

☞ *The object of this work is to enable those who consult it to help themselves to the law; and thereby to dispense, as far as possible, with professional assistance and advice. There are many wrongs and grievances which persons submit to from time to time through not knowing how or where to apply for redress; and many persons have as great a dread of a lawyer's office as of a lion's den. With this book at hand it is believed that many a* SIX-AND-EIGHTPENCE *may be saved; many a wrong redressed; many a right reclaimed; many a law suit avoided; and many an evil abated. The work has established itself as the standard legal adviser of all classes, and has also made a reputation for itself as a useful book of reference for lawyers residing at a distance from law libraries, who are glad to have at hand a work embodying recent decisions and enactments.*

⁂ OPINIONS OF THE PRESS.

"It is a complete code of English Law, written in plain language, which all can understand.... Should be in the hands of every business man, and all who wish to abolish lawyers' bills."—*Weekly Times.*

"A useful and concise epitome of the law, compiled with considerable care."—*Law Magazine.*

"A complete digest of the most useful facts which constitute English law."—*Globe.*

"This excellent handbook.... Admirably done, admirably arranged, and admirably cheap."—*Leeds Mercury.*

"A concise, cheap and complete epitome of the English law. So plainly written that he who runs may read, and he who reads may understand."—*Figaro.*

"A dictionary of legal facts well put together. The book is a very useful one."—*Spectator.*

"A work which has long been wanted, which is thoroughly well done, and which we most cordially recommend."—*Sunday Times.*

"The latest edition of this popular book ought to be in every business establishment, and on every library table."—*Sheffield Post.*

"A complete epitome of the law; thoroughly intelligible to non-professional readers."—*Bell's Life.*

Legal Guide for Pawnbrokers.

THE PAWNBROKERS', FACTORS' AND MERCHANTS' GUIDE TO THE LAW OF LOANS AND PLEDGES. With the Statutes and a Digest of Cases. By H. C. FOLKARD, Esq., Barrister-at-Law. Fcap. 8vo, 3s. 6d. cloth.

The Law of Contracts.

LABOUR CONTRACTS: A Popular Handbook on the Law of Contracts for Works and Services. By DAVID GIBBONS. Fourth Edition, Appendix of Statutes by T. F. UTTLEY, Solicitor. Fcap. 8vo, 3s. 6d. cloth.

The Factory Acts.

SUMMARY OF THE FACTORY AND WORKSHOP ACTS (1878-1891). For the Use of Manufacturers and Managers. By EMILE GARCKE and J. M. FELLS. (Reprinted from "FACTORY ACCOUNTS.") Crown 8vo, 6d. sewed.

OGDEN, SMALE AND CO. LIMITED, PRINTERS, GREAT SAFFRON HILL, E.C.

Weale's Rudimentary Series.

LONDON,
1862.
THE PRIZE
MEDAL
Was awarded to the
Publishers of
**"WEALE'S
SERIES."**

A NEW LIST OF

WEALE'S SERIES

RUDIMENTARY SCIENTIFIC, EDUCATIONAL, AND CLASSICAL.

Comprising nearly <u>Three Hundred and Fifty</u> distinct works in almost every department of Science, Art, and Education, recommended to the notice of <u>Engineers, Architects, Builders, Artisans, and Students generally</u>, as well as to those interested in <u>Workmen's Libraries, Literary and Scientific Institutions, Colleges, Schools, Science Classes</u>, &c., &c.

☞ "WEALE'S SERIES includes Text-Books on almost every branch of Science and Industry, comprising such subjects as Agriculture, Architecture and Building, Civil Engineering, Fine Arts, Mechanics and Mechanical Engineering, Physical and Chemical Science, and many miscellaneous Treatises. The whole are constantly undergoing revision, and new editions, brought up to the latest discoveries in scientific research, are constantly issued. The prices at which they are sold are as low as their excellence is assured." — *American Literary Gazette.*

"Amongst the literature of technical education, WEALE'S SERIES has ever enjoyed a high reputation, and the additions being made by Messrs. CROSBY LOCKWOOD & SON render the series more complete, and bring the information upon the several subjects down to the present time." — *Mining Journal.*

"It is not too much to say that no books have ever proved more popular with, or more useful to, young engineers and others than the excellent treatises comprised in WEALE'S SERIES." — *Engineer.*

"The excellence of WEALE'S SERIES is now so well appreciated, that it would be wasting our space to enlarge upon their general usefulness and value." — *Builder.*

"The volumes of WEALE'S SERIES form one of the best collections of elementary technical books in any language." — *Architect.*

"WEALE'S SERIES has become a standard as well as an unrivalled collection of treatises in all branches of art and science." —*Public Opinion.*

PHILADELPHIA, 1876.
THE PRIZE MEDAL
Was awarded to the Publishers for
Books: Rudimentary, Scientific,
"WEALE'S SERIES," ETC.

CROSBY LOCKWOOD & SON,
7, STATIONERS' HALL COURT, LUDGATE HILL, LONDON, E.C.

WEALE'S RUDIMENTARY SCIENTIFIC SERIES.

∗∗ The volumes of this Series are freely Illustrated with Woodcuts, or otherwise, where requisite. Throughout the following List it must be understood that the books are bound in limp cloth, unless otherwise stated; *but the volumes marked with a ‡ may also be had strongly bound in cloth boards for 6d. extra.*

N.B.—*In ordering from this List it is recommended, as a means of facilitating business and obviating error, to quote the numbers affixed to the volumes, as well as the titles and prices.*

CIVIL ENGINEERING, SURVEYING, ETC.

No.
31. *WELLS AND WELL-SINKING.* By JOHN GEO. SWINDELL, A.R.I.B.A., and G. R. BURNELL, C.E. Revised Edition. With a New Appendix on the Qualities of Water. Illustrated. 2s.
35. *THE BLASTING AND QUARRYING OF STONE,* for Building and other Purposes. By Gen. Sir J. BURGOYNE, Bart. 1s. 6d.
43. *TUBULAR, AND OTHER IRON GIRDER BRIDGES,* particularly describing the Britannia and Conway Tubular Bridges. By G. DRYSDALE DEMPSEY, C.E. Fourth Edition. 2s.
44. *FOUNDATIONS AND CONCRETE WORKS,* with Practical Remarks on Footings, Sand, Concrete, Béton, Pile-driving, Caissons, and Cofferdams, &c. By E. DOBSON. Seventh Edition, 1s. 6d.
60. *LAND AND ENGINEERING SURVEYING.* By T. BAKER, C.E. Fifteenth Edition, revised by Professor J. R. YOUNG. 2s.‡
80*. *EMBANKING LANDS FROM THE SEA.* With examples and Particulars of actual Embankments, &c. By J. WIGGINS, F.G.S. 2s.
81. *WATER WORKS,* for the Supply of Cities and Towns. With a Description of the Principal Geological Formations of England as influencing Supplies of Water, &c. By S. HUGHES, C.E. New Edition. 4s.‡
118. *CIVIL ENGINEERING IN NORTH AMERICA,* a Sketch of. By DAVID STEVENSON, F.R.S.E., &c. Plates and Diagrams. 3s.
167. *IRON BRIDGES, GIRDERS, ROOFS, AND OTHER WORKS.* By FRANCIS CAMPIN, C.E. 2s. 6d.‡
197. *ROADS AND STREETS.* By H. LAW, C.E., revised and enlarged by D. K. CLARK, C.E., including pavements of Stone, Wood, Asphalte, &c. 4s. 6d.‡
203. *SANITARY WORK IN THE SMALLER TOWNS AND IN VILLAGES.* By

C. SLAGG, A.M.I.C.E. Revised Edition. 3s.‡
212. *GAS-WORKS, THEIR CONSTRUCTION AND ARRANGEMENT*; and the Manufacture and Distribution of Coal Gas. Originally written by SAMUEL HUGHES, C.E. Re-written and enlarged by WILLIAM RICHARDS, C.E. Eighth Edition, with important additions. 5s. 6d.‡
213. *PIONEER ENGINEERING*. A Treatise on the Engineering Operations connected with the Settlement of Waste Lands in New Countries. By EDWARD DOBSON, Assoc. Inst. C.E. 4s. 6d.‡
216. *MATERIALS AND CONSTRUCTION*; A Theoretical and Practical Treatise on the Strains, Designing, and Erection of Works of Construction. By FRANCIS CAMPIN, C.E. Second Edition, revised. 3s.‡
219. *CIVIL ENGINEERING*. By HENRY LAW, M.Inst. C.E. Including HYDRAULIC ENGINEERING by GEO. R. BURNELL, M.Inst. C.E. Seventh Edition, revised, with large additions by D. KINNEAR CLARK, M.Inst. C.E. 6s. 6d., Cloth boards, 7s. 6d.
268. *THE DRAINAGE OF LANDS, TOWNS, & BUILDINGS*. By G. D. DEMPSEY, C.E. Revised, with large Additions on Recent Practice in Drainage Engineering, by D. KINNEAR CLARK, M.I.C.E. Second Edition, Corrected. 4s. 6d.‡

MECHANICAL ENGINEERING, ETC.

33. *CRANES*, the Construction of, and other Machinery for Raising Heavy Bodies. By JOSEPH GLYNN, F.R.S. Illustrated. 1s. 6d.
34. *THE STEAM ENGINE*. By Dr. LARDNER. Illustrated. 1s. 6d.
59. *STEAM BOILERS*: their Construction and Management. By R. ARMSTRONG, C.E. Illustrated. 1s. 6d.
82. *THE POWER OF WATER*, as applied to drive Flour Mills, and to give motion to Turbines, &c. By JOSEPH GLYNN, F.R.S. 2s.‡
98. *PRACTICAL MECHANISM*, the Elements of; and Machine Tools. By T. BAKER, C.E. With Additions by J. NASMYTH, C.E. 2s. 6d.‡
139. *THE STEAM ENGINE*, a Treatise on the Mathematical Theory of, with Rules and Examples for Practical Men. By T. BAKER, C.E. 1s. 6d.
164. *MODERN WORKSHOP PRACTICE*, as applied to Steam Engines, Bridges, Ship-building, &c. By J. G. WINTON. New Edition. 3s. 6d.‡
165. *IRON AND HEAT*, exhibiting the Principles concerned in the Construction of Iron Beams, Pillars, and Girders. By J. ARMOUR. 2s. 6d.‡
166. *POWER IN MOTION*: Horse-Power, Toothed-Wheel Gearing, Long and Short Driving Bands, and Angular Forces. By J. ARMOUR, 2s.‡
171. *THE WORKMAN'S MANUAL OF ENGINEERING DRAWING*. By J. MAXTON. 7th Edn. With 7 Plates and 350 Cuts. 3s. 6d.‡
190. *STEAM AND THE STEAM ENGINE*, Stationary and Portable. By J. SEWELL and D. K. CLARK, C.E. 3s. 6d.‡
200. *FUEL*, its Combustion and Economy. By C. W. WILLIAMS. With Recent Practice in the Combustion and Economy of Fuel—Coal, Coke, Wood, Peat, Petroleum, &c.—by D. K. CLARK, M.I.C.E. 3s. 6d.‡
202. *LOCOMOTIVE ENGINES*. By G. D. DEMPSEY, C.E.; with large additions by D. KINNEAR CLARK, M.I.C.E. 3s.‡
211. *THE BOILERMAKER'S ASSISTANT* in Drawing, Templating, and Calculating Boiler and Tank Work. By JOHN COURTNEY, Practical Boiler

Maker. Edited by D. K. CLARK, C.E. 100 Illustrations. 2s.

217. *SEWING MACHINERY*: Its Construction, History, &c., with full Technical Directions for Adjusting, &c. By J. W. URQUHART, C.E. 2s.‡

223. *MECHANICAL ENGINEERING*. Comprising Metallurgy, Moulding, Casting, Forging, Tools, Workshop Machinery, Manufacture of the Steam Engine, &c. By FRANCIS CAMPIN, C.E. Second Edition. 2s. 6d.‡

236. *DETAILS OF MACHINERY*. Comprising Instructions for the Execution of various Works in Iron. By FRANCIS CAMPIN, C.E. 3s.‡

237. *THE SMITHY AND FORGE*; including the Farrier's Art and Coach Smithing. By W. J. E. CRANE. Illustrated. 2s. 6d.‡

238. *THE SHEET-METAL WORKER'S GUIDE*; a Practical Hand-book for Tinsmiths, Coppersmiths, Zincworkers, &c. With 94 Diagrams and Working Patterns. By W. J. E. CRANE. Second Edition, revised. 1s. 6d.

251. *STEAM AND MACHINERY MANAGEMENT*: with Hints on Construction and Selection. By M. POWIS BALE, M.I.M.E. 2s. 6d.‡

254. *THE BOILERMAKER'S READY-RECKONER*. By J. COURTNEY. Edited by D. K. CLARK, C.E. 4s.

⁂ Nos. 211 and 254 in One Vol., half-bound, entitled "THE BOILERMAKER'S READY-RECKONER AND ASSISTANT." By J. Courtney and D. K. CLARK. 7s.

255. *LOCOMOTIVE ENGINE-DRIVING*. A Practical Manual for Engineers in charge of Locomotive Engines. By MICHAEL REYNOLDS, M.S.E. Eighth Edition. 3s. 6d., limp; 4s. 6d. cloth boards.

256. *STATIONARY ENGINE-DRIVING*. A Practical Manual for Engineers in charge of Stationary Engines. By MICHAEL REYNOLDS, M.S.E. Fourth Edition. 3s. 6d. limp; 4s. 6d. cloth boards.

260. *IRON BRIDGES OF MODERATE SPAN*: their Construction and Erection. By HAMILTON W. PENDRED, C.E. 2s.

MINING, METALLURGY, ETC.

4. *MINERALOGY*, Rudiments of; a concise View of the General Properties of Minerals. By A. RAMSAY, F.G.S., F.R.G.S., &c. Third Edition, revised and enlarged. Illustrated. 3s. 6d.‡

117. *SUBTERRANEOUS SURVEYING*, with and without the Magnetic Needle. By T. FENWICK and T. BAKER, C.E. Illustrated. 2s. 6d.‡

135. *ELECTRO-METALLURGY*; Practically Treated. By ALEXANDER WATT. Ninth Edition, enlarged and revised, with additional Illustrations, and including the most recent Processes. 3s. 6d.‡

172. *MINING TOOLS*, Manual of. For the Use of Mine Managers, Agents, Students, &c. By WILLIAM MORGANS. 2s. 6d.

172*. *MINING TOOLS, ATLAS* of Engravings to Illustrate the above, containing 235 Illustrations, drawn to Scale. 4to. 4s. 6d.

176. *METALLURGY OF IRON*. Containing History of Iron Manufacture, Methods of Assay, and Analyses of Iron Ores, Processes of Manufacture of Iron and Steel, &c. By H. BAUERMAN, F.G.S. Sixth Edition, revised and enlarged. 5s.‡

180. *COAL AND COAL MINING*. By the late Sir WARINGTON W. SMYTH, M.A., F.R.S. Seventh Edition, revised. 3s. 6d.‡

195. *THE MINERAL SURVEYOR AND VALUER'S COMPLETE GUIDE*. By W. LINTERN, M.E. Third Edition, including Magnetic and Angular Surveying. With Four Plates. 3s. 6d.‡

214. *SLATE AND SLATE QUARRYING*, Scientific, Practical, and Commercial. By D. C. DAVIES, F.G.S., Mining Engineer, &c. 3s.‡

264. *A FIRST BOOK OF MINING AND QUARRYING*, with the Sciences connected therewith, for Primary Schools and Self Instruction. By J. H. COLLINS, F.G.S. Second Edition, with additions. 1s. 6d.

ARCHITECTURE, BUILDING, ETC.

16. *ARCHITECTURE—ORDERS*—The Orders and their Æsthetic Principles. By W. H. LEEDS. Illustrated. 1s. 6d.

17. *ARCHITECTURE—STYLES*—The History and Description of the Styles of Architecture of Various Countries, from the Earliest to the Present Period. By T. TALBOT BURY, F.R.I.B.A., &c. Illustrated. 2s.

∴ ORDERS AND STYLES OF ARCHITECTURE, *in One Vol., 3s. 6d.*

18. *ARCHITECTURE—DESIGN*—The Principles of Design in Architecture, as deducible from Nature and exemplified in the Works of the Greek and Gothic Architects. By E. L. GARBETT, Architect. Illustrated. 2s. 6d.

∴ *The three preceding Works, in One handsome Vol., half-bound, entitled* "MODERN ARCHITECTURE," *price 6s.*

22. THE ART OF BUILDING, Rudiments of. General Principles of Construction, Materials used in Building, Strength and Use of Materials, Working Drawings, Specifications, and Estimates. By E. DOBSON, 2s.‡

25. *MASONRY AND STONECUTTING*: Rudimentary Treatise on the Principles of Masonic Projection and their application to Construction. By EDWARD DOBSON, M.R.I.B.A., &c. 2s. 6d.‡

42. *COTTAGE BUILDING*. By C. BRUCE ALLEN, Architect. Eleventh Edition, revised and enlarged. With a Chapter on Economic Cottages for Allotments, by EDWARD E. ALLEN, C.E. 2s.

45. *LIMES, CEMENTS, MORTARS, CONCRETES, MASTICS, PLASTERING*, &c. By G. R. BURNELL, C.E. Fourteenth Edition. 1s. 6d.

57. *WARMING AND VENTILATION*. An Exposition of the General Principles as applied to Domestic and Public Buildings, Mines, Lighthouses, Ships, &c. By C. TOMLINSON, F.R.S., &c. Illustrated. 3s.

111. *ARCHES, PIERS, BUTTRESSES*, &c.: Experimental Essays on the Principles of Construction. By W. BLAND. Illustrated. 1s. 6d.

116. *THE ACOUSTICS OF PUBLIC BUILDINGS*; or, The Principles of the Science of Sound applied to the purposes of the Architect and Builder. By T. ROGER SMITH, M.R.I.B.A., Architect. Illustrated. 1s. 6d.

127. *ARCHITECTURAL MODELLING IN PAPER*, the Art of. By T. A. RICHARDSON, Architect. Illustrated. 1s. 6d.

128. *VITRUVIUS—THE ARCHITECTURE OF MARCUS VITRUVIUS POLLO*. In Ten Books. Translated from the Latin by JOSEPH GWILT, F.S.A., F.R.A.S. With 23 Plates. 5s.

130. *GRECIAN ARCHITECTURE*, An Inquiry into the Principles of Beauty in; with an Historical View of the Rise and Progress of the Art in Greece. By the EARL OF ABERDEEN. 1s.

∴ *The two preceding Works in One handsome Vol., half bound, entitled* "ANCIENT ARCHITECTURE," *price 6s.*

132. *THE ERECTION OF DWELLING-HOUSES*. Illustrated by a Perspective View, Plans, Elevations, and Sections of a pair of Semi-detached Villas, with the Specification, Quantities, and Estimates, &c. By S. H. BROOKS. New Edition, with Plates. 2s. 6d.‡

156. *QUANTITIES & MEASUREMENTS* in Bricklayers', Masons', Plasterers', Plumbers', Painters', Paperhangers', Gilders', Smiths', Carpenters' and Joiners' Work. By A. C. BEATON, Surveyor. Ninth Edition. 1s. 6d.

175. *LOCKWOOD'S BUILDER'S PRICE BOOK FOR 1893*. A Comprehensive Handbook of the Latest Prices and Data for Builders, Architects, Engineers, and Contractors. Re-constructed, Re-written, and further Enlarged. By FRANCIS T. W. MILLER, A.R.I.B.A. 700 pages. 3s. 6d.; cloth boards, 4s.

[Just Published.

182. *CARPENTRY AND JOINERY*—THE ELEMENTARY PRINCIPLES OF CARPENTRY. Chiefly composed from the Standard Work of THOMAS TREDGOLD, C.E. With a TREATISE ON JOINERY by E. WYNDHAM TARN, M.A. Fifth Edition, Revised. 3s. 6d.‡

182*. *CARPENTRY AND JOINERY. ATLAS* of 35 Plates to accompany the above. With Descriptive Letterpress. 4to. 6s.

185. *THE COMPLETE MEASURER*; the Measurement of Boards, Glass, &c.; Unequal-sided, Square-sided, Octagonal-sided, Round Timber and Stone, and Standing Timber, &c. By RICHARD HORTON. Fifth Edition. 4s.; strongly bound in leather, 5s.

187. *HINTS TO YOUNG ARCHITECTS*. By G. WIGHTWICK. New Edition. By G. H. GUILLAUME. Illustrated. 3s. 6d.‡

188. *HOUSE PAINTING, GRAINING, MARBLING, AND SIGN WRITING*: with a Course of Elementary Drawing for House-Painters, Sign-Writers, &c., and a Collection of Useful Receipts. By ELLIS A. DAVIDSON. Sixth Edition. With Coloured Plates. 5s. cloth limp; 6s. cloth boards.

189. *THE RUDIMENTS OF PRACTICAL BRICKLAYING*. In Six Sections: General Principles; Arch Drawing, Cutting, and Setting; Pointing; Paving, Tiling, Materials; Slating and Plastering; Practical Geometry, Mensuration, &c. By ADAM HAMMOND. Seventh Edition. 1s. 6d.

191. *PLUMBING*. A Text-Book to the Practice of the Art or Craft of the Plumber. With Chapters upon House Drainage and Ventilation. Sixth Edition. With 380 Illustrations. By W. P. BUCHAN. 3s. 6d.‡

192. *THE TIMBER IMPORTER'S, TIMBER MERCHANT'S, and BUILDER'S STANDARD GUIDE*. By R. E. GRANDY. 2s.

206. *A BOOK ON BUILDING, Civil and Ecclesiastical*, including CHURCH RESTORATION. With the Theory of Domes and the Great Pyramid, &c. By Sir EDMUND BECKETT, Bart., LL.D., Q.C., F.R.A.S. 4s. 6d.‡

226. *THE JOINTS MADE AND USED BY BUILDERS* in the Construction of various kinds of Engineering and Architectural Works. By WYVILL J. CHRISTY, Architect. With upwards of 160 Engravings on Wood. 3s.‡

228. *THE CONSTRUCTION OF ROOFS OF WOOD AND IRON*. By E. WYNDHAM TARN, M.A., Architect. Second Edition, revised. 1s. 6d.

229. *ELEMENTARY DECORATION*: as applied to the Interior and Exterior Decoration of Dwelling-Houses, &c. By J. W. FACEY. 2s.

257. *PRACTICAL HOUSE DECORATION*. A Guide to the Art of Ornamental Painting. By JAMES W. FACEY. 2s. 6d.

∴ *The two preceding Works, in One handsome Vol., half-bound, entitled* "HOUSE DECORATION, ELEMENTARY AND PRACTICAL," *price 5s.*

230. *A PRACTICAL TREATISE ON HANDRAILING*. Showing New and Simple Methods. By G. COLLINGS. Second Edition, Revised, including A TREATISE ON STAIRBUILDING. Plates. 2s. 6d.

247. *BUILDING ESTATES*: a Rudimentary Treatise on the Development, Sale, Purchase, and General Management of Building Land. By FOWLER MAITLAND, Surveyor. Second Edition, revised. 2s.

248. *PORTLAND CEMENT FOR USERS.* By HENRY FAIJA, Assoc. M. Inst. C.E. Third Edition, corrected. Illustrated. 2s.

252. *BRICKWORK*: a Practical Treatise, embodying the General and Higher Principles of Bricklaying, Cutting and Setting, &c. By F. WALKER. Third Edition, Revised and Enlarged. 1s. 6d.

23. *THE PRACTICAL BRICK AND TILE BOOK.* Comprising:

189. 265. BRICK AND TILE MAKING, by E. DOBSON, A.I.C.E.; PRACTICAL BRICKLAYING by A. HAMMOND; BRICKCUTTING AND SETTING, by A. HAMMOND. 534 pp. with 270 Illustrations. 6s. Strongly half-bound.

253. *THE TIMBER MERCHANT'S, SAW-MILLER'S, AND IMPORTER'S FREIGHT-BOOK AND ASSISTANT.* By WM. RICHARDSON. With Additions by M. POWIS BALE, A.M.Inst.C.E. 3s.‡

258. *CIRCULAR WORK IN CARPENTRY AND JOINERY.* A Practical Treatise on Circular Work of Single and Double Curvature. By GEORGE COLLINGS. Second Edition, 2s. 6d.

259. *GAS FITTING*: A Practical Handbook treating of every Description of Gas Laying and Fitting. By JOHN BLACK. 2s. 6d.‡

261. *SHORING AND ITS APPLICATION*: A Handbook for the Use of Students. By GEORGE H. BLAGROVE. 1s. 6d.

265. *THE ART OF PRACTICAL BRICK CUTTING & SETTING.* By ADAM HAMMOND. With 90 Engravings. 1s. 6d.

267. *THE SCIENCE OF BUILDING*: An Elementary Treatise on the Principles of Construction. By E. WYNDHAM TARN, M.A. Lond. Third Edition, Revised and Enlarged. 3s. 6d.‡

271. *VENTILATION*: a Text-book to the Practice of the Art of Ventilating Buildings. By W. P. BUCHAN, R.P., Sanitary Engineer, Author of "Plumbing," &c. 3s. 6d.‡

272. *ROOF CARPENTRY*; Practical Lessons in the Framing of Wood Roofs. For the Use of Working Carpenters. By GEO. COLLINGS, Author of "Handrailing and Stairbuilding," &c. 2s.

[*Just published.*]

273. *THE PRACTICAL PLASTERER*: A Compendium of Plain and Ornamental Plaster Work. By WILFRED KEMP. 2s.

[*Just published.*]

SHIPBUILDING, NAVIGATION, ETC.

51. *NAVAL ARCHITECTURE.* An Exposition of the Elementary Principles. By J. PEAKE. Fifth Edition, with Plates. 3s. 6d.‡

53*. *SHIPS FOR OCEAN & RIVER SERVICE*, Elementary and Practical Principles of the Construction of. By H. A. SOMMERFELDT. 1s. 6d.

53**. *AN ATLAS OF ENGRAVINGS* to Illustrate the above. Twelve large folding plates. Royal 4to, cloth. 7s. 6d.

54. *MASTING, MAST-MAKING, AND RIGGING OF SHIPS.* Also Tables of Spars, Rigging, Blocks; Chain, Wire, and Hemp Ropes, &c., relative to every class of vessels. By ROBERT KIPPING, N.A. 2s.

54*. *IRON SHIP-BUILDING.* With Practical Examples and Details. By JOHN GRANTHAM, C.E. Fifth Edition. 4s.

55. *THE SAILOR'S SEA BOOK*: a Rudimentary Treatise on Navigation. By

JAMES GREENWOOD, B.A. With numerous Woodcuts and Coloured Plates. New and enlarged edition. By W. H. ROSSER. 2s. 6d.‡

80. *MARINE ENGINES AND STEAM VESSELS.* By ROBERT MURRAY, C.E. Eighth Edition, thoroughly Revised, with Additions by the Author and by GEORGE CARLISLE, C.E. 4s. 6d. limp; 5s. cloth boards.

83*bis*. *THE FORMS OF SHIPS AND BOATS.* By W. BLAND. Eighth Edition, Revised, with numerous Illustrations and Models, 1s. 6d.

99. *NAVIGATION AND NAUTICAL ASTRONOMY,* in Theory and Practice. By Prof. J. R. YOUNG. New Edition. 2s. 6d.

106. *SHIPS' ANCHORS,* a Treatise on. By G. COTSELL, N.A. 1s. 6d.

149. *SAILS AND SAIL-MAKING.* With Draughting, and the Centre of Effort of the Sails; Weights and Sizes of Ropes: Masting, Rigging, and Sails of Steam Vessels, &c. 12th Edition. By R. KIPPING. N.A. 2s. 6d.‡

155. *ENGINEER'S GUIDE TO THE ROYAL & MERCANTILE NAVIES.* By a PRACTICAL ENGINEER. Revised by D. F. M'CARTHY. 3s.

55 & 204. *PRACTICAL NAVIGATION.* Consisting of The Sailor's Sea-Book. By JAMES GREENWOOD and W. H. ROSSER. Together with the requisite Mathematical and Nautical Tables for the Working of the Problems. By H. LAW, C.E., and Prof. J. R. YOUNG. 7s. Half-bound.

AGRICULTURE, GARDENING, ETC.

61*. *A COMPLETE READY RECKONER FOR THE ADMEASUREMENT OF LAND, &c.* By A. ARMAN. Third Edition, revised and extended by C. NORRIS, Surveyor, Valuer, &c. 2s.

131. *MILLER'S, CORN MERCHANT'S, AND FARMER'S READY RECKONER.* Second Edition, with a Price List of Modern Flour-Mill Machinery, by W. S. HUTTON, C.E. 2s.

140. *SOILS, MANURES, AND CROPS.* (Vol. 1. OUTLINES OF MODERN FARMING.) By R. SCOTT BURN. Woodcuts. 2s.

141. *FARMING & FARMING ECONOMY,* Notes, Historical and Practical, on. (Vol. 2. OUTLINES OF MODERN FARMING.) By R. SCOTT BURN. 3s.

142. *STOCK; CATTLE, SHEEP, AND HORSES.* (Vol. 3. OUTLINES OF MODERN FARMING.) By R. SCOTT BURN. Woodcuts. 2s. 6d.

145. *DAIRY, PIGS, AND POULTRY,* Management of the. By R. SCOTT BURN. (Vol. 4. OUTLINES OF MODERN FARMING.) 2s.

146. *UTILIZATION OF SEWAGE, IRRIGATION, AND RECLAMATION OF WASTE LAND.* (Vol. 5. OUTLINES OF MODERN FARMING.) By R. SCOTT BURN. Woodcuts. 2s. 6d.

∴ Nos. 140-1-2-5-6, in One Vol., handsomely half-bound, entitled "OUTLINES OF MODERN FARMING." By ROBERT SCOTT BURN. Price 12s.

177. *FRUIT TREES,* The Scientific and Profitable Culture of. From the French of DU BREUIL. Revised by GEO. GLENNY. 187 Woodcuts. 3s. 6d.‡

198. *SHEEP; THE HISTORY, STRUCTURE, ECONOMY, AND DISEASES OF.* By W. C. SPOONER, M.R.V.C., &c. Fifth Edition, enlarged, including Specimens of New and Improved Breeds. 3s. 6d.‡

201. *KITCHEN GARDENING MADE EASY.* By GEORGE M. F. GLENNY. Illustrated, 1s. 6d.‡

207. *OUTLINES OF FARM MANAGEMENT, and the Organization of Farm Labour.* By R. SCOTT BURN. 2s. 6d.‡

208. *OUTLINES OF LANDED ESTATES MANAGEMENT.* By R. SCOTT BURN.

2s. 6d.

❖ Nos. 207 & 208 in One Vol., handsomely half-bound, entitled "OUTLINES OF LANDED ESTATES AND FARM MANAGEMENT." By R. SCOTT BURN. Price 6s.

209. *THE TREE PLANTER AND PLANT PROPAGATOR*. A Practical Manual on the Propagation of Forest Trees, Fruit Trees, Flowering Shrubs, Flowering Plants, &c. By SAMUEL WOOD. 2s.

210. *THE TREE PRUNER*. A Practical Manual on the Pruning of Fruit Trees, including also their Training and Renovation; also the Pruning of Shrubs, Climbers, and Flowering Plants. By SAMUEL WOOD. 1s. 6d.

❖ Nos. 209 & 210 in One Vol., handsomely half-bound, entitled "THE TREE PLANTER, PROPAGATOR, AND PRUNER." By SAMUEL WOOD. Price 3s. 6d.

218. *THE HAY AND STRAW MEASURER*: Being New Tables for the Use of Auctioneers, Valuers, Farmers, Hay and Straw Dealers, &c. By JOHN STEELE. Fifth Edition. 2s.

222. *SUBURBAN FARMING*. The Laying-out and Cultivation of Farms, adapted to the Produce of Milk, Butter, and Cheese, Eggs, Poultry, and Pigs. By Prof. JOHN DONALDSON and R. SCOTT BURN. 3s. 6d.‡

231. *THE ART OF GRAFTING AND BUDDING*. By CHARLES BALTET. With Illustrations. 2s. 6d.‡

232. *COTTAGE GARDENING*; or, Flowers, Fruits, and Vegetables for Small Gardens. By E. HOBDAY, 1s. 6d.

233. *GARDEN RECEIPTS*. Edited by CHARLES W. QUIN. 1s. 6d.

234. *MARKET AND KITCHEN GARDENING*. By C. W. SHAW, late Editor of "Gardening Illustrated." 3s.‡

239. *DRAINING AND EMBANKING*. A Practical Treatise, embodying the most recent experience in the Application of Improved Methods. By JOHN SCOTT, late Professor of Agriculture and Rural Economy at the Royal Agricultural College, Cirencester. With 68 Illustrations, 1s. 6d.

240. *IRRIGATION AND WATER SUPPLY*. A Treatise on Water Meadows, Sewage Irrigation, and Warping; the Construction of Wells, Ponds, and Reservoirs, &c. By Prof. JOHN SCOTT. With 34 Illus. 1s. 6d.

241. *FARM ROADS, FENCES, AND GATES*. A Practical Treatise on the Roads, Tramways, and Waterways of the Farm; the Principles of Enclosures; and the different kinds of Fences, Gates, and Stiles. By Professor JOHN SCOTT. With 75 Illustrations, 1s. 6d.

242. *FARM BUILDINGS*. A Practical Treatise on the Buildings necessary for various kinds of Farms, their Arrangement and Construction, with Plans and Estimates. By Prof. JOHN SCOTT. With 105 Illus. 2s.

243. *BARN IMPLEMENTS AND MACHINES*. A Practical Treatise on the Application of Power to the Operations of Agriculture; and on various Machines used in the Threshing-barn, in the Stock-yard, and in the Dairy, &c. By Prof. J. SCOTT. With 123 Illustrations. 2s.

244. *FIELD IMPLEMENTS AND MACHINES*. A Practical Treatise on the Varieties now in use, with Principles and Details of Construction, their Points of Excellence, and Management. By Professor JOHN SCOTT. With 138 Illustrations. 2s.

245. *AGRICULTURAL SURVEYING*. A Practical Treatise or Land Surveying, Levelling, and Setting-out; and on Measuring and Estimating Quantities, Weights, and Values of Materials, Produce, Stock, &c. By Prof. JOHN SCOTT. With 62 Illustrations, 1s. 6d.

❖ Nos. 239 to 245 in One Vol., handsomely half-bound, entitled "THE COMPLETE TEXT-BOOK OF FARM ENGINEERING." By Professor JOHN SCOTT. Price 12s.

250. *MEAT PRODUCTION.* A Manual for Producers, Distributors, &c. By John Ewart. 2s. 6d.‡

266. *BOOK-KEEPING FOR FARMERS & ESTATE OWNERS.* By J. M. Woodman, Chartered Accountant. 2s. 6d. cloth limp; 3s. 6d. cloth boards.

MATHEMATICS, ARITHMETIC, ETC.

32. *MATHEMATICAL INSTRUMENTS,* a Treatise on; Their Construction, Adjustment, Testing, and Use concisely Explained. By J. F. Heather, M.A. Fourteenth Edition, revised, with additions, by A. T. Walmisley, M.I.C.E., Fellow of the Surveyors' Institution. Original Edition, in 1 vol., Illustrated. 2s.‡

∴ In ordering the above, be careful to say, "Original Edition" (No. 32), to distinguish it from the Enlarged Edition in 3 vols. (Nos. 168-9-70.)

76. *DESCRIPTIVE GEOMETRY,* an Elementary Treatise on; with a Theory of Shadows and of Perspective, extracted from the French of G. Monge. To which is added, a description of the Principles and Practice of Isometrical Projection. By J. F. Heather, M.A. With 14 Plates. 2s.

178. *PRACTICAL PLANE GEOMETRY:* giving the Simplest Modes of Constructing Figures contained in one Plane and Geometrical Construction of the Ground. By J. F. Heather, M.A. With 215 Woodcuts. 2s.

83. *COMMERCIAL BOOK-KEEPING.* With Commercial Phrases and Forms in English, French, Italian, and German. By James Haddon, M.A., Arithmetical Master of King's College School, London, 1s. 6d.

84. *ARITHMETIC,* a Rudimentary Treatise on: with full Explanations of its Theoretical Principles, and numerous Examples for Practice. By Professor J. R. Young. Eleventh Edition, 1s. 6d.

84*. A Key to the above, containing Solutions in full to the Exercises, together with Comments, Explanations, and Improved Processes, for the Use of Teachers and Unassisted Learners. By J. R. Young, 1s. 6d.

85. *EQUATIONAL ARITHMETIC,* applied to Questions of Interest, Annuities, Life Assurance, and General Commerce; with various Tables by which all Calculations may be greatly facilitated. By W. Hipsley. 2s.

86. *ALGEBRA,* the Elements of. By James Haddon, M.A. With Appendix, containing miscellaneous Investigations, and a Collection of Problems in various parts of Algebra. 2s.

86*. A Key and Companion to the above Book, forming an extensive repository of Solved Examples and Problems in Illustration of the various Expedients necessary in Algebraical Operations. By J. R. Young, 1s. 6d.

88. 89. *EUCLID,* The Elements of: with many additional Propositions and Explanatory Notes: to which is prefixed, an Introductory Essay on Logic. By Henry Law, C.E. 2s. 6d.‡

∴ Sold also separately, viz.:—

88. Euclid, The First Three Books. By Henry Law, C.E. 1s. 6d.
89. Euclid, Books 4, 5, 6, 11, 12. By Henry Law, C.E. 1s. 6d.

90. *ANALYTICAL GEOMETRY AND CONIC SECTIONS,* By James Hann. A New Edition, by Professor J. R. Young. 2s.‡

91. *PLANE TRIGONOMETRY,* the Elements of. By James Hann, formerly Mathematical Master of King's College, London, 1s. 6d.

92. *SPHERICAL TRIGONOMETRY,* the Elements of. By James Hann. Revised by Charles H. Dowling, C.E. 1s.

∴ *Or with "The Elements of Plane Trigonometry," in One Volume, 2s. 6d.*

93. *MENSURATION AND MEASURING.* With the Mensuration and Levelling of Land for the Purposes of Modern Engineering. By T. BAKER, C.E. New Edition by E. NUGENT, C.E. Illustrated. 1s. 6d.

101. *DIFFERENTIAL CALCULUS,* Elements of the. By W. S. B. WOOLHOUSE, F.R.A.S., &c. 1s. 6d.

102. *INTEGRAL CALCULUS,* Rudimentary Treatise on the. By HOMERSHAM COX, B.A. Illustrated. 1s.

136. *ARITHMETIC,* Rudimentary, for the Use of Schools and Self-Instruction. By JAMES HADDON, M.A. Revised by A. ARMAN, 1s. 6d.

137. A KEY TO HADDON'S RUDIMENTARY ARITHMETIC. By A. ARMAN, 1s. 6d.

168. *DRAWING AND MEASURING INSTRUMENTS.* Including —I. Instruments employed in Geometrical and Mechanical Drawing, and in the Construction, Copying, and Measurement of Maps and Plans. II. Instruments used for the purposes of Accurate Measurement, and for Arithmetical Computations. By J. F. HEATHER, M.A. Illustrated. 1s. 6d.

169. *OPTICAL INSTRUMENTS.* Including (more especially) Telescopes, Microscopes, and Apparatus for producing copies of Maps and Plans by Photography. By J. F. HEATHER, M.A. Illustrated. 1s. 6d.

170. *SURVEYING AND ASTRONOMICAL INSTRUMENTS.* Including —I. Instruments Used for Determining the Geometrical Features of a portion of Ground. II. Instruments Employed in Astronomical Observations. By J. F. HEATHER, M.A. Illustrated. 1s. 6d.

∴ *The above three volumes form an enlargement of the Author's original work "Mathematical Instruments." (See No. 32 in the Series.)*

168. 169. 170. *MATHEMATICAL INSTRUMENTS.* By J. F. HEATHER, M.A. Enlarged Edition, for the most part entirely re-written. The 3 Parts as above, in One thick Volume. With numerous Illustrations. 4s. 6d.‡

158. *THE SLIDE RULE, AND HOW TO USE IT;* containing full, easy, and simple Instructions to perform all Business Calculations with unexampled rapidity and accuracy. By CHARLES HOARE, C.E. Sixth Edition. With a Slide Rule in tuck of cover. 2s. 6d.‡

196. *THEORY OF COMPOUND INTEREST AND ANNUITIES;* with Tables of Logarithms for the more Difficult Computations of Interest, Discount, Annuities, &c. By FÉDOR THOMAN. Fourth Edition. 4s.‡

199. *THE COMPENDIOUS CALCULATOR;* or, Easy and Concise Methods of Performing the various Arithmetical Operations required in Commercial and Business Transactions; together with Useful Tables. By D. O'GORMAN. Twenty-seventh Edition, carefully revised by C. NORRIS. 2s 6d., cloth limp; 3s. 6d., strongly half-bound in leather.

204. *MATHEMATICAL TABLES,* for Trigonometrical, Astronomical, and Nautical Calculations; to which is prefixed a Treatise on Logarithms. By HENRY LAW, C.E. Together with a Series of Tables for Navigation and Nautical Astronomy. By Prof. J. R. YOUNG. New Edition. 4s.

204*. *LOGARITHMS.* With Mathematical Tables for Trigonometrical, Astronomical, and Nautical Calculations. By HENRY LAW, M.Inst.C.E. New and Revised Edition. (Forming part of the above Work). 3s.

221. *MEASURES, WEIGHTS, AND MONEYS OF ALL NATIONS,* and an Analysis of the Christian, Hebrew, and Mahometan Calendars. By W. S. B. WOOLHOUSE, F.R.A.S., F.S.S. Seventh Edition, 2s. 6d.‡

227. *MATHEMATICS AS APPLIED TO THE CONSTRUCTIVE ARTS.* Illustrating the various processes of Mathematical Investigation, by means

of Arithmetical and Simple Algebraical Equations and Practical Examples. By Francis Campin. C.E. Second Edition. 3s.‡

PHYSICAL SCIENCE, NATURAL PHILOSOPHY, ETC.

1. *CHEMISTRY.* By Professor George Fownes, F.R.S. With an Appendix on the Application of Chemistry to Agriculture, 1s.
2. *NATURAL PHILOSOPHY,* Introduction to the Study of. By C. Tomlinson. Woodcuts. 1s. 6d.
6. *MECHANICS,* Rudimentary Treatise on. By Charles Tomlinson. Illustrated. 1s. 6d.
7. *ELECTRICITY*; showing the General Principles of Electrical Science, and the purposes to which it has been applied. By Sir W. Snow Harris, F.R.S., &c. With Additions by R. Sabine, C.E., F.S.A. 1s. 6d.
7*. *GALVANISM.* By Sir W. Snow Harris. New Edition by Robert Sabine, C.E., F.S.A. 1s. 6d.
8. *MAGNETISM*; being a concise Exposition of the General Principles of Magnetical Science. By Sir W. Snow Harris. New Edition, revised by H. M. Noad, Ph.D. With 165 Woodcuts. 3s. 6d.‡
11. *THE ELECTRIC TELEGRAPH*; its History and Progress; with Descriptions of some of the Apparatus. By R. Sabine, C.E., F.S.A. 3s.
12. *PNEUMATICS,* including Acoustics and the Phenomena of Wind Currents, for the Use of Beginners. By Charles Tomlinson, F.R.S. Fourth Edition, enlarged. Illustrated. 1s. 6d.
72. *MANUAL OF THE MOLLUSCA*; a Treatise on Recent and Fossil Shells. By Dr. S. P. Woodward, A.L.S. Fourth Edition. With Plates and 300 Woodcuts. 7s. 6d., cloth.
96. *ASTRONOMY.* By the late Rev. Robert Main, M.A. Third Edition, by William Thynne Lynn, B.A., F.R.A.S. 2s.
97. *STATICS AND DYNAMICS,* the Principles and Practice of; embracing also a clear development of Hydrostatics, Hydrodynamics, and Central Forces. By T. Baker, C.E. Fourth Edition. 1s. 6d.
173. *PHYSICAL GEOLOGY,* partly based on Major-General Portlock's "Rudiments of Geology." By Ralph Tate, A.L.S., &c. Woodcuts. 2s.
174. *HISTORICAL GEOLOGY,* partly based on Major-General Portlock's "Rudiments." By Ralph Tate, A.L.S., &c. Woodcuts. 2s. 6d.
173 & 174. *RUDIMENTARY TREATISE ON GEOLOGY,* Physical and Historical. Partly based on Major-General Portlock's "Rudiments of Geology." By Ralph Tate, A.L.S., F.G.S., &c. In One Volume. 4s. 6d.‡
183 & 184. *ANIMAL PHYSICS,* Handbook of. By Dr. Lardner, D.C.L., formerly Professor of Natural Philosophy and Astronomy in University College, Lond. With 520 Illustrations. In One Vol. 7s. 6d., cloth boards.

⁂ *Sold also in Two Parts, as follows:—*
183. *Animal Physics.* By Dr. Lardner. Part I., Chapters I.—VII. 4s.
184. *Animal Physics.* By Dr. Lardner. Part II., Chapters VIII.—XVIII. 3s.

269. *LIGHT*: an Introduction to the Science of Optics for the Use of Students of Architecture, Engineering, and other Applied Sciences. By E. Wyndham Tarn, M.A. 1s. 6d.

[Just published.

FINE ARTS.

20. *PERSPECTIVE FOR BEGINNERS.* Adapted to Young Students and Amateurs in Architecture, Painting, &c. By George Pyne. 2s.
40. *GLASS STAINING, AND THE ART OF PAINTING ON GLASS.* From the German of Dr. Gessert and Emanuel Otto Fromberg. With an Appendix on The Art of Enamelling. 2s. 6d.
69. *MUSIC,* A Rudimentary and Practical Treatise on. With numerous Examples. By Charles Child Spencer. 2s. 6d.
71. *PIANOFORTE,* The Art of Playing the. With numerous Exercises & Lessons from the Best Masters. By Charles Child Spencer. 1s. 6d.
69-71. *MUSIC & THE PIANOFORTE.* In one vol. Half-bound, 5s.
181. *PAINTING POPULARLY EXPLAINED,* including Fresco, Oil, Mosaic, Water Colour, Water-Glass, Tempera, Encaustic, Miniature, Painting on Ivory, Vellum, Pottery, Enamel, Glass, &c. With Historical Sketches of the Progress of the Art by Thomas John Gullick, assisted by John Timbs, F.S.A. Sixth Edition, revised and enlarged. 5s.‡
186. *A GRAMMAR OF COLOURING,* applied to Decorative Painting and the Arts. By George Field. New Edition, enlarged and adapted to the Use of the Ornamental Painter and Designer. By Ellis A. Davidson. With two new Coloured Diagrams, &c. 3s.‡
246. *A DICTIONARY OF PAINTERS, AND HANDBOOK FOR PICTURE AMATEURS;* including Methods of Painting, Cleaning, Relining and Restoring, Schools of Painting, &c. With Notes on the Copyists and Imitators of each Master. By Philippe Daryl. 2s. 6d.‡

INDUSTRIAL AND USEFUL ARTS.

23. *BRICKS AND TILES,* Rudimentary Treatise on the Manufacture of. By E. Dobson, M.R.I.B.A. Illustrated, 3s.‡
67. *CLOCKS, WATCHES, AND BELLS,* a Rudimentary Treatise on. By Sir Edmund Beckett, LL.D., Q.C. Seventh Edition, revised and enlarged. 4s. 6d. limp; 5s. 6d. cloth boards.
83**. *CONSTRUCTION OF DOOR LOCKS.* Compiled from the Papers of A. C. Hobbs, and Edited by Charles Tomlinson. F.R.S. 2s. 6d.
162. *THE BRASS FOUNDER'S MANUAL;* Instructions for Modelling, Pattern-Making, Moulding, Turning, Filing, Burnishing, Bronzing, &c. With copious Receipts, &c. By Walter Graham. 2s.‡
205. *THE ART OF LETTER PAINTING MADE EASY.* By J.G. Badenoch. Illustrated with 12 full-page Engravings of Examples, 1s. 6d.
215. *THE GOLDSMITH'S HANDBOOK,* containing full Instructions for the Alloying and Working of Gold. By George E. Gee, 3s.‡
225. *THE SILVERSMITH'S HANDBOOK,* containing full Instructions for the Alloying and Working of Silver. By George E. Gee. 3s.‡

∴ *The two preceding Works, in One handsome Vol., half-bound, entitled* "The Goldsmith's & Silversmith's Complete Handbook," 7s.

249. *THE HALL-MARKING OF JEWELLERY PRACTICALLY CONSIDERED.* By George E. Gee. 3s.‡
224. *COACH BUILDING,* A Practical Treatise, Historical and Descriptive. By J. W. Burgess. 2s. 6d.‡
235. *PRACTICAL ORGAN BUILDING.* By W. E. Dickson, M.A., Precentor of

Ely Cathedral. Illustrated. 2s. 6d.‡

262. *THE ART OF BOOT AND SHOEMAKING.* By JOHN BEDFORD LENO. Numerous Illustrations. Third Edition. 2s.

263. *MECHANICAL DENTISTRY*: A Practical Treatise on the Construction of the Various Kinds of Artificial Dentures, with Formulæ, Tables, Receipts, &c. By CHARLES HUNTER. Third Edition. 3s.‡

270. *WOOD ENGRAVING*: A Practical and Easy Introduction to the Study of the Art. By W. N. BROWN, 1s. 6d.

MISCELLANEOUS VOLUMES.

36. *A DICTIONARY OF TERMS used in ARCHITECTURE, BUILDING, ENGINEERING, MINING, METALLURGY, ARCHÆOLOGY, the FINE ARTS, &c.* By JOHN WEALE. Sixth Edition. Revised by ROBERT HUNT, F.R.S. Illustrated. 5s. limp; 6s. cloth boards.

50. *LABOUR CONTRACTS.* A Popular Handbook on the Law of Contracts for Works and Services. By DAVID GIBBONS. Fourth Edition, Revised, with Appendix of Statutes by T. F. UTTLEY, Solicitor, 3s. 6d. cloth.

112. *MANUAL OF DOMESTIC MEDICINE.* By R. GOODING, B.A., M.D. A Family Guide in all Cases of Accident and Emergency. 2s.

112*. *MANAGEMENT OF HEALTH.* A Manual of Home and Personal Hygiene. By the Rev. JAMES BAIRD, B.A. 1s.

150. *LOGIC*, Pure and Applied. By S. H. EMMENS. 1s. 6d.

153. *SELECTIONS FROM LOCKE'S ESSAYS ON THE HUMAN UNDERSTANDING.* With Notes by S. H. EMMENS. 2s.

154. *GENERAL HINTS TO EMIGRANTS.* 2s.

157. *THE EMIGRANT'S GUIDE TO NATAL.* By R. MANN. 2s.

193. *HANDBOOK OF FIELD FORTIFICATION.* By Major W. W. KNOLLYS, F.R.G.S. With 163 Woodcuts. 3s.‡

194. *THE HOUSE MANAGER*: Being a Guide to Housekeeping, Practical Cookery, Pickling and Preserving, Household Work, Dairy Management, &c. By AN OLD HOUSEKEEPER. 3s. 6d.‡

194, 112 & 112*. *HOUSE BOOK (The).* Comprising:—I. THE HOUSE MANAGER. By an OLD HOUSEKEEPER. II. DOMESTIC MEDICINE. By R. GOODING, M.D. III. MANAGEMENT OF HEALTH. By J. BAIRD. In One Vol., half-bound, 6s.

☞ *The ‡ indicates that these vols may be had strongly bound at 6d. extra.*

EDUCATIONAL AND CLASSICAL SERIES.
HISTORY.

1. **England, Outlines of the History of**; more especially with reference to the Origin and Progress of the English Constitution. By WILLIAM DOUGLAS HAMILTON, F.S.A., of Her Majesty's Public Record Office. 4th Edition, revised. 5s.; cloth boards, 6s.

5. **Greece, Outlines of the History of**; in connection with the Rise of the Arts and Civilization in Europe. By W. DOUGLAS HAMILTON, of University

College, London, and EDWARD LEVIEN, M.A., of Balliol College, Oxford. 2s. 6d.; cloth boards, 3s. 6d.
7. **Rome, Outlines of the History of:** from the Earliest Period to the Christian Era and the Commencement of the Decline of the Empire. By EDWARD LEVIEN, of Balliol College, Oxford. Map, 2s. 6d.; cl. bds. 3s. 6d.
9. **Chronology of History, Art, Literature, and Progress,** from the Creation of the World to the Present Time. The Continuation by W. D. HAMILTON, F.S.A. 3s.; cloth boards, 3s. 6d.

ENGLISH LANGUAGE AND MISCELLANEOUS.

11. **Grammar of the English Tongue,** Spoken and Written. With an Introduction to the Study of Comparative Philology. By HYDE CLARKE, D.C.L. Fifth Edition, 1s. 6d.
12. **Dictionary of the English Language,** as Spoken and Written. Containing above 100,000 Words. By HYDE CLARKE, D.C.L. 3s. 6d.; cloth boards, 4s. 6d.; complete with the GRAMMAR, cloth bds., 5s. 6d.
48. **Composition and Punctuation,** familiarly Explained for those who have neglected the Study of Grammar. By JUSTIN BRENAN. 18th Edition, 1s. 6d.
49. **Derivative Spelling-Book:** Giving the Origin of Every Word from the Greek, Latin, Saxon, German, Teutonic, Dutch, French, Spanish, and other Languages; with their present Acceptation and Pronunciation. By J. ROWBOTHAM, F.R.A.S. Improved Edition, 1s. 6d.
51. **The Art of Extempore Speaking:** Hints for the Pulpit, the Senate, and the Bar. By M. BAUTAIN, Vicar-General and Professor at the Sorbonne. Translated from the French. 8th Edition, carefully corrected. 2s. 6d.
53. **Places and Facts in Political and Physical Geography,** for Candidates in Examinations. By the Rev. EDGAR RAND, B.A. 1s.
54. **Analytical Chemistry,** Qualitative and Quantitative, a Course of. To which is prefixed, a Brief Treatise upon Modern Chemical Nomenclature and Notation. By WM. W. PINK and GEORGE E. WEBSTER. 2s.

THE SCHOOL MANAGERS' SERIES OF READING BOOKS,
Edited by the Rev. A. R. GRANT, Rector of Hitcham, and Honorary Canon of Ely; formerly H.M. Inspector of Schools.

INTRODUCTORY PRIMER, 3*d.*

	s.	d.		s.	d.
First Standard	0	6	Fourth Standard	1	2
Second "	0	10	Fifth "	1	6
Third "	1	0	Sixth "	1	6

LESSONS FROM THE BIBLE. Part 1. Old Testament, 1s.
LESSONS FROM THE BIBLE. Part II. New Testament, to which is added THE GEOGRAPHY OF THE BIBLE, for very young Children. By Rev. C. THORNTON FORSTER, 1s. 2d. ⁂ Or the Two Parts in One Volume. 2s.

FRENCH.

24. **French Grammar.** With Complete and Concise Rules on the Genders of

French Nouns. By G. L. STRAUSS, Ph.D. 1s. 6d.
25. **French-English Dictionary.** Comprising a large number of New Terms used in Engineering, Mining, &c. By ALFRED ELWES. 1s. 6d.
26. **English-French Dictionary.** By ALFRED ELWES. 2s.
25, 26. **French Dictionary** (as above). Complete, in One Vol., 3s.; cloth boards, 3s. 6d. ∴ Or with the GRAMMAR, cloth boards, 4s. 6d.
47. **French and English Phrase Book**: containing Introductory Lessons, with Translations, several Vocabularies of Words, a Collection of suitable Phrases, and Easy Familiar Dialogues, 1s. 6d.

GERMAN.

39. **German Grammar.** Adapted for English Students, from Heyse's Theoretical and Practical Grammar, by Dr. G. L. STRAUSS, 1s. 6d.
40. **German Reader**: A Series of Extracts, carefully culled from the most approved Authors of Germany; with Notes, Philological and Explanatory. By G. L. STRAUSS, Ph.D. 1s.
41-43. **German Triglot Dictionary.** By N. E. S. A. HAMILTON. In Three Parts. Part I. German-French-English. Part II. English-German-French. Part III. French-German-English. 3s., or cloth boards, 4s.
41-43 & 39. **German Triglot Dictionary** (as above), together with German Grammar (No. 39), in One Volume, cloth boards, 5s.

ITALIAN.

27. **Italian Grammar**, arranged in Twenty Lessons, with a Course of Exercises. By ALFRED ELWES. 1s. 6d.
28. **Italian Triglot Dictionary**, wherein the Genders of all the Italian and French Nouns are carefully noted down. By ALFRED ELWES. Vol. 1. Italian-English-French. 2s. 6d.
30. **Italian Triglot Dictionary.** By A. ELWES. Vol. 2. English-French-Italian. 2s. 6d.
32. **Italian Triglot Dictionary.** By ALFRED ELWES. Vol. 3. French-Italian-English. 2s. 6d.
28, 30, 32. **Italian Triglot Dictionary** (as above). In One Vol., 7s. 6d. Cloth boards.

SPANISH AND PORTUGUESE.

34. **Spanish Grammar**, in a Simple and Practical Form. With a Course of Exercises. By ALFRED ELWES. 1s. 6d.
35. **Spanish-English and English-Spanish Dictionary.** Including a large number of Technical Terms used in Mining, Engineering, &c. with the proper Accents and the Gender of every Noun. By ALFRED ELWES 4s.; cloth boards, 5s. ∴ Or with the GRAMMAR, cloth boards, 6s.
55. **Portuguese Grammar**, in a Simple and Practical Form. With a Course of Exercises. By ALFRED ELWES. 1s. 6d.
56. **Portuguese-English and English-Portuguese Dictionary.** Including a large number of Technical Terms used in Mining, Engineering, &c., with the proper Accents and the Gender of every Noun. By ALFRED ELWES. Second Edition, Revised, 5s.; cloth boards, 6s. ∴ Or with the GRAMMAR, cloth boards, 7s.

HEBREW.

46*. **Hebrew Grammar.** By Dr. BRESSLAU. 1s. 6d.
44. **Hebrew and English Dictionary,** Biblical and Rabbinical; containing the Hebrew and Chaldee Roots of the Old Testament Post-Rabbinical Writings. By Dr. BRESSLAU. 6s.
46. **English and Hebrew Dictionary.** By Dr. BRESSLAU. 3s.
44, 46. 46*. **Hebrew Dictionary** (as above), in Two Vols., complete, with the GRAMMAR, cloth boards, 12s.

LATIN.

19. **Latin Grammar.** Containing the Inflections and Elementary Principles of Translation and Construction. By the Rev. THOMAS GOODWIN, M.A., Head Master of the Greenwich Proprietary School. 1s. 6d.
20. **Latin-English Dictionary.** By the Rev. THOMAS GOODWIN, M.A. 2s.
22. **English-Latin Dictionary;** together with an Appendix of French and Italian Words which have their origin from the Latin. By the Rev. THOMAS GOODWIN, M.A. 1s. 6d.
20, 22. **Latin Dictionary** (as above). Complete in One Vol., 3s. 6d. cloth boards, 4s. 6d. ∴ Or with the GRAMMAR, cloth boards, 5s. 6d.

LATIN CLASSICS. With Explanatory Notes in English.

1. **Latin Delectus.** Containing Extracts from Classical Authors, with Genealogical Vocabularies and Explanatory Notes, by H. YOUNG. 1s. 6d.
2. **Cæsaris** Commentarii de Bello Gallico. Notes, and a Geographical Register for the Use of Schools, by H. YOUNG. 2s.
3. **Cornelius Nepos.** With Notes. By H. YOUNG. 1s.
4. **Virgilii** Maronis Bucolica et Georgica. With Notes on the Bucolics by W. RUSHTON, M.A., and on the Georgics by H. YOUNG. 1s. 6d.
5. **Virgilii** Maronis Æneis. With Notes, Critical and Explanatory, by H. YOUNG. New Edition, revised and improved. With copious Additional Notes by Rev. T. H. L. LEARY, D.C.L., formerly Scholar of Brasenose College, Oxford. 3s.
5* ————— Part 1 Books i.—vi., 1s. 6d.
5** ———— Part 2. Books vii.-xii., 2s.
6. **Horace;** Odes, Epode, and Carmen Sæculare. Notes by H. YOUNG. 1s. 6d.
7. **Horace;** Satires, Epistles, and Ars Poetica. Notes by W. BROWNRIGG SMITH, M.A., F.R.G.S. 1s. 6d.
8. **Sallustii** Crispi Catalina et Bellum Jugurthinum. Notes, Critical and Explanatory, by W. M. DONNE, B.A., Trin. Coll., Cam. 1s. 6d.
9. **Terentii** Andria et Heautontimorumenos. With Notes, Critical and Explanatory, by the Rev. JAMES DAVIES, M.A. 1s. 6d.
10. **Terentii** Adelphi, Hecyra, Phormio. Edited, with Notes, Critical and Explanatory, by the Rev. JAMES DAVIES, M.A. 2s.
11. **Terentii** Eunuchus, Comœdia. Notes, by Rev. J. DAVIES, M.A. 1s. 6d.
12. **Ciceronis** Oratio pro Sexto Roscio Amerino. Edited, with an Introduction, Analysis, and Notes, Explanatory and Critical, by the Rev. JAMES DAVIES, M.A. 1s. 6d.
13. **Ciceronis** Orationes in Catilinam, Verrem, et pro Archia. With Introduction, Analysis, and Notes, Explanatory and Critical, by Rev. T. H. L. LEARY,

D.C.L. formerly Scholar of Brasenose College, Oxford. 1s. 6d.

14. **Ciceronis** Cato Major, Lælius, Brutus, sive de Senectute, de Amicitia, de Claris Oratoribus Dialogi. With Notes by W. BROWNRIGG SMITH, M.A., F.R.G.S. 2s.

16. **Livy:** History of Rome. Notes by H. YOUNG and W. B. SMITH, M.A. Part 1. Books i., ii., 1s. 6d.

16*. — — — — Part 2. Books iii., iv., v., 1s. 6d.

17. — — — — Part 3. Books xxi., xxii., 1s. 6d.

19. **Latin Verse Selections**, from Catullus, Tibullus, Propertius, and Ovid. Notes by W. B. DONNE, M.A., Trinity College, Cambridge. 2s.

20. **Latin Prose Selections**, from Varro, Columella, Vitruvius, Seneca, Quintilian, Florus, Velleius Paterculus, Valerius Maximus Suetonius, Apuleius, &c. Notes by W. B. DONNE, M.A. 2s.

21. **Juvenalis** Satiræ. With Prolegomena and Notes by T. H. S. ESCOTT, B.A, Lecturer on Logic at King's College, London. 2s.

GREEK.

14. **Greek Grammar**, in accordance with the Principles and Philological Researches of the most eminent Scholars of our own day. By HANS CLAUDE HAMILTON. 1s. 6d.

15, 17. **Greek Lexicon.** Containing all the Words in General Use, with their Significations, Inflections, and Doubtful Quantities. By HENRY R. HAMILTON. Vol. 1. Greek-English, 2s. 6d.; Vol. 2. English-Greek, 2s. Or the Two Vols. in One, 4s. 6d.; cloth boards, 5s.

14, 15. 17. **Greek Lexicon** (as above). Complete, with the GRAMMAR, in One Vol., cloth boards, 6s.

GREEK CLASSICS. With Explanatory Notes in English.

1. **Greek Delectus.** Containing Extracts from Classical Authors, with Genealogical Vocabularies and Explanatory Notes, by H. YOUNG. New Edition, with an improved and enlarged Supplementary Vocabulary, by JOHN HUTCHISON, M.A., of the High School, Glasgow. 1s. 6d.

2, 3. **Xenophon's Anabasis**; or, The Retreat of the Ten Thousand. Notes and a Geographical Register, by H. YOUNG. Part 1. Books i. to iii., 1s. Part 2. Books iv. to vii., 1s.

4. **Lucian's Select Dialogues.** The Text carefully revised, with Grammatical and Explanatory Notes, by H. YOUNG. 1s. 6d.

5-12. **Homer**, The Works of. According to the Text of BAEUMLEIN. With Notes, Critical and Explanatory, drawn from the best and latest Authorities, with Preliminary Observations and Appendices, by T. H. L. LEARY, M.A., D.C.L.

THE ILIAD:	Part 1. Books i. to vi., 1s. 6d.	Part 3. Books xiii. to xviii., 1s. 6d.
	Part 2. Books vii. to xii., 1s.6d.	Part 4. Books xix. to xxiv., 1s. 6d.
THE ODYSSEY:	Part 1. Books i. to vi., 1s. 6d	Part 3. Books xiii. to xviii., 1s. 6d.

| | Part 2. Books vii. to xii., 1s. 6d. | Part 4. Books xix. to xxiv., and Hymns, 2s. |

13. **Plato's Dialogues**: The Apology of Socrates, the Crito, and the Phædo. From the Text of C. F. HERMANN. Edited with Notes, Critical and Explanatory, by the Rev. JAMES DAVIES, M.A. 2s.

14-17. **Herodotus**, The History of, chiefly after the Text of GAISFORD. With Preliminary Observations and Appendices, and Notes, Critical and Explanatory, by T. H. L. LEARY, M.A., D.C.L.

Part 1.	Books i., ii. (The Clio and Euterpe), 2s.
Part 2.	Books iii., iv. (The Thalia and Melpomene), 2s.
Part 3.	Books v.-vii. (The Terpsichore, Erato, and Polymnia), 2s.
Part 4.	Books viii., ix. (The Urania and Calliope) and Index, 1s. 6d.

18. **Sophocles**: Œdipus Tyrannus. Notes by H. YOUNG. 1s.
20. **Sophocles**: Antigone. From the Text of DINDORF. Notes, Critical and Explanatory, by the Rev. JOHN MILNER, B.A. 2s.
23. **Euripides**: Hecuba and Medea. Chiefly from the Text of DINDORF. With Notes, Critical and Explanatory, by W. BROWNRIGG SMITH, M.A., F.R.G.S. 1s. 6d.
26. **Euripides**: Alcestis. Chiefly from the Text of DINDORF. With Notes, Critical and Explanatory, by JOHN MILNER, B.A. 1s. 6d.
30. **Æschylus**: Prometheus Vinctus: The Prometheus Bound. From the Text of DINDORF. Edited, with English Notes, Critical and Explanatory, by the Rev. JAMES DAVIES, M.A. 1s.
32. **Æschylus**: Septem Contra Thebes: The Seven against Thebes. From the Text of DINDORF. Edited, with English Notes, Critical and Explanatory, by the Rev. JAMES DAVIES, M.A. 1s.
40. **Aristophanes**: Acharnians. Chiefly from the Text of C. H. WEISE. With Notes, by C. S. T. TOWNSHEND, M.A. 1s. 6d.
41. **Thucydides**: History of the Peloponnesian War. Notes by H. YOUNG. Book 1. 1s. 6d.
42. **Xenophon's Panegyric on Agesilaus.** Notes and Introduction by LL. F. W. JEWITT. 1s. 6d.
43. **Demosthenes.** The Oration on the Crown and the Philippics. With English Notes. By Rev. T. H. L. LEARY, D.C.L., formerly Scholar of Brasenose College, Oxford. 1s. 6d.

LONDON: CROSBY LOCKWOOD AND SON,

7, STATIONERS' HALL COURT, LUDGATE HILL, E.C.